AF595082

Advances in Sensors, Big Data and Machine Learning in Intelligent Animal Farming

Advances in Sensors, Big Data and Machine Learning in Intelligent Animal Farming

Editors

Yongliang Qiao
Lilong Chai
Dongjian He
Daobilige Su

MDPI • Basel • Beijing • Wuhan • Barcelona • Belgrade • Manchester • Tokyo • Cluj • Tianjin

Editors
Yongliang Qiao
University of Sydney
Australia

Lilong Chai
College of Agricultural and Environmental Sciences
Athens Campus
USA

Dongjian He
Northwest A&F University
China

Daobilige Su
China Agricultural University
China

Editorial Office
MDPI
St. Alban-Anlage 66
4052 Basel, Switzerland

This is a reprint of articles from the Special Issue published online in the open access journal *Sensors* (ISSN 1424-8220) (available at: https://www.mdpi.com/journal/sensors/special_issues/animal_farming).

For citation purposes, cite each article independently as indicated on the article page online and as indicated below:

LastName, A.A.; LastName, B.B.; LastName, C.C. Article Title. *Journal Name* **Year**, *Volume Number*, Page Range.

ISBN 978-3-0365-4035-1 (Hbk)
ISBN 978-3-0365-4036-8 (PDF)

Cover image courtesy of Judah Estrada

Contents

About the Editors

Yongliang Qiao (Research Associate) received B.S.in Electrical Engineering and Automation and M.S. degree in Agricultural Electrification and Automation from Northwest A&F University, Yangling, China, in 2010 and 2013 respectively. He received Ph.D. degree in computer science from the University of Technology of Belfort-Montbéliard, France in 2017. He was a research associate at the Australian Centre for Field Robotics, the University of Sydney, Australia. Since 2021, he has been on the youth editorial board of the journal of Smart Agriculture. His research interests include computer vision, agricultural robots, deep learning, multi-sensor fusion, and pattern recognition.

Lilong Chai (Assistant Professor) is a tenure-track Assistant Professor & Engineering Specialist in the Department of Poultry Science, College of Agricultural and Environmental Sciences at the University of Georgia (UGA) in Athens, USA. Chai is an affiliated member of Phenomics and Plant Robotics Center & Integrative Precision Agriculture team at UGA. He is a leading researcher in the areas of animal environmental engineering & precision poultry farming. Chai's current projects at UGA focus on innovation in machine vision-based methods for monitoring behaviors, health, and welfare indicators of layers and broilers and air emissions mitigation from poultry houses. Prior to joining UGA, Chai was a postdoc in the Department of Agricultural and Biosystems Engineering at Iowa State University (ISU). At ISU, Chai led the field study on commercial farms for two USDA-NIFA programs for improving animal welfare and air quality of commercial poultry houses. Chai's contributions include 110 scientific publications, PI/Co-PI of over 20 grants/contracts, and 20+ awards/honors including the 2021 Sunkist Young Designer Award from American Society of Agricultural and Biological Engineers (ASABE). Chai is currently the Coordinator of Georgia Precision Poultry Farming Conference, and the Vice Chair of ASABE-Environmental Air Quality Committee.

Dongjian He (Professor) received his PhD from College of Mechanical and Electronic Engineering, Northwest A&F University in 1998 and worked as PhD Adviser and Professor in College of Mechanical and Electronic Engineering, Northwest A&F University from 1998 to 2000. He also worked as Senior visiting research fellow at Agric. Info. Technol, Tsukuba University, Japan from 1999 to 2000. Between 2000 and 2009, he worked as Dean, PhD Adviser and Professor of the College of Information Engineering, Northwest A&F University. From 2009 to 2015, he works as Dean, PhD Adviser and Professor of the College of Mechanical and Electronic Engineering, Northwest A&F University and Managing Dean of the Yangling International Academy of Modern Agriculture. His research interests are image analysis and recognition; intelligent detecting and control; agricultural information technology; digital plant and Virtual technology; animal health and welfare.

Daobilige Su (Associate Professor) received his B. Eng. in Mechatronic Engineering from Zhejiang University, China in 2010, M. Eng. in Automation and Robotics from Warsaw University of Technology, Poland and M.Eng. in Automation from University of Genova, Italy through European Master on Advanced Robotics (EMARO) program in 2012, and Ph. D. in robotics at Centre for Autonomous System (CAS), University of Technology Sydney (UTS), Australia in 2017. He was a post-doctoral research associate at Australian Centre for Field Robotics (ACFR), The University of Sydney from 2017 to 2020. He is currently an Associate Professor at College of Engineering, China Agricultural University, China. His current research areas include field robotics, SLAM, computer vision, robot audition and machine learning.

Preface to "Advances in Sensors, Big Data and Machine Learning in Intelligent Animal Farming"

Animal production (e.g., milk, meat, and eggs) provides valuable protein for human beings and animals. However, animal production is facing a number of challenges worldwide such as environmental impacts and animal welfare/health concerns. Maintaining the good health and welfare of livestock and poultry is very important in terms of production efficiency, social economy, and sustainability. In livestock and poultry farming operations, accurate and efficient monitoring of livestock and poultry information can help us to analyze the health and welfare status of animals. Early detection of sick or abnormal individuals can help reduce economic losses and protect animal welfare. In recent years, there has been growing interest in animal welfare. At present, livestock and poultry farming mainly relies on manual observation to obtain animal information, but the method is labor intensive and subjective to human errors. The contact method of implanting devices/sensors into animals to monitor animals' physiological conditions has been tested widely. The concern of this contact method is causing animal stress responses and impacting animal wellbeing. Noninvasive monitoring technologies of computer vision systems can reduce or avoid the impact of observers on animals and related stress response to animals in the monitoring of animal behaviors and welfare, while there is a lack of artificial intelligent strategies, e.g., machine learning or deep learning, that can track animals and extract welfare indicators accurately and quantitatively. Therefore, innovating engineering strategies, such as computer vision-based systems, to identify issues related to animal health and welfare automatically in real-time are critical for enhancing animal production efficiency and welfare. This book therefore aims to gather information and updated research on "Advances in Sensors, Big Data and Machine Learning in Intelligent Animal Farming".

Yongliang Qiao, Lilong Chai, Dongjian He, and Daobilige Su
Editors

Article

Evaluation of an Active LF Tracking System and Data Processing Methods for Livestock Precision Farming in the Poultry Sector

Camille Marie Montalcini [1,*], Bernhard Voelkl [2], Yamenah Gómez [1], Michael Gantner [3] and Michael J. Toscano [1]

1 Center for Proper Housing: Poultry and Rabbits (ZTHZ), Division of Animal Welfare, VPH Institute, University of Bern, Burgerweg 22, 3052 Zollikofen, Switzerland; yamenah.gomez@vetsuisse.unibe.ch (Y.G.); michael.toscano@vetsuisse.unibe.ch (M.J.T.)
2 Division of Animal Welfare, VPH Institute, University of Bern, Längassstrasse 120, 3012 Bern, Switzerland; bernhard.voelkl@vetsuisse.unibe.ch
3 Gantner Pigeon Systems GmbH, 6780 Schruns, Austria; michael.gantner@gantnersolutions.com
* Correspondence: camille.montalcini@vetsuisse.unibe.ch

Abstract: Tracking technologies offer a way to monitor movement of many individuals over long time periods with minimal disturbances and could become a helpful tool for a variety of uses in animal agriculture, including health monitoring or selection of breeding traits that benefit welfare within intensive cage-free poultry farming. Herein, we present an active, low-frequency tracking system that distinguishes between five predefined zones within a commercial aviary. We aimed to evaluate both the processed and unprocessed datasets against a "ground truth" based on video observations. The two data processing methods aimed to filter false registrations, one with a simple deterministic approach and one with a tree-based classifier. We found the unprocessed data accurately determined birds' presence/absence in each zone with an accuracy of 99% but overestimated the number of transitions taken by birds per zone, explaining only 23% of the actual variation. However, the two processed datasets were found to be suitable to monitor the number of transitions per individual, accounting for 91% and 99% of the actual variation, respectively. To further evaluate the tracking system, we estimated the error rate of registrations (by applying the classifier) in relation to three factors, which suggested a higher number of false registrations towards specific areas, periods with reduced humidity, and periods with reduced temperature. We concluded that the presented tracking system is well suited for commercial aviaries to measure individuals' transitions and individuals' presence/absence in predefined zones. Nonetheless, under these settings, data processing remains a necessary step in obtaining reliable data. For future work, we recommend the use of automatic calibration to improve the system's performance and to envision finer movements.

Keywords: low-frequency tracking; commercial aviary; laying hens; false registrations; tree-based classifier; animal behaviour

Citation: Montalcini, C.M.; Voelkl, B.; Gómez, Y.; Gantner, M.; Toscano, M.J. Evaluation of an Active LF Tracking System and Data Processing Methods for Livestock Precision Farming in the Poultry Sector. *Sensors* **2022**, *22*, 659. https://doi.org/10.3390/s22020659

Academic Editors: Yongliang Qiao, Lilong Chai, Dongjian He and Daobilige Su

Received: 14 October 2021
Accepted: 13 January 2022
Published: 15 January 2022

1. Introduction

Tracking technologies generate sequences of chronologically ordered location data and offer a way to monitor movement of many individuals over long time periods with minimal disturbances. Tracking technologies have become valuable for detecting health issues in farm animals at an early stage [1–4] and in cage-free poultry farming, for their potential to select breeding traits that benefit welfare within cage-free systems [5,6] as well as to provide scientific information for optimal management [7]. However, cage-free housings are uniquely complex and may introduce numerous challenges for tracking technologies. For instance, cage-free housings of laying hens often contain a relatively high concentration of material that can interfere with tracking signals, including metal hardware (e.g., perches, floor, feeding lines) and multiple stacked horizontal levels that prevent direct lines of sight require by some automated tracking technologies (e.g., video tracking, infrared). Furthermore, compared to most other livestock, laying hens are relatively small animals that can be housed in large groups at very high densities, which would likely alter ultra-high frequency (UHF) radio signals [8]. Compared to most other commonly tracked livestock (e.g., swine, cattle), laying hens move differently (e.g., flying, jumping between horizontal tiers) and often faster. These challenges might induce measurement errors (as defined by the difference between a measured quantity and its true value), both of random and systematic natures [9]. Random errors are often inevitable and unpredictable, but their effects can be minimized, for example, by increasing the sample size. On the other hand, systematic errors are often predictable with consistent causes (e.g., environmental interference, improper calibration), but their effects are harder to compensate for and can lead to biases if not appropriately addressed during analysis.

Tracking systems have already been used to examine laying hens within the interior of a commercial system [10–12]; these tracking systems had to overcome the housing complexities described above. However, measurement errors were primarily evaluated within less complex settings (e.g., in small interior or outdoor settings) than commercial aviaries but focusing on movements of greater precision (i.e., individual location) than the current effort (transitions between predefined zones). For instance, using an ultra-wide band (UWB) system, Rodenburg et al. [6] reported an accuracy of 85% in detecting individuals' location, and Stadig et al. [13] reported an error of less than 50 cm in 80% of measurements. These results present great potential for tracking systems to represent individual positions within free-range areas, as well as a margin to refine the data. Systematic errors were also investigated, although only within settings less complex than commercial aviaries. For instance, comparing registrations generated by a UWB system against video observations, Sluis et al. [14] observed an average overestimation of 40% of in the distance of broilers moving less than 15 m and an average underestimation of 15% in the distance of broilers moving more than 30 m. Furthermore, Stadig et al. [13] observed a larger error in certain areas of the experimental field and a negative influence of rain on the percentage of successful registrations. These results suggest that various factors, such as the individual level of activity, specific areas, and weather conditions, could cause errors in measurement. Although tracking systems within cage-free housing systems are becoming more popular, they still have challenges to overcome. We therefore studied long-term tracking in commercial aviaries at the level of visited zones (with five zones) instead of precise individual locations. In the current study, we used active tags with low-frequency (LF) tracking and UHF communication that distinguished five zones with key resources, including the three stacked tiers of a commercial aviary (top floor, nest box, lower floor), the littered floor underneath, and an outside covered winter garden. This tracking system is comparable to UWB tracking systems with lower frequencies, with the aim of reducing possible interactions with the environment, such as liquid and metallic materials [15].

To overcome measurement errors, some studies have mentioned novel placement of tracking system components [13,16], filtering of registrations that are not possible [13], or filtering of individual positions that do not move more than the 95% confidence interval of the system's positioning errors [17]. When modifying the configuration of the tracking system is not an option, data processing may be the only alternative to refine and, in some cases, obtain validated data. Furthermore, tracking data often contain metadata associated to each registration, which could be used to detect false registrations and increase accuracy. Due to a potentially large number of available features and interaction effects, manually defining a rule-based algorithm can be time-consuming and suboptimal, whereas machine learning may offer a valuable solution for filtering false registrations. Despite potential for data refinement, there are only a few studies on UWB systems and related technologies that scrutinize data-processing methods, particularly within the unique settings of housings of cage-free laying hens. In the current study, we aimed to contribute to the collective effort of evaluating tracking systems for laying-hen farming, with a focus on the interior of a commercial aviary system. To achieve this aim, two analysis steps were involved. First, two data-processing approaches were applied to filter false registrations, including a simple deterministic approach that filters stays of short durations (SD method) and a machine learning approach (ML method) based on a tree-based classifier. The two processed datasets and the unprocessed dataset were compared against video-observation results (our gold standard). This evaluation was conducted in terms of the number of transitions per individual within the predefined zones and individuals' presence/absence in each zone every second. Secondly, to better evaluate the tracking system, we studied the effect of filtering false registrations based on the ML method over a two-month period on 144 tracked animals under three potential influencing factors: different areas of the aviary, external temperature, and external humidity. We selected these factors because they have already shown to be associated, to some extent, with tracking-system performance and could introduce biases in our own work and that of others using comparable technology if associated with false-registrations.

2. Materials and Methods

2.1. Ethical Statement

The study was conducted according to the cantonal and federal regulations for the ethical treatment of experimentally used animals and approved by the Bern Cantonal Veterinary Office (BE-45/20).

2.2. Animals and Housing

As part of a larger study examining effects of on-farm hatching, approximately 4800 chicks were reared in an Inauen Natura rearing barn previously described by Stratmann et al. [18] and located at the Aviforum facility in Zollikofen, Switzerland. At seven days of age, focal animals were selected, and at approximately 16 weeks of age, all animals were transferred to an on-site commercial laying barn containing a Bolegg Vencomatic Terrace aviary. The aviary system is split into 20 identical pens separated by a vertical grid, with each pen containing 225 animals and an outside, covered winter garden that can be accessed through a pop hole (illustrated in Figure 1). Eight of the 20 pens were used for the current study, with 18 focal animals per pen (a total of 144). On the same day as the transfer to the laying barn, we mounted a tracking tag enclosed within a cloth backpack (mass: 15.6 g; height: 14.5 cm; width: 13 cm) on the back of each focal hen. These backpacks were identifiable from video cameras based on their unique colour combination.

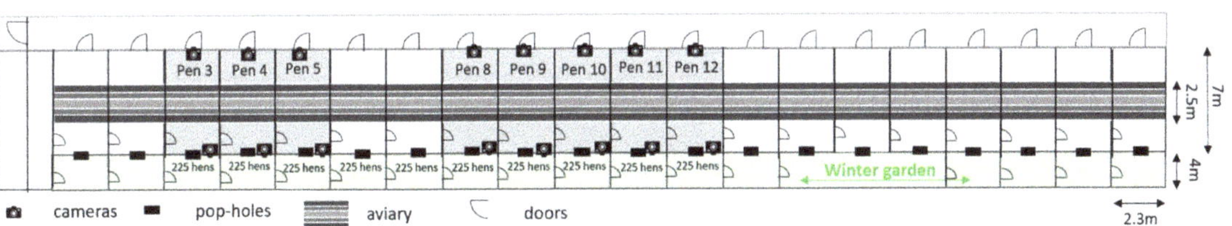

Figure 1. Housing setup, including the pens and aviary location in the barn, winter-garden zone, pop holes, and cameras.

2.3. Tracking System

To track hens across different areas within a pen, we distinguished five zones with key resources, including the three stacked tiers of a commercial aviary (top, nest box, lower), the littered floor underneath, and the winter garden, as illustrated in Figure 2A. During the laying phase, transitions between the five zones were assessed continuously for each focal hen by means of a customized tracking system. For this, three identical stations of a low-power, active tracking system (®Gantner Solutions GmbH, Schruns, Austria) were installed within the laying barn, each covering either two or three pens (Stations 3–5: pens 3, 4, 5; Stations 8–9: pens 8, 9; Stations 10–12: pens 10, 11, 12). Each station involved several components, including five markers (1 per zone) emitting signals through a cable (creating separately enclosed fields for each zone; Figure 2B); active tags (mass: 28.1 g) that can receive signals; and lastly, a reader that communicates through UHF (868 MHz), with the tags and a dedicated computer.

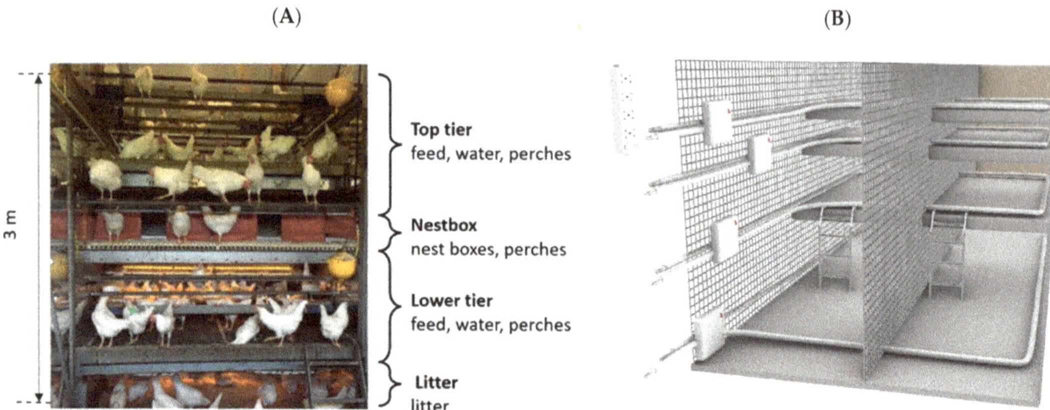

Figure 2. (**A**) Side view of the aviary within one pen with its three aviary zones (top tier, nest box, lower tier) and the littered floor; (**B**) simplified 3D model of the tracking system covering two pens, including the four markers of the four indoor zones and their associated cables.

The receiving strength of the LF signal (RSS) is used to determine theoretical distance to the antenna loop. At almost every position in any zone, the tag can receive the signal of multiple markers. This signal last for 50 ms. It is important that during that time, the tag only receives the signal of one marker; otherwise, signals would overlap and might not be valid. Therefore, the markers send at different transmission intervals (varying from 1.6 to 2.1 s depending on the zone) a fixed low carrier frequency signal of 0.125 MHz (LF-signal) that is modulated to allow markers to be differentiated. Within a 10-s interval, a tag could theoretically receive between five and six signals per marker, but this number will often be lower, as every marker has a maximum range of only two to three metres. Every time a tag receives an LF-signal, an algorithm (tag-algorithm) is applied to the registered LF signals received within the past 10 s to evaluate whether the tagged hen has transitioned to a new zone. The tag algorithm reports a new transition when a tag receives the absolute strongest signal value from the same marker twice within 10 s and if the associated zone differs from the last registered zone (pseudo-code in the Supplementary Text S1). Following the installation of the tracking-system stations, each pen was calibrated under field conditions to ensure a correct interpretation of information obtained by the devices. More specifically, a tracking tag was positioned in each of the 44 predefined critical locations per pen (e.g., where two zones border one another) to evaluate RSS against observed distance to the antenna loop and to adjust the LF signal of specific markers as necessary.

Individual transitions to a zone registered by the tracking system are hereafter called registrations. More specifically, we will refer to correct registrations (CR) for registered zones where the animal is located (i.e., true zone as determined by video) and to false registrations (FR) for registered zones not consistent with the true zone for the bird (FR). Among CRs, we distinguish two types of registrations: (1) registrations that are not associated with a true transition (corrected registrations) and (2) registrations associated with a true transition (transitional registrations). Our goal was to obtain only transitional registrations, and data processing was used towards this objective.

2.4. Video Observations to Detect False Registrations

Two cameras per pen were placed within the indoor portion of each pen in such a manner that each location where an animal could transition between any of the three indoor zones was visible. The view did not cover the interior of the pop hole nor the winter garden and thus did not allow transitions to the winter garden to be filmed. For the generation of the video-based tracking data as a gold standard, video data were collected over the third and fourth weeks for an 11-day period simultaneously with the collection of the tracking data. Single animals were visually tracked by two trained observers independent of one another in order to classify each registration as FR or CR. An inter-rater reliability test between the two observers for 137 registrations, including four random hens and four different days, resulted in an inter-rater reliability of perfect agreement, with all recordings classified correctly by both observers.

For the evaluation of the two processing methods (SD and ML) and the unprocessed tracking data against video-based tracking data as the gold standard, two sets of registrations were analysed through video, generating two datasets: (1) the training dataset used to develop the ML method; and (2) the test dataset used to evaluate the two processing methods, as well as the unprocessed tracking data, against video-based tracking data. As described in the Section 2.5, the training dataset was used in a cross-validation process to split the data into validation and training sets and select for the optimal models.

The training dataset was composed of 4274 registrations classified as FR or CR by means of 241 h of video observations divided into 79 batches, varying from 0.5 h to 7 h, involving 44 tracking tags over 11 days. The batches were selected based on the visual representations of individuals' movement across all days to ensure a broad variation of movement sequences and a reasonable number of observations across zones, stations, and tracking tags. To avoid introducing noise in model training, the training dataset did not contain registrations from the winter-garden zone due to the limited camera view in the pop holes described earlier. The training dataset comprised 13% FR and 87% CR.

The test dataset was composed of 865 registrations classified as FR or CR by means of 96 h of video observation. More specifically, 48 batches (six/pen) of 2-h video (including 47 randomly selected tracking tags) were randomly chosen over six days and reduced to 42 batches due to technical issues (e.g., backpacks not visible from the cameras). As the test dataset was used to evaluate two processing methods, including one that did not require training, the test dataset contained registrations from each of the five zones, including the winter garden. However, as the classifiers can only be tested on classes included in the training process, all registrations from the winter garden were processed solely by the SD method. Registrations in the winter garden were retained in the evaluation of both processing approaches for two main reasons: first, to avoid any bias towards poorer/greater performance of the SD method, if that zone would be more easily/laboriously detected by the tracking system compared to the other zones; second, even if the winter-garden zone is processed by the SD method when evaluating the ML method, its performance is still influenced by the ML method, typically when the ML method filters a registration to the litter zone reported between registrations in the winter-garden zone (as there would be one less transition to the winter garden). When a registration to the winter-garden zone could not be clearly classified through video observation (i.e., animal could be either in the pop hole or the winter garden), CR was used for biological relevance. We decided to define the pop-hole area (illustrated in Figure 1) as part of the winter-garden zone (and not the litter zone), as exposure to natural light in the pop hole is more similar to the winter-garden zone than the litter zone. To better evaluate the tracking system, in addition to the tracking system's registrations, the test dataset contained all true transitions observed during video observations that were not reported by the tracking system (missed transitions). Missed transitions represented 0.6% of the test dataset. The test dataset comprised 5% FR and 95% CR.

2.5. Evaluation of the Two Data Processing Methods

As the tracking system used in this study evaluated the location of a tag every time the tag received an LF-signal, longer records have more opportunities for self-correction and therefore are more likely to be accurately record the location. Therefore, an intuitive and simple way to process the data is to filter all registrations that last for less than a certain threshold (SD method). We used a one-minute threshold with the objective of minimizing loss of actual transitions while maintaining a good representation of the true data.

To account for more of the available information during data processing, we used a machine learning approach (ML method) based on decision-trees, which, in addition to the registration duration used by the SD method, employed 13 features of the registrations (detailed in Table 1), including the RSS, the zone, and the station identities. The zone identities of the previous and next registrations (of the same tag) were also included to account for the movement sequence. The durations of the previous and next registrations were also included, as we expected the duration to be the most important feature for detecting FR. Our goal was to build a model to process (clean) the data rather than generate predictions about hen movement patterns. We aimed to isolate the true signal of hen movement, which can be used in future research to evaluate the drivers of hen behaviour. As such, our model is independent of external factors that could be of potential interest for future investigation (e.g., weather). Three classifiers (random forest, gradient boosting, CatBoost) based on decision-trees [19] were used to account for potential non-linearity and interaction effects [20]. The gradient-boosting classifier is a greedy algorithm that sequentially trains a shallow decision tree in order to correct the errors of the previously trained tree [21], and the CatBoost model is a recently developed gradient-boosting algorithm [22,23] that was selected in this study for its ability to process categorical features during training (algorithms of the classifiers further detailed in Supplementary Text S2). Following hyperparameter selection through a 3-fold cross-validated grid search (detailed in Table S1 of the Supplementary Materials) and model training on the training dataset, the performances of the classifiers were evaluated on the held-out test dataset using three common classifier performance measures [24]: (1) accuracy, defined as the fraction of predictions correctly classified by the model; (2) precision of class X, defined as the proportion of the predicted class X that is correctly classified by the model; and (3) recall of class X, defined as the proportion of the observed (true) class X, that is correctly classified by the model. To better contrast predictionsof the three tree-based classifiers on the test dataset in order to select one for the ML method, we used McNemar's non-parametric test for pairwise binary classifier comparison [25] to test the null hypothesis that two models have similar proportions of errors. The normalized importance of features was generated for the selected model to understand the model's reliance on each feature when producing its predictions. Finally, the ML method used the selected classifier to classify registrations as FR and CR and then filtered FR from the unprocessed data. However, due to the limitations of video in covering the pop-hole area, the SD method was applied here to filter registrations in the winter-garden zone.

Table 1. Record features used to train the model and the normalized importance of features in the final CatBoost model.

Feature Name	Description
previous zone; zone; next zone	zone identity of the previous/considered/next registered record with the strongest LF signal, indicating the zone where the individual has transitioned/is transitioning/will transition to
RSS	a measurement of the power present in the strongest received LF signal (dB)
tracking system ID	identity of the tracking-system copy
previous duration; duration; next duration	reported time of stay in the zone from the previous/considered/next registered record
zone2	second zone identity with the strongest LF signal
RSS of zone2	a measurement of the power present in the second strongest received LF signal (dB)
zone3exist	binary feature that equals 1 if the tag registers a signal of at least three different zones during the last 10 s, and otherwise equals 0
next2zone = zone; previous2zone = zone	binary feature that equals 1 if the registered second zone from the next/previous record is the same as the occurring zone, and otherwise equals 0

We contrasted the two data-processing approaches by applying them to the unprocessed test dataset (i.e., including CR and FR). The resulting two processed datasets (ML and SD datasets), as well as the unprocessed test dataset, were then compared against the respective gold-standard dataset (i.e., registrations identified as CR through video observation). In each case, we evaluated two things: (1) the animal's location (or more specifically, their presence/absence in each zone) and (2) the animal's movement. To evaluate how well these datasets represented individuals' presence/absence in each zone at each second, we compared their associated categorical time series (containing five categories, one for each zone). The performance was evaluated in terms of accuracy, macro-averaged recall, and macro-averaged precision (where the macro-averaged recall/precision is the average of the recall/precision across each zone). To evaluate how well these datasets represented individuals' movement, we compared the total number of individual transitions per batch, per zone in each case. Performance was evaluated with the explained variance score (EV) and the mean absolute error (MAE), defined as:

$$EV = 1 - \frac{variance\{y_{GS} - \hat{y}\}}{variance\{y_{GS}\}}, \quad MAE = \frac{1}{n_{samples}} \sum_{i=0}^{n_{samples}-1} \left| y_{GSi} - \hat{y}_i \right|$$

where $\hat{y}$ contains information from a processed dataset and y_{GS} contains the respective gold-standard information. The EV is used to measure the magnitude of the expected effect on the number of transitions [26]. The MAE is used to measure, in an unambiguous and natural manner, the magnitude of the expected average error [27] in terms of the number of transitions (for a two-hour batch). This analysis was performed with Python version 3.8.5 using the SciKit Learn package [28] for the performance measures and the CatBoost package [22] for the CatBoost classifier.

2.6. Investigation of Influencing Factors

When comparing large datasets with thousands of hours of tracking per animal, comparison with video recordings as a gold standard becomes impractical. Therefore, to further evaluate the tracking system, we used the tree-based classifier from the ML method to identify FR (IFR) and studied the estimated error rate, defined as the number of IFRs against the total number of registrations, in relation to specific factors. The estimated error rate had a value of one when all records were filtered by the ML method and a value of zero when none was filtered. This approach has some limitations due to probable FRs not being detected or some being falsely detected. However, by removing the limitation on the number of days and individuals used, a broader investigation of the systems' performance can be conducted. Data processing with the ML method is shown in the Results section to filter most of the true FRs (recall of class FR: 93%) and to filter mostly true FRs (precision of class FR: 84%). Therefore, IFRs should highlight most of the FRs from the unprocessed data and should be composed mainly of FRs. We applied the ML method over a two-month period, involving 144 animals, during which the hens were kept under similar management conditions every day, including 15 h of artificial light and six hours with access to the winter garden. To avoid biasing the data towards a greater error rate when the winter garden was closed, we excluded all registrations of transitions to the winter garden for periods when it was closed. We evaluated the estimated error rate in relation to different areas by reporting the mean $\pm$ SD of the estimated error rate across individuals for each of the five zones in each of the eight pens (40 pen-zone areas). We evaluated the estimated error rate in relation to external weather variables by fitting a mixed-effects logistic regression (link function: logit, R package "lme4" [29]) on the ratio of IFR to the total registrations minus IFR (per hour), with pen identity nested in station identity as a random term and hourly external humidity (%) and temperature (°C) as explanatory variables. External humidity was rescaled by dividing its values by 10. To control for variations barn management and animal behaviour throughout the day, the hour of the day was also added as a fixed effect. External humidity and temperature were obtained from the LSZB weather station (~12 km from the barn) and accessed via the Wolfram alpha API in Python.

3. Results

3.1. Evaluation of the Two Data Processing Methods

On the test dataset, the three classifiers showed stable (over 100 random seeds) accuracy, recall, and precision (Figure S1 of the Supplementary Materials), and the McNemar's test showed a similar proportion of errors between each classifier ($p > 0.05$). Thus, with an accuracy of 99%, we selected the CatBoost algorithm for the ML method because of its ability to handle categorical variables in Python. Additionally, 84% of the time that the model identified an FR, the model prediction was correct (precision of class FR). and 100% of the time that the model identified a CR, the model prediction was correct (precision of class CR). Additionally, 93% of the FR observations were classified by the model as FR (recall of class FR), and 99% of the CR observations were classified by the model as CR (recall of class CR). The zone identity, RSS, and the previous registrations' zone identity were the three most important features, accounting for 21%, 19%, and 13% of the overall importance of the features, respectively, while duration accounted for 7% (Figure 3A). To further illustrate the importance of the features, Figure 3B show the RSS and duration of the test dataset's registrations, split into CR and FR (from video observations). The receiving strength of the LF signal was generally higher for the correct registrations of all indoor zones. We also observed longer duration of stay to be more frequent among the correct registrations, with the exception of registrations in the lower perch zone, where no difference in the duration of stay was observed between correct and false registrations.

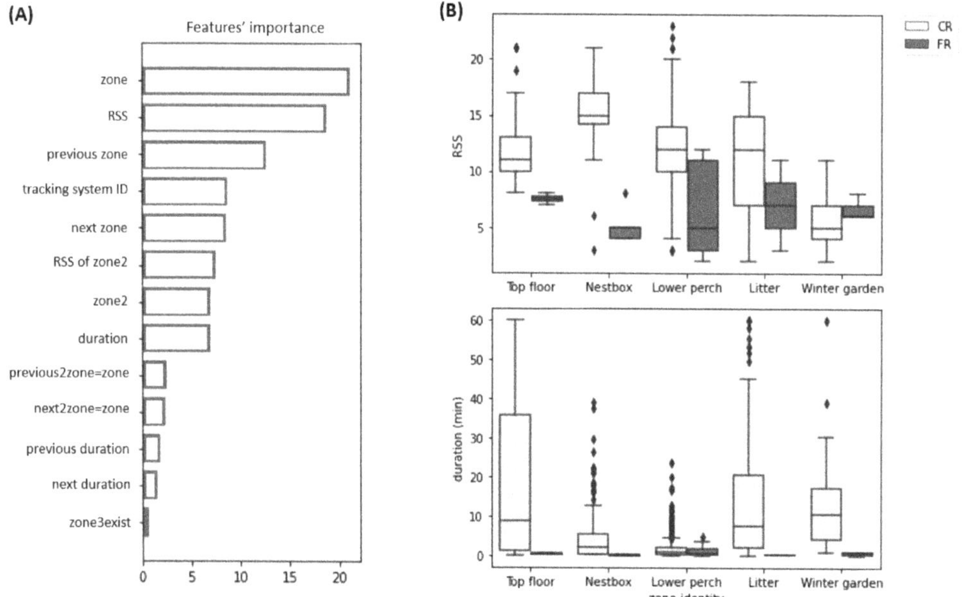

Figure 3. (**A**) Normalized importance of features for the selected CatBoost model. (**B**) Box plots of the RSS (top) and duration of stay (bottom) of the test dataset registrations, split into CR (light grey) and FR (grey), as produced by video observations, are displayed for each zone.

The unprocessed, SD and ML datasets all determined an individual's zone (every second), with an accuracy of 99%, 98%, and 100%, respectively, and displayed the same values (99%, 98%, and 100%, respectively) for the macro-averaged precision and macro-averaged recall. We found the ML and the SD datasets to underestimate the number of transitions by an average 0.27 and 0.06 transitions per zone, respectively, for a two-hour batch, in contrast to the unprocessed dataset, which overestimated the number of transitions by approximately 0.5 transitions per zone, on average, for a two-hour batch (average number of transitions per batch, per zone by video observation was 1.8). The percentage of variance of the ground-truth data recovered by the unprocessed, SD and ML datasets was 23%, 91%, and 99%, respectively, which is further illustrated in Figure 4.

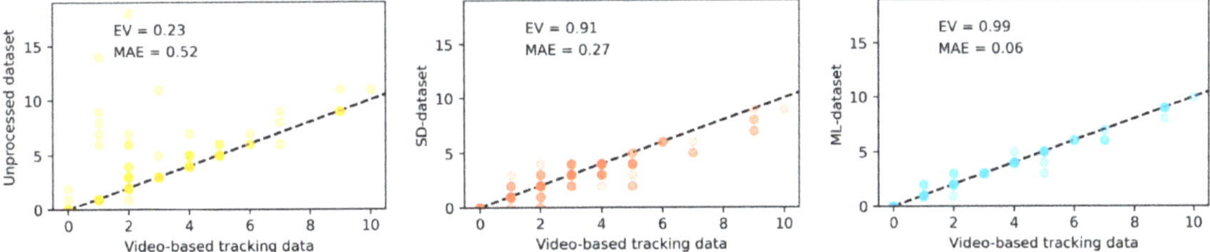

Figure 4. Number of transitions per individual (per batch, per zone) for the unprocessed, SD and ML datasets against video-based tracking data and associated *EV* and *MAE* scores. Overlapping data points are represented by darker shading.

3.2. Investigation of Influencing Factors

The estimated error rate across pen-zone areas varies from 0.0 ± 0.0 (e.g., litter area within each pen of Stations 10–12) to 0.5 ± 0.19 for Pen 8, suggesting that half of the registrations in the winter garden from Pen 8 were filtered by the ML method. The estimated error rate per pen-zone area is further detailed in Figure 5. Furthermore, we found a negative effect of humidity ($p = 0.003$) on the estimated error rate, with an odds ratio of 0.96 (95%-CI [0.94, 0.99]), indicating a 4% lower likelihood of obtaining a false registration with an increase in humidity of 10%. Additionally, we found a negative effect of temperature ($p < 0.001$) on the estimated error rate, with an odds ratio of 0.97 (95%-CI [0.96–0.98]), indicating a 3% lower likelihood of obtaining a false registration with an increase in temperature of 1 °C (for further details, see Table S2 of the Supplementary Materials). The difference between the unprocessed and the processed data (by the ML-method) is further illustrated in Figure 6 through a visual representation of an animal's transitions over eight consecutive days. For instance, observed several transitions filtered by the ML method between the lower-perch and top-floor zones.

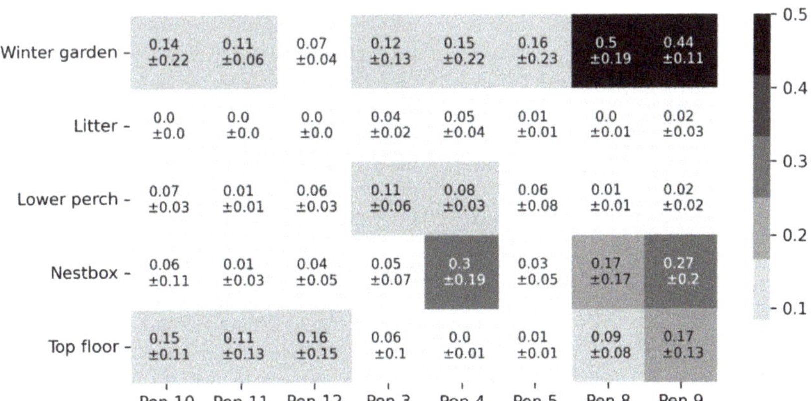

Figure 5. Estimated error rate (mean ± SD) for pen-zone areas.

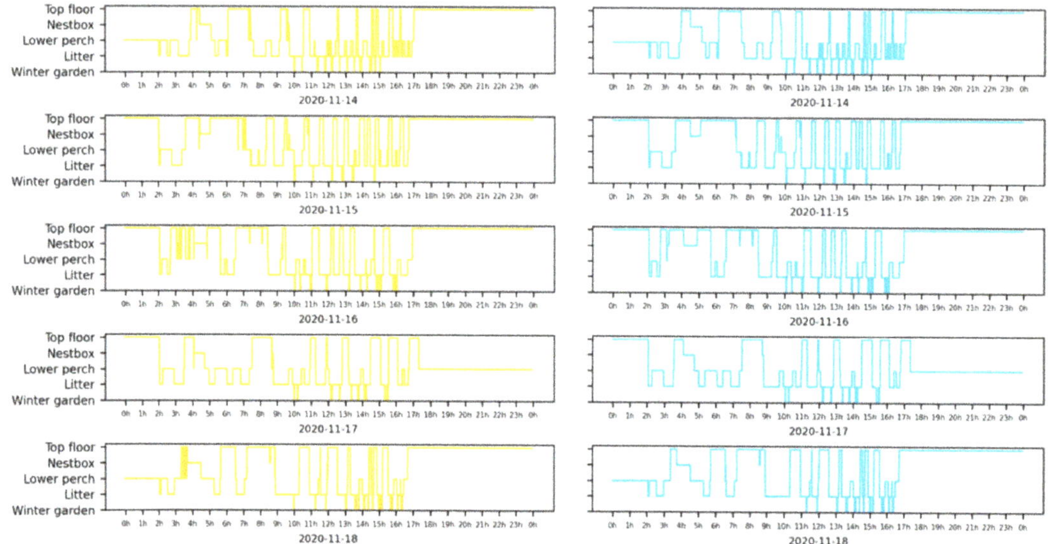

Figure 6. Unprocessed data (yellow) and processed data by the ML method (blue) of a single animal over five consecutive days. Each row represents a single 24-hour day, with the zone identities represented on the y-axis (the indoor zones are ordered following the aviary order, and the winter garden is represented below the indoor zones).

4. Discussion

We found the presented LF tracking system accurately determined the presence of animals in a given zone (at the second level), with macro-averaged precision, and macro-averaged recall of 99% when compared against video observations of the test dataset. This good performance might be explained by the tag algorithm, which searches for new transitions, on average, every 0.5 s (i.e., each time a tag receives an LF signal), thus regularly providing opportunities for correctional records. However, the number of transitions in a zone generated by the tracking system was overestimated and only explained 23% of the true variance (as observed by video). Therefore, the unprocessed tracking data did not constitute a good representation of individual transitions between the five zones, which could be emphasized by the observed differences in the estimated error rate within specific pen-zone areas. On the one hand, we observed clear differences in the estimated error rate of a given zone across different stations (e.g., winter-garden zones in Stations 10–12, Stations 3–5, and Stations 8–9 had a mean estimated error rate varying, across their respective pens, between 0.07 and 0.14, 0.12 and 0.16, and 0.44 and 0.5, respectively). On the other hand, we observed differences within pens of the same station (e.g., nest-box zone in Stations 3–4 had an estimated error rate of 0.05 ± 0.07 in Pen 3, 0.3 ± 0.19 in Pen 4, and 0.03 ± 0.05 in Pen 5). The observed differences in the estimated error rate across different pen-zone areas aligned well with locations described through anecdotal notes made during video observations describing precise locations where a tracking tag generated a high amount of FR (by repeatedly switching between two, sometimes non-neighbouring zones) while the animal was immobile (weak spots). An explanation for the existence of weak spots may be the pen furnishing blocking the line of sight between tags and signal cables, which is known to cause signal interference in UWB systems [16]. More specifically, metallic materials can absorb the signal and distort the electromagnetic field, which could either block or enhance the signal, rendering RSS a poor representation of the distance to the signal cable, possibly explain errors between non-neighbouring zones. Our tracking system was designed to use a lower frequency than a common UWB system in order to avoid possible interactions with metallic materials, although signals may still be affected.

Furthermore, the existence of weak spots may be attributed to the calibration process, a manual and time-consuming step performed independently for each station and iteratively through each pen. When a tracking tag was detected in an incorrect zone during the calibration process, the LF values of specific markers were adjusted. As the pens are steel cages, the LF field generated by each marker can be slightly inhomogeneous. As a result, when an LF signal value is adjusted, all measurements must be repeated to ensure that the change in the LF value did not lead to further detection errors. The difficulty lies in setting the LF values of the markers in such a way that the correct zone is detected in all locations of the tracking system. In particular, the nest-box zone is a small zone located between two zones (Figure 2), and a change in the LF value of the marker had a greater effect on the neighbouring zones because it quickly led to the tracking tag being detected in the incorrect zone. Therefore, we recommend the use of an automatic calibration process to improve the system's performance. To achieve this, within each zone, several tags would be placed at predefined locations of critical measurement points. Each signal strength received by any tag from any marker would be registered. An algorithm would be executed every 10 s (ensuring enough time for adjustment of LF signal values to take place in the field) on all RSS registered within the past 10 s. This algorithm would identify the most problematic zone, defined, for example, by the zone with the smallest dB difference in relation to another zone (across all tags in that zone). If this difference does not exceed the limit of 1 dB in relation to another zone, the LF signal value of the associated marker is automatically adjusted. As soon as 100 consecutive runs induce no adjustment of an LF signal, the calibration is complete. An automatic calibration would save time as only one person would work on the calibration. This would also offer new opportunities, such as smaller zones, allowing for registration of finer movements. For instance, in our settings, it might be possible to differentiate between the nest boxes and the balcony in front the nest boxes (currently, both are registered as the nest-box zone). Furthermore, automatic calibration would ensure more homogeneous LF values across markers from the same zone across all pens, and consequently, more comparable datasets across different stations and pens would be generated.

Our tracking system's poor performance in representing individual transitions highlights the importance of processing automatically generated datasets. Relevant data-processing studies are lacking, although they could help to standardize this process to generate comparable datasets across different studies. The benefit of this work is most essential in light of rapid development in technology in order to manage and improve the welfare of animals within commercial livestock systems [1,2,30–35]. We showed that the data processed by a simple filtering of registrations associated with short durations (<1 min) of stays was suitable for monitoring the number of transitions per individual per zone, accounting for 91% of the actual variance (as observed by video). We further reported a gain in performance using a tree-based classifier to filter false registrations, accounting for 99% of the true variance in the number of transitions per individuals, which could partly be explained by the additional information provided to the ML method. Indeed, zone identity and RSS were the two main features upon which the tree-based classifier based its predictions, while the SD method was based solely on the duration of the stay. Interestingly, this also suggests that our expectation of the record's duration being the most important feature to detect FR was incorrect when other features are included. The current study did not allow for this comparison when a single feature is used; however, further studies using a simple rule-based approach should consider the RSS addition to the records' duration. The importance of features further suggests that the zone identities of the previous and next registered record are of greater importance than the duration of stay from the previous and next registrations. Results concerning the importance of can offer direction on how to improve similar tracking systems, for instance, by including a threshold of RSS values for each zone based, for example, on the result of an automatic calibration. Another possibility would be to include the SD method as part of the tag algorithm, although this would eliminate the possibility of registering fast transitions between two zones (<1 min).

The ML method required additional efforts, for instance, more video observations, compared to the SD method and statistical modelling in R or Python. Therefore, the choice between both methods relies on a compromise between time accorded in data processing and the performance of the processed data. In the current study, the SD method recovered 8% less of the true variance in the number of transitions per individuals than to the ML-method. To put this value in context, we used simulated sampling to estimate the impact of a comparable loss on the effect size (measured by the Pearson correlation) of a simulated movement variable, M, on a simulated health variable, H. The two simulated variables (M and H) followed a standard normal distribution, with a Pearson correlation coefficient varying from 0.15 to 0.40 to cover potentially interesting ranges of effect sizes when studying movements in relation to health [36,37] and heritability of behaviour [38–40]. By adding noise to M (and calling the result M'), we estimated (over 10,000 simulations, per sample size) the percentage of cases where significance would be lost ($p > 0.05$), depending on the initial effect size and sample size. Our estimations suggest that a change in percentage of the initial variance explained by M′ from 0.99 to 0.91 would change the significance of a critical test in 26% (or 25%) of cases when applying a sample size of 80 (or 120) and an initial effect size of 0.25 (or 0.2), respectively (see details in Figure S2 of the Supplementary Materials). Therefore, using a tree-based classifier to filter false registrations can be greater value for studies with small sample and effect sizes (e.g., $n = 120$, effect size of 0.2) than the filtering approach using stays of short duration as threshold. For large sample sizes or samples with strong correlations between the measured movement and the trait of interest, the SD method might produce equally reliable results as the ML method.

Our results further reported a marginal effect of periods of time characterized by higher humidity or higher temperature, associated with a lower estimated error rate of transitions to the winter garden. Because air is our medium of signal transmission, when humidity is changes, the magnetic field is also expected to change. As calibration was conducted in August 2020, the performance of the tracking system may be optimized for a period with higher temperature than average. Additionally, as Richards et al. [41] reported, associations between daily weather conditions and mean pop-hole usage in laying hens, including an increase in mean pop-hole usage associated with an increase in temperature, and the influence of the weather conditions on animal behaviour may be explanatory. In spite of these results, external environmental factors cannot be controlled for and are part of the experiment. However, these results can aid in interpretation and awareness of possible limitations for subsequent analyses of these or similar tracking data.

5. Conclusions

The active LF tracking system evaluated in this study determined the presence/absence of birds in a zone with an accuracy of 99% but overestimated the number of transitions by birds per zone, explaining only 23% of the true variation (as observed by videos). However, we showed that filtering stays of short durations rendered the data suitable for monitoring the number of transitions per individual, explaining 91% of the true variation, and that the use of a tree-based classifier to filter false registrations recovered an additional 8% of the true variation. Simulations further suggested that a machine learning approach for data processing could be of greater value than a simple deterministic approach in studies with small sample and small effect sizes. Results also suggest that filtering false registrations may reduce the effect of systematic errors towards certain pen-zone areas and towards periods of time characterized by lower humidity or temperature values. However, results also suggest that these factors might, to some extent, remain in the processed data and should be considered properly in subsequent analyses. In conclusion, this tracking system is well suited for complex indoor housing (similar to commercial aviaries) to measure the transitions of individuals and the presence/absence of birds in predefined zones (thus, duration of stays in zones). Nonetheless, under these settings, data processing remains a necessary step in obtaining reliable tracking data. For future work, we recommend the use of automatic calibration to improve the system's performance and to envision finer movements.

Supplementary Materials: The following are available online at https://www.mdpi.com/article/10.3390/s22020659/s1, Table S1. Selected grid-search parameters for the random forest, gradient boosting, and CatBoost classifiers. When a parameter is not applicable for a specific classifier, the "-" notation is used. Table S2. Output of the logistic regression with the proportion of IFR to the number of registrations minus IFR as response variable and the humidity, temperature, and hour of day as fixed effects. Figure S1. Precision per class (0: FR; 1: CR), recall per class (0: FR; 1: CR), and accuracy of the three classifiers over 100 random seeds. Figure S2. Percentage of simulations that lost significance ($p > 0.05$) of associated initial effect size (measure by Pearson correlation between two simulated samples from a normal distribution: M′ and H) after a change in percentage of the true variance recovered by M′ from 0.99 to 0.91, depending on the initial effect size (varying from 0.16 to 0.4) and sample size (varying from 80 to 280).

Author Contributions: Conceptualization, C.M.M., B.V., Y.G. and M.J.T.; methodology, C.M.M., B.V. and Y.G.; formal analysis, C.M.M.; investigation, C.M.M. and M.J.T.; writing—original draft preparation, C.M.M. and M.J.T.; writing—review and editing, C.M.M., B.V., Y.G., M.G. and M.J.T.; visualization, C.M.M.; supervision, M.J.T.; funding acquisition, M.J.T. All authors have read and agreed to the published version of the manuscript.

Funding: This research was funded by the Swiss National Science Foundation, grant number 310030_189056. The APC was funded by the University of Bern.

Institutional Review Board Statement: The animal study protocol was approved by the Canton of Bern (BE45/20, date of approval: 28 April 2020).

Informed Consent Statement: Not applicable.

Data Availability Statement: The full code and video observations results (test and training datasets) can be found on a public GitHub repository (https://github.com/cam4ani/PhD-AnimalWelfare/tree/main/Chapter0-GantnerCleaning (accessed on 12 October 2021)).

Acknowledgments: We would like to thank Masha Marincek and Doriana Sportelli for conducting all video observations, as well as Abdulsatar Abdel Rahman for technical support throughout the project and assistance with catching and handling of hens. We also are grateful to Aviforum staff for providing animal care.

Conflicts of Interest: The authors declare no conflict of interest. The funders had no role in the design of the study; in the collection, analyses, or interpretation of data; in the writing of the manuscript; or in the decision to publish the results.

References

1. Schillings, J.; Bennett, R.; Rose, D.C. Exploring the Potential of Precision Livestock Farming Technologies to Help Address Farm Animal Welfare. *Front. Anim. Sci.* **2021**, *2*. [CrossRef]
2. Berckmans, D. Precision Livestock Farming Technologies for Welfare Management in Intensive Livestock Systems. *Rev. Sci. Tech. Off. Int. Epiz* **2014**, *33*, 189–196. [CrossRef]
3. Herlin, A.; Brunberg, E.; Hultgren, J.; Högberg, N.; Rydberg, A.; Skarin, A. Animal Welfare Implications of Digital Tools for Monitoring and Management of Cattle and Sheep on Pasture. *Animals* **2021**, *11*, 829. [CrossRef]
4. Weary, D.M.; Huzzey, J.M.; von Keyserlingk, M.A.G. Board-Invited Review: Using Behavior to Predict and Identify Ill Health in Animals. *J. Anim. Sci.* **2009**, *87*, 770–777. [CrossRef]
5. Ellen, E.D.; van der Sluis, M.; Siegford, J.; Guzhva, O.; Toscano, M.J.; Bennewitz, J.; van der Zande, L.E.; van der Eijk, J.A.J.; Haas, E.N.; Norton, T.; et al. Review of Sensor Technologies in Animal Breeding: Phenotyping Behaviors of Laying Hens to Select Against Feather Pecking. *Animals* **2019**, *9*, 108. [CrossRef]
6. Rodenburg, T.B.; Bennewitz, J.; de Haas, E.N.; Košťál, L.; Pichová, K.; Piette, D.; Tetens, J.; van der Eijk, J.; Visser, B.; Ellen, E.D. The Use of Sensor Technology and Genomics to Breed for Laying Hens That Show Less Damaging Behaviour. In Proceedings of the 8th European Conference on Precision Livestock Farming, Nantes, France, 12–14 September 2017.
7. Li, L.; Zhao, Y.; Oliveira, J.; Verhoijsen, W.; Liu, K.; Xin, H. A UHF RFID System for Studying Individual Feeding and Nesting Behaviors of Group-Housed Laying Hens. *Trans. ASABE* **2017**, *60*, 1337–1347. [CrossRef]
8. Triguero-Ocaña, R.; Vicente, J.; Acevedo, P. Performance of Proximity Loggers under Controlled Field Conditions: An Assessment from a Wildlife Ecological and Epidemiological Perspective. *Anim. Biotelem.* **2019**, *7*, 24. [CrossRef]
9. Dodge, Y. *The Concise Encyclopedia of Statistics*; Springer Science: Berlin/Heidelberg, Germany, 2008.
10. Rufener, C.; Abreu, Y.; Asher, L.; Berezowski, J.A.; Maximiano Sousa, F.; Stratmann, A.; Toscano, M.J. Keel Bone Fractures Are Associated with Individual Mobility of Laying Hens in an Aviary System. *Appl. Anim. Behav. Sci.* **2019**, *217*, 48–56. [CrossRef]

11. Rufener, C.; Berezowski, J.; Maximiano Sousa, F.; Abreu, Y.; Asher, L.; Toscano, M.J. Finding Hens in a Haystack: Consistency of Movement Patterns within and across Individual Laying Hens Maintained in Large Groups. *Sci. Rep.* **2018**, *8*, 12303. [CrossRef] [PubMed]
12. Sibanda, T.Z.; Walkden-Brown, S.W.; Kolakshyapati, M.; Dawson, B.; Schneider, D.; Welch, M.; Iqbal, Z.; Cohen-Barnhouse, A.; Morgan, N.K.; Boshoff, J.; et al. Flock Use of the Range Is Associated with the Use of Different Components of a Multi-Tier Aviary System in Commercial Free-Range Laying Hens. *Br. Poult. Sci.* **2019**, *61*, 97–106. [CrossRef]
13. Stadig, L.M.; Ampe, B.; Rodenburg, T.B.; Reubens, B.; Maselyne, J.; Zhuang, S.; Criel, J.; Tuyttens, F.A.M. An Automated Positioning System for Monitoring Chickens' Location: Accuracy and Registration Success in a Free-Range Area. *Appl. Anim. Behav. Sci.* **2018**, *201*, 31–39. [CrossRef]
14. Van der Sluis, M.; Klerk, B.; Ellen, E.D.; Haas, Y.; Hijink, T.; Rodenburg, T.B. Validation of an Ultra-Wideband Tracking System for Recording Individual Levels of Activity in Broilers. *Animals* **2019**, *9*, 580. [CrossRef] [PubMed]
15. Gharat, V.; Colin, E.; Baudoin, G.; Richard, D. Impact of Ferromagnetic Obstacles on LF-RFID Based Indoor Positioning Systems. In Proceedings of the 2017 IEEE International Conference on RFID Technology and Application, RFID-TA 2017, Warsaw, Poland, 20–22 September 2017; pp. 284–289. [CrossRef]
16. Liu, H.; Darabi, H.; Banerjee, P.; Liu, J. Survey of Wireless Indoor Positioning Techniques and Systems. *IEEE Trans. Syst. Man Cybern. Part C (Appl. Rev.)* **2007**, *37*, 1067–1080. [CrossRef]
17. Nakayama, S.; Laskowski, K.L.; Klefoth, T.; Arlinghaus, R. Between- and within-Individual Variation in Activity Increases with Water Temperature in Wild Perch. *Behav. Ecol.* **2016**, *27*, 1676–1683. [CrossRef]
18. Stratmann, A.; Fröhlich, E.K.F.; Gebhardt-Henrich, S.G.; Harlander-Matauschek, A.; Würbel, H.; Toscano, M.J. Modification of Aviary Design Reduces Incidence of Falls, Collisions and Keel Bone Damage in Laying Hens. *Appl. Anim. Behav. Sci.* **2015**, *165*, 112–123. [CrossRef]
19. Rokach, L.; Maimon, O. Decision Trees. In *Data Mining and Knowledge Discovery Handbook*; Maimon, O., Rokach, L., Eds.; Springer: New York, NY, USA, 2005; pp. 165–192, ISBN 978-0-387-25465-4.
20. Ebiele, F.M.J.; Atemkeng, M. Conventional Machine Learning Based on Feature Engineering for Detecting Pneumonia from Chest X-Rays. In Proceedings of the Conference of the South African Institute of Computer Scientists and Information Technologists, Cape Town, South Africa, 14–16 September 2020; pp. 149–155. [CrossRef]
21. Friedman, J.H. Greedy function approximation: A gradient boosting machine. *Ann. Stat.* **2001**, *29*, 1189–1232. [CrossRef]
22. Dorogush, A.V.; Ershov, V.; Gulin, A. CatBoost: Gradient Boosting with Categorical Features Support. *arXiv* **2018**, arXiv:1810.11363.
23. Prokhorenkova, L.; Gusev, G.; Vorobev, A.; Dorogush, A.V.; Gulin, A. CatBoost: Unbiased Boosting with Categorical Features. *arXiv* **2017**, arXiv:1706.09516.
24. Gorunescu, F. *Data Mining: Concepts, Models and Techniques*; Springer: Heidelberg, Germany, 2011.
25. McNemar, Q. Note on the Sampling Error of the Difference between Correlated Proportions or Percentages. *Psychometrika* **1947**, *12*, 153–157. [CrossRef]
26. Good, R.; Fletcher, H.J. Reporting Explained Variance. *J. Res. Sci. Teach.* **1981**, *18*, 1–7. [CrossRef]
27. Willmott, C.J.; Matsuura, K. Advantages of the Mean Absolute Error (MAE) over the Root Mean Square Error (RMSE) in Assessing Average Model Performance. *Clim. Res.* **2005**, *30*, 79–82. [CrossRef]
28. Pedregosa, F.; Varoquaux, G.; Gramfort, A.; Michel, V.; Thirion, B.; Grisel, O.; Blondel, M.; Prettenhofer, P.; Weiss, R.; Dubourg, V.; et al. Scikit-Learn: Machine Learning in Python. *J. Mach. Learn. Res.* **2011**, *12*, 2825–2830.
29. Bates, D.; Mächler, M.; Bolker, B.M.; Walker, S.C. Fitting Linear Mixed-Effects Models Using Lme4. *arXiv* **2014**, arXiv:1406.5823.
30. Li, N.; Ren, Z.; Li, D.; Zeng, L. Review: Automated Techniques for Monitoring the Behaviour and Welfare of Broilers and Laying Hens: Towards the Goal of Precision Livestock Farming. *Animal* **2020**, *14*, 617–625. [CrossRef]
31. Halachmi, I.; Guarino, M.; Bewley, J.; Pastell, M. Smart Animal Agriculture: Application of Real-Time Sensors to Improve Animal Well-Being and Production. *Annu. Rev. Anim. Biosci.* **2019**, *7*, 403–425. [CrossRef] [PubMed]
32. Rowe, E.; Dawkins, M.S.; Gebhardt-Henrich, S.G. A Systematic Review of Precision Livestock Farming in the Poultry Sector: Is Technology Focussed on Improving Bird Welfare? *Animals* **2019**, *9*, 614. [CrossRef] [PubMed]
33. Ahmed, G.; Malick, R.A.S.; Akhunzada, A.; Zahid, S.; Sagri, M.R.; Gani, A. An Approach towards IoT-Based Predictive Service for Early Detection of Diseases in Poultry Chickens. *Sustainability* **2021**, *13*, 13396. [CrossRef]
34. Gómez, Y.; Stygar, A.H.; Boumans, I.J.M.M.; Bokkers, E.A.M.; Pedersen, L.J.; Niemi, J.K.; Pastell, M.; Manteca, X.; Llonch, P. A Systematic Review on Validated Precision Livestock Farming Technologies for Pig Production and Its Potential to Assess Animal Welfare. *Front. Vet. Sci.* **2021**, *8*, 660565. [CrossRef]
35. Odintsov Vaintrub, M.; Levit, H.; Chincarini, M.; Fusaro, I.; Giammarco, M.; Vignola, G. Review: Precision Livestock Farming, Automats and New Technologies: Possible Applications in Extensive Dairy Sheep Farming. *Animal* **2021**, *15*, 100143. [CrossRef]
36. Fogsgaard, K.K.; Røntved, C.M.; Sørensen, P.; Herskin, M.S. Sickness Behavior in Dairy Cows during Escherichia Coli Mastitis. *J. Dairy Sci.* **2012**, *95*, 630–638. [CrossRef]
37. Tizard, I. Sickness Behavior, Its Mechanisms and Significance. *Anim. Health Res. Rev.* **2008**, *9*, 87–99. [CrossRef] [PubMed]
38. Dochtermann, N.A.; Schwab, T.; Anderson Berdal, M.; Dalos, J.; Royauté, R. The Heritability of Behavior: A Meta-Analysis. *J. Hered.* **2019**, *110*, 403–410. [CrossRef]
39. Stirling, D.G.; Réale, D.; Roff, D.A. Selection, Structure and the Heritability of Behaviour. *J. Evol. Biol.* **2002**, *15*, 277–289. [CrossRef]

40. Dingemanse, N.J.; Both, C.; Drent, P.J.; van Oers, K.; van Noordwijk, A.J. Repeatability and Heritability of Exploratory Behaviour in Great Tits from the Wild. *Anim. Behav.* **2002**, *64*, 929–938. [CrossRef]
41. Richards, G.J.; Wilkins, L.J.; Knowles, T.G.; Booth, F.; Toscano, M.J.; Nicol, C.J.; Brown, S.N.; Richards, G.J.; Wilkins, L.J.; Knowles, T.G.; et al. Pop Hole Use by Hens with Different Keel Fracture Status Monitored throughout the Laying Period. *Vet. Rec.* **2012**, *170*, 494. [CrossRef]

Article

Modular E-Collar for Animal Telemetry: An Animal-Centered Design Proposal

Marta Siguín [1], Teresa Blanco [1,2], Federico Rossano [3] and Roberto Casas [1,*]

1 Howlab (Human Openware Research Lab) Research Group, I3A (Aragon Institute of Engineering Research), University of Zaragoza, 50009 Zaragoza, Spain; msiguin@unizar.es (M.S.); tblanco@unizar.es (T.B.)
2 GeoSpatium Lab S.L., Carlos Marx 6, 50015 Zaragoza, Spain
3 CCL (Comparative Cognition Lab), University of California, San Diego, CA 92093, USA; frossano@ucsd.edu
* Correspondence: rcasas@unizar.es

Abstract: Animal telemetry is a subject of great potential and scientific interest, but it shows design-dependent problems related to price, flexibility and customization, autonomy, integration of elements, and structural design. The objective of this paper is to provide solutions, from the application of design, to cover the niches that we discovered by reviewing the scientific literature and studying the market. The design process followed to achieve the objective involved a development based on methodologies and basic design approaches focused on the human experience and also that of the animal. We present a modular collar that distributes electronic components in several compartments, connected, and powered by batteries that are wirelessly recharged. Its manufacture is based on 3D printing, something that facilitates immediacy in adaptation and economic affordability. The modularity presented by the proposal allows for adapting the size of the modules to the components they house as well as selecting which specific modules are needed in a project. The homogeneous weight distribution is transferred to the comfort of the animal and allows for a better integration of the elements of the collar. This device substantially improves the current offer of telemetry devices for farming animals, thanks to an animal-centered design process.

Keywords: wearables design; animal farming; animal-centered design; animal telemetry; modularity; smart collar; design contributions; additive manufacturing

Citation: Siguín, M.; Blanco, T.; Rossano, F.; Casas, R. Modular E-Collar for Animal Telemetry: An Animal-Centered Design Proposal. *Sensors* **2022**, *22*, 300. https://doi.org/10.3390/s22010300

Academic Editors: Yongliang Qiao, Lilong Chai, Dongjian He and Daobilige Su

Received: 29 November 2021
Accepted: 28 December 2021
Published: 31 December 2021

1. Introduction

Telemetry combines the use of different sensors and wireless communications to perform physical and/or chemical measurements remotely. Applied to the study of animals, it allows for the acquisition of animal life data through a device placed on the animal that sends signals to a receptor [1]. In this way, different issues related to the individual and their environment can be monitored in a much less invasive way, without having to come into direct contact with them, except for the placement of the device.

Since the 1960s, radiotelemetry has been used as an instrument to track the position of animals and study their behavior [2–4]. In the last few years, the development of telemetry devices in animal studies has provided noteworthy advances in the direction of increasing the batteries' lifetime, improving the precision and functionality of systems, miniaturizing devices, increasing the variety and novelty of data collected, and research in data processing as well as the use of eco-friendly materials and renewable energy sources [5–10].

Recently, small sensors from mobile and communication technologies and location systems have gradually been integrated in animal telemetry: accelerometers, magnetometers, cameras, temperature sensors, pressure sensors, etc. They have been combined and placed in global positioning system (GPS) collars, allowing for the study of ecological issues around migration, foraging behavior, physiological performance, habitat selection and social interaction, particularly, of medium and large terrestrial mammals [4,11].

Therefore, telemetry has become a very powerful tool in the study of animal life for the purposes of evolutionary, behavioral, and veterinary research; for the monitoring of animals in their environment; and for the conservation of fauna. It has brought greater efficiency and objectivity, and has allowed professionals to work with animals that would have been unthinkable to study in their habitat just a few years ago. In addition, the automation of animal monitoring ensures the continuity of data collection, which becomes an obstacle in extreme situations such as the one we currently experience due to *Coronavirus Disease 2019* (COVID-19) [12,13].

In the field of smart livestock, monitoring has led to the improvement in animal protection and welfare through the monitoring of their behavioral and physiological states, which represents a step toward the responsible production and consumption of animal materials. Monitoring allows (i) to improve the traceability of animal welfare; (ii) facilitate decision-making to which the farmer submits; and (iii) favor the management of the exploitation [6,9,13–15].

The application of telemetry in animals presents difficulties, mainly related to (i) weight distribution; (ii) autonomy; (iii) flexibility in design; and (iv) cost of the devices. These problems have become apparent from the analysis of three types of sources: scientific publications referring to the physical design of animal telemetry devices [16–24]; scientific publications referring to the evaluation of animal telemetry [13,25–27] as well as commercial solutions currently available [28–36].

Most of these problems are highly dependent on design. Design is a process capable of connecting technology with the real requirements of users, providing the market with products and services that respond to the diverse cultural and social context in which we live, which currently requires an indispensable technological adaptation. On one hand, a good design strategy brings innovation to the processes (i) of contextual and user research; (ii) detection of needs and definition of functionalities and requirements; (iii) of ideation and conceptualization; and (iv) finally evaluation. This can help to solve problems meshing the product, user, and environment as well as planning and formulating multidisciplinary strategies thanks to the holistic training of the designer, accustomed to working in a team and in various areas not related to their discipline [37]. On the other hand, design can also contribute to technical aspects of product development, defining it structurally and formally, and analyzing and making a good choice of materials and manufacturing processes.

This paper describes the design and evaluation of a low-cost telemetric device for the study of the behavior of medium and large mammals such as farm animals, which provides solutions to the problems detected in current devices. We hope to contribute both at a methodological level and in order to facilitate and extend the use of animal telemetry for the study of animals and their environment.

2. Materials and Methods

2.1. Design Concept

The structure of a typical telemetry collar is mainly composed of the following parts (Figure 1):

1. Strap: structural element on which the portability of the device is based;
2. Electronic module: envelope that contains the active part of the device inside. In general, collars have a single electronic module that is placed in the lower part of the animal's neck, allowing the antenna to be correctly oriented thanks to the action of gravity;
3. Antenna: it is the component that allows the transmission of the information collected by the electronic device. The antenna can be external, a wire rope or a more sophisticated independent element such as the one in Figure 1, or internal, integrated into a printed circuit board (PCB);
4. Coating: sometimes electronic modules and/or antennae are covered with plastic materials to protect them;

5. Unions: the connection of the external elements with the strap is usually carried out by rivets or bolt–nut unions;
6. Drop-Off: this is a mechanism used in the field of wildlife to be able to recover the device without having to recapture the animal that carries it. These devices can be electronic or mechanical. The latter are based on the degradation of the material that composes them; when the material has degraded in the expected time, the collar falls off and can be recovered by scientists;
7. Closure system: the safest closures are made by means of two bolt–nut connections.

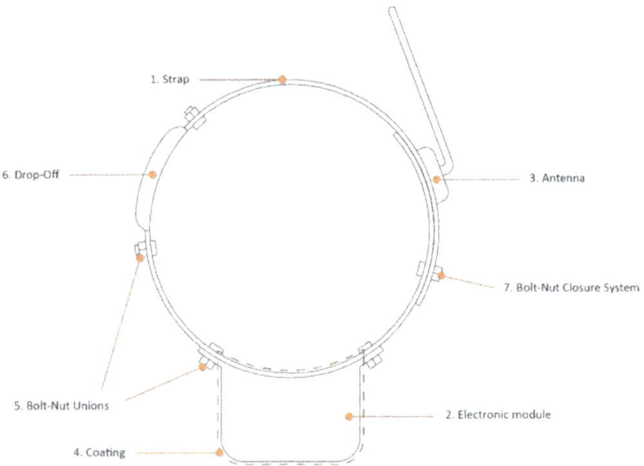

Figure 1. Main parts of a telemetry collar.

2.2. Design Methodology

The design process followed for the development of this telematic device (Figure 2) was established based on basic design methodologies and approaches: (i) the Double Diamond of the British Design Council [38]; (ii) the ideology of People-Centered Design of IDEO [39]; (iii) the Design Thinking process of the D. School [40]; and (iv) of the design applied to IoT: Cosica [41,42]. These methodologies have been oriented to the experience of the animal and the human and to the design of wearables, and have been adapted to the context and ecosystem in which the project was developed.

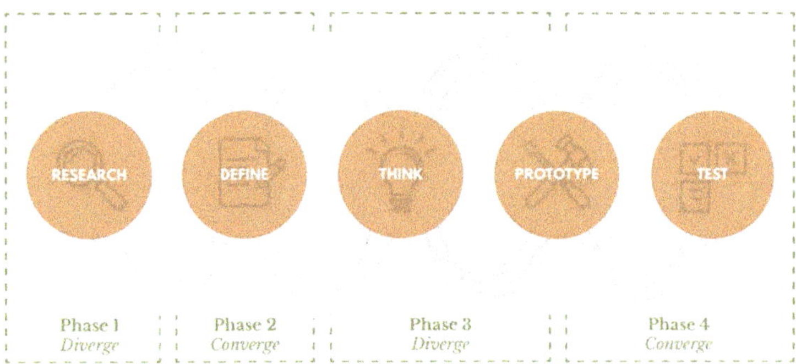

Figure 2. Methodological process of the investigation.

A collaborative design process was carried out, in which the following had participated: designers, electronic engineers, and telecommunications engineers as well as veterinary experts, technical personnel who work in continuous contact with animals, behavioral researchers, and coordinators of animal centers.

2.2.1. Phase 1–Research (Diverge)

The objective of this phase is to collect as much information as possible from users and the context, in order to subsequently define requirements based on their real needs. To achieve this, the following strategy is proposed (Table 1).

Table 1. Design strategy followed for the development of Phase 1.

Objective	Analysis & Source	Method
Obtain information from the telemetric context in which the project will be developed; and compile recommendations and solutions resulting from the investigative exercise.	State of the art from papers in scientific journals.	Literature review.
Extract considerations to take into account in relation to environmental conditions and use cases.	Analysis of the environment of use through semi-structured interviews with users.	Synthesis of the information according to cases of use and location of the animals.
Define user profiles; and detect their needs.	User analysis through semi-structured interviews with users.	Person method (modeling the characteristics of the different groups of users). Quotes (collect literary phrases that express the wishes or concerns of users). Team meetings to synthesize the information.
Define the morphometric measurements of the animals that determine the design of the device; and decide what type of device is going to be developed according to its placement.	Morphometric analysis by semi-structured interviews with users and morphometric tables.	Information synthesis.
Know what is currently being offered in terms of telemetric devices and what market niches or problems currently exist in it.	Market study and structural analysis from market offer and scientific papers.	Search for products by manufacturers. Synthesis of the characteristics belonging to the elements that make up a standard collar.
Decide what functions are going to be implemented; and define the electronic components that the wearable must have.	Functional analysis through meetings with users and with the team.	Information synthesis.
Define a manufacturing strategy that reduces production costs.	Manufacturing Context studied from papers in scientific journals.	Literature review.

2.2.2. Phase 2–Define (Converge)

After the divergence of the previous phase, where a large amount of information has been collected, in phase 2, it is intended to define the problem and divide it, in order to tackle it more easily. To do this, in the first place, the information collected is synthesized to highlight the most relevant, which helps to correctly focus the ideation process. This is done through *Clustering*, grouping the most important revelations in relation to the problems detected and the design requirements to be considered. Subsequently, designers, electronic engineers, and telecommunications engineers work together to propose and define technical solutions to the problems detected and to define the electronic components that the device will have in order to consider them in the design process. Finally, the defined problem is divided by presenting 12 design challenges that will guide the next creative stages.

The structural proposed challenges are:

- Challenge 1 (Body—Strap): Choice of materials and strap size;

- Challenge 2 (Body—Modules): How can a watertight and flexible union of modules be made?
- Challenge 3 (Body—Modules): How can a cover–body joint of the module be watertight and safe?
- Challenge 4 (Body—Modules): How can the union or registration between the modules and the strap be made?
- Challenge 5 (Body—Modules): What should the shape of the modules be so that they do not cause discomfort to the animal?
- Challenge 6 (Body—Cover): How should the modules be protected?
- Challenge 7 (Drop-Off): Adaptation and development of a Drop-Off system based on the degradation of latex tubes;
- Challenge 8 (Closure): Proposal of a rapid closing system; and
- Challenge 9 (Distribution): Proposal of a correct distribution of the elements along the strap.

These challenges must be carried out taking into account three transversal challenges:

- Challenge 10 (Environmental conditions): Design of a collar resistant to environmental conditions;
- Challenge 11 (Integration): Design of a compact collar whose parts are integrated; and
- Challenge 12 (Impact): Design a collar that has the least impact on the animal.

2.2.3. Phase 3–Think and Prototype (Diverge)

This phase aims to solve the challenges defined in the previous phase. The process carried out to achieve this phase has a highly iterative component (Figure 3). The system of challenges defined in the previous phase is followed to tackle the problems to be solved in a structured and defined way. As ideas are generated, they are prototyped and/or evaluated with users and the team to rule out options or to validate them.

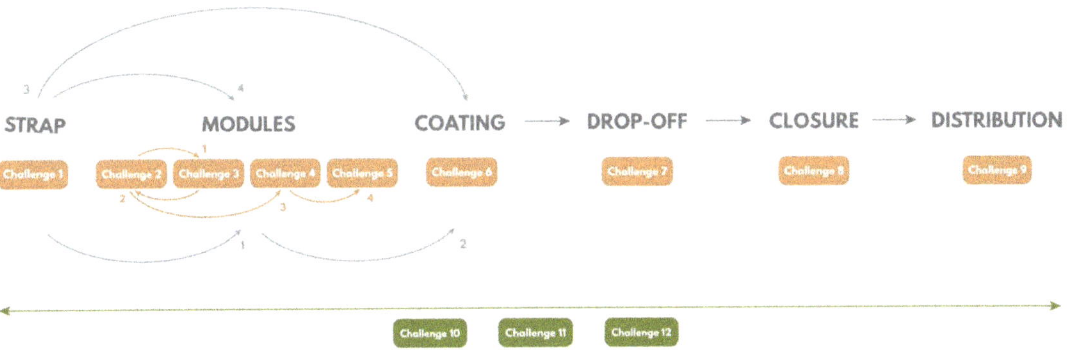

Figure 3. Iterative temporal development that was performed during the ideation process.

2.2.4. Phase 4–Prototype and Test (Converge)

In order to test the proposals: two final prototypes of the collar were made, where all the variations are represented, and a final evaluation was structured.

Evaluation is one of the key points of any process. The literature on the design of animal monitoring devices always structures its discourse taking into account a final evaluation that assesses the results of the use of the new proposed product. However, in most cases, these results focus only on technological deployment (autonomy, signal range, failed devices, etc.), and in the case of evaluating the physical design of the device, these studies serve few and very superficial objectives (for example, the device has been broken, has caused injuries to animals, or the mortality rate) [16–24]. In addition, these evaluations are merely quantitative, not giving value to the experience of the professionals.

Our complete evaluation of the design elements and of the overall design of the collar was carried out using a mixed methods approach [43–45].

In such a particular context as the one in which we find ourselves, the use of both quantitative and qualitative evaluation methods can help us (i) to complement the information when collecting data on dimensions that have only been evaluated by a single method (the adequacy of the force applied to close the collar or autonomy); (ii) to generate a more complete concept of the objectives through the combination (iterative methods with users have been combined with semi-structured interviews, so that the information obtained through the iteration has helped us to identify key points to deal in interviews); and (iii) and to refine the results by triangulating information on the same dimension (weight, ease of use).

Sources

The collar was evaluated through various sources:

- Experts in the Environment (EE): Managers of animal centers and workers, who act as potential clients and animal experts. They work with the animals and put the collar on them;
- Research Experts (RE): Behavioral researchers, also acting as potential clients and experts in animal interaction in a context of behavioral research;
- Engineers (E), who evaluate technical specifications of the collar in the laboratory; and
- Current Offer (CO), which allows the proposed collar to be evaluated against current designs.

Methods

- Laboratory Experiments (LE): Laboratory tests were carried out at different times in the process to evaluate technical issues such as tightness. Rapid prototyping techniques were also used to evaluate the physical designs of the parts and the distribution of weights;
- Iterative Methods with users (IM): Regular contact with experts was maintained. Through various methods such as meetings, open interviews, small product presentations, sending samples, etc., information was extracted on their opinions and judgments. These methods guided the design process and allowed us to detect elements that should be emphasized in future evaluations;
- Focus Group (FG): A focus group was held with four experts in behavioral science with extensive experience with animals. The objective of the focus group was to gain the opinions that research experts have in relation to the proposals and what they can contribute to their work;
- Real-Life Testing (RLT): The prototypes were evaluated with animals, which allowed the designers to observe how they relate to the morphometry of the animal. On the other hand, experts in the environment also observed the behavior of animals in relation to the collars. The collars were tested on sheep (rasa aragonesa and roya bilbilitana), goats (murciano granadina and mestiza de Florida), and horses (hispano-bretón) under the approval of the Ethical Committee of the University of Zaragoza (PI55/20, 28 October 2020);
- Semi-Structured Interviews (SSI): Semi-structured interviews were carried out with the experts in the environment to evaluate the alternatives reflected in the prototypes, which are detailed later; and
- Document Analysis (DA): To evaluate the proposals against the current panorama on animal telemetry, a table was compiled in which the characteristics of different collars on the market were compared.

Table 2 shows the objectives evaluated in relation to the sources and the methods used for their evaluation.

Table 2. Objectives evaluated in relation to the sources and the methods used for their evaluation.

		Objectives to Evaluate	Source [1]	Methods [2]
Elements	Strap	Material (malleability; and resistance to environmental conditions)	EE	IM + RLT + SSI
		Width (adaptation to the morphometry of the animal)	EE	IM + RLT + SSI
		Length (adaptation to the morphometry of the animal)	EE	IM + RLT + SSI
	Modules—Body	Shape (comfort for the animal)	EE + E	IM + RLT + SSI + LE
	Modules—Lid	Shape (comfort for the animal)	EE + E	IM + RLT + SSI + LE
	Coating	Adaptation to the collar	E	LE
		Resistance (to be worn)	EE	IM + RLT + SSI
	Drop-Off	Robustness	EE + E	IM + RLT + SSI + LE
		Structure (change proposed)	EE + E	IM + RLT + SSI + LE
	Closure	Structure (new design)	EE + E	IM + RLT + SSI + LE
		Ease of use	EE + RE + E	IM + FG + RLT + SSI + LE
		Force applied (required for handling)	EE	IM + RLT + SSI
	Unions	Weight reduction	CO	DA
		Body—Lid sealing	E	LE
		Module—Module sealing	E	LE
Composition	Collar	Weight	EE + RE + CO	RLT + SSI + DA + IM
		Weight distribution	EE + RE + CO + E	RLT + SSI + DA + IM + FG + LE
		Integration of elements and formal and aesthetic adaptation	EE + RE + CO	RLT + SSI + DA + IM + FG
		Autonomy	CO	DA
		Design flexibility	CO	DA
		Comfort for the animal	EE + RE	RLT + SSI + IM + FG
		Ease of use	EE + RE + CO	RLT + SSI + DA + IM + FG
		Interaction	EE + RE	RLT + SSI + IM + FG

[1] Source abbreviations: EE (Experts in the Environment); E (Engineers); RE (Research Experts); CO (Current Offer). [2] Methods abbreviations: IM (Iterative Methods with experts); RLT (Real-Life Testing); SSI (Semi-Structured Interviews); LE (Laboratory Experiments); FG (Focus Group); DA (Document Analysis).

3. Results

As stated before, we aimed to solve four main problems detected in animal telemetry devices: weight distribution, autonomy, flexibility in design, and price. We proposed the design of a modular collar that distributes the electronic components in several compartments, connected and powered by rechargeable batteries. The manufacturing of the device was based on 3D printing.

The distribution of the elements must bear in mind two main premises: (i) distributing the weight as evenly as possible along the collar so that the animal does not suffer; and (ii) that the antennae are always in the most convenient position of the collar to allow proper communications (e.g., GPS/satellite must be at upper position pointing to the sky). A low-level prototype was created to check and adjust the distribution of the elements, which was based on the balance of weights. The device needs at least six modules.

However, to demonstrate the modularity and customization of the proposal, we decided to prototype a collar with seven modules, where four of them are batteries. This results in the composition depicted in Figure 4.

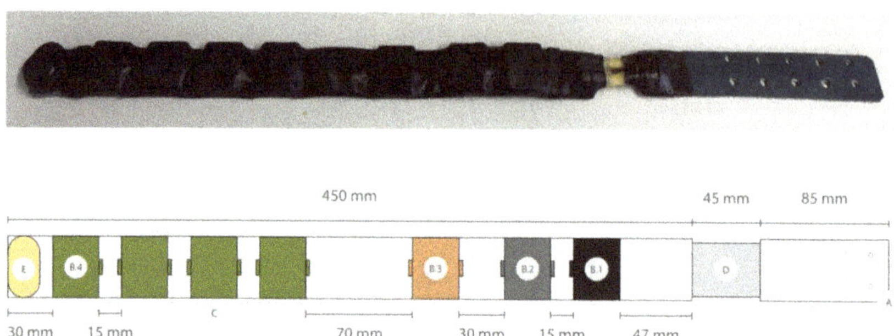

Figure 4. Bounded distribution of the elements of the collar (A Strap; B.1 GPS module; B.2 Communications module; B.3 Sensors module; B.4 Battery modules; C Coating; D Drop-Off; E Closure).

As seen in Figure 5, the main structure of the collars can be divided into (A) strap; (B) modules, which are divided into body and lid; (C) coating; (D) drop-off; (E) and closure.

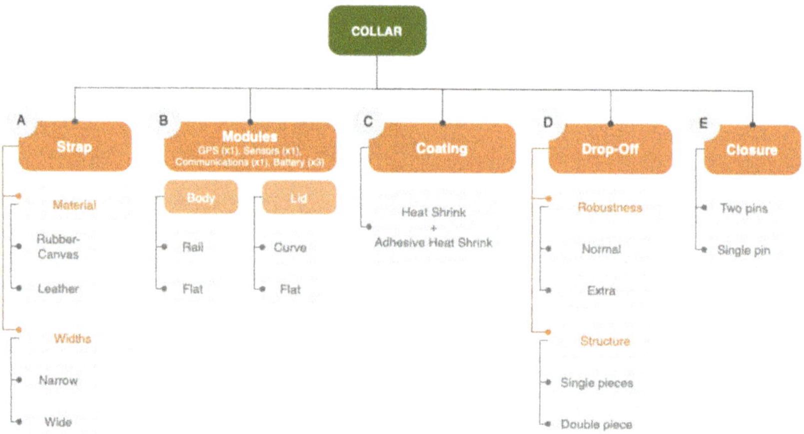

Figure 5. Block diagram of collar structure and design options.

Several options are proposed for each of the structural elements in order to evaluate them (Figure 4).

3.1. Strap

The strap (Figure 4A) is the structural element on which the rest of the components are mounted. The choice of the material of the strap is a decision that revolves around its rigidity and its response to handling and weather conditions, since it will not come into contact with the animal's skin because of the coating.

The use of two different materials (Figure 5A) was considered: a rubber–canvas composite material with several interleaved layers and natural leather. The leather strap was less rigid, more malleable, and adaptable to the movement of the animal, which can be more comfortable for the animal but also less resistant to pulling or biting, while the rubber–canvas strap was more rigid and helps define the shape of the collar, reversing the advantages and disadvantages compared to leather.

Various strap widths (Figure 5A) were also assessed, one equal to the height of the modules containing the electronics, so that the collar is more compact and is more protected from the action of animals; and another a little lower, with the idea of reducing the material to make it more flexible and comfortable for the animal.

3.2. Modules

3.2.1. Shell

To favor the fractionation of the electronics, they were housed in independent but interconnected modules. These modules must be watertight in all of their joints and have a shape that is comfortable for the animal.

The modules (Figures 4B and 6) consist of two pieces: the body and the lid; these are manufactured by 3D printing in acrylonitrile butadiene styrene (ABS). The body is the element that keeps its measurements constant, while the lid varies in height depending on what the module contains. This allows different components to be accommodated just by changing one measure. In our case, the gap that houses the electronics always maintains its height (32 mm) and its width (22 mm), combining its depths between 5.4 mm for electronic modules and 16.4 mm for battery modules.

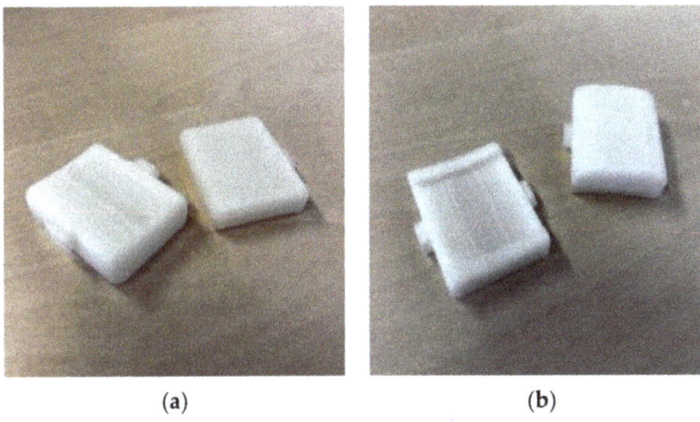

(**a**) (**b**)

Figure 6. Modules: (**a**) flat body with flat lid; (**b**) rail body with curved lid.

In terms of shape, two types of body (Figure 5B) were tested depending on the width of the strap: a flat body simply attached on the wide strap, and a rail body that embraces the narrow strap, generating battlements between module and module. To enhance the integration of the modules on the strap, the base of the body is provided with a curvature that accompanies the circumference of the animal's neck.

As for the lids, a flat version and a more curved one (Figure 5B) are proposed. The flat lid reduces the material and thickness of the modules to the minimum, while the curved lid follows the curvature of the collar and generates fewer edges, although the thickness of the modules increases.

The union between the body and the lid is carried out by means of the adhesive and sealing of both parts by the chemical reaction that occurs between ABS and acetone in a tongue and groove that runs along its perimeter. Both body options have a ledge on each side where a heat shrink tube adheres and compresses. This allows, on one hand, for the protection of the connection of the modules that is made by wires, and on the other, to join the modules together. Modules were attached on the strap using double-sided tape.

3.2.2. Electronics

Building blocks of the electronics inside the collar vary according to the target animal and the monitoring features required (Table 3).

Table 3. Electronic features depending on the type of monitoring.

	Intensive Monitoring	Remote Monitoring
Target animals	Small-medium sized mammals (sheep, goat, etc.)	Medium and large sized mammals (horse, cow, etc.)
Monitoring scenario	Animals are estabulated or in confined facilities that allow periodic check in.	Animals are free and move in large areas not seen for months.
Electronic blocks	1 block with: Movement and magnetic sensor SD card for massive sensor datalogging, Bluetooth for communication and proximity sensing (Figure 7).	3 blocks with: Movement sensor with smart analysis to extract activity. GPS (including antenna). Lora communication.
Battery	6000 mA·h Li-Ion battery made up by 6 pieces of 1 A·h	4000 mA·h Li-Ion battery made up by 4 pieces of 1 A·h
Energy expenditure	Low power (when no movement detected): 11.2 J/day Sensor datalogging (5′ proximity scan and 16 h of movement recorded): 663.5 J/day Data downloading (30′ once per day): 46.8 J/day	Low power (when no movement detected): 33.7 J/day Sensor data logging (1 h proximity scan and 2 h of movement recorded): 77.7 J/day GPS data logging (24 locations/day) and activity sending (1 h periodicity): 62.6 J/day GPS data logging (4 locations/day): 10.4 J/day
Device lifetime	91 days	Smart mode (24 gps/day) + Activity + BLE –> 236 days Smart mode (4 gps/day) –> 5.5 years

- Sensing, computing, and datalogging: these were implemented using a microcontroller (to manage data and rest of the hardware) and small sensors measuring linear and angular acceleration, sound, magnetic field, etc.
- Communications: these were implemented using different communication modules depending on the required range, data throughput, and antenna size (e.g., Bluetooth (short range, high throughput, smallest antenna), VHF (very long range, very little throughput, large antenna) and Lora (long range, low throughput, small antenna)).
- Location: this can be undertaken using a global navigation satellite system (GNSS) module for precise and global location or using wireless communication modules for rough positioning.
- Energy: battery is required to run the electronics and its technology and size defines the system's lifetime by dividing the energy available inside the battery by the energy required by the electronics (calculated as the sum of the products of the power required by each electronic block inside the device times the time this piece is running).

$$\text{lifetime} = \frac{\text{battery\_energy}}{\sum_{\text{electronic\_blocks}}(\text{running\_power} \times \text{time\_running})} \quad (1)$$

We designed two different electronics that fit inside the collar; all of them fulfilled the dimensional restrictions of the maximum area of 20 mm × 30 mm.

Figure 7. Electronic block for intensive monitoring (20 × 30 mm).

3.3. Coating

The coating of the collar (Figure 4C) must have a double function, reinforce it against climatic conditions and resist the manipulation of the various elements that compose it. For this reason, conventional heat shrink tubing was used to cover the collar and its ends were sealed with adhesive heat shrink strips (Figure 5C). Conventional heat shrink was also used to hide mechanical joints that have shiny elements, in order not to attract the attention of the animal, its companions or other species.

3.4. Drop-Off

Drop-off systems, typical in the study of wildlife, can be used as a security system in case the animal is trapped because of the collar, which can be an interesting element to also incorporate in intelligent farming. The drop-off system (Figures 4D and 8) that the collar had is similar to that of Telonics commercial solutions [46], since we considered it to be a successful method as a safety system against hanging. This was based on the degradation of latex against the action of the environment. This system consists of two latex tubes that are joined by means of nylon thread to the connecting ends of some pieces fixed to the strap thanks to a nut–screw connection. Nevertheless, certain novelties that improve the design of the commercial models on several levels have been introduced in regard to fasteners. The redesign joins the two elements of the commercial model in a single piece in such a way that its assembly is facilitated and its robustness is increased. Moreover, it offers the possibility of combining or choosing between the two structure options, single or double, in order to adapt the collar to each context and animal species (in some cases the double option could improve its resistance) (Figure 5D).

3.5. Closure

The closures that are currently used in market devices make the placing of the collar a complex and time-consuming activity. The most difficult issue that surrounds this element is that it must attend to the needs of two main types of user: animal (it must resist its force and have a mechanism that is difficult for them to open) and the veterinarian (it must be easy and quick to open for veterinarians).

To improve this problem, we developed a closing system (Figures 4E and 9) based on a magnetic head that locks and unlocks on a pin thanks to the action of a neodymium magnet. Two concepts were designed (Figure 5E): a version composed of (i) a base with two pins, a first pin that allows closure and a second pin that keeps the collar fixed without allowing it to rotate; and (ii) an upper piece that contains the magnetic mechanism and guides the second pin; and a second version with a single pin to allow us to learn whether the use of the second pin is really necessary or if, morphometrically, there is not enough clearance for the strap to rotate too much on the animal's neck. This latest version has a top piece that contains the magnetic mechanism and a base with a single pin.

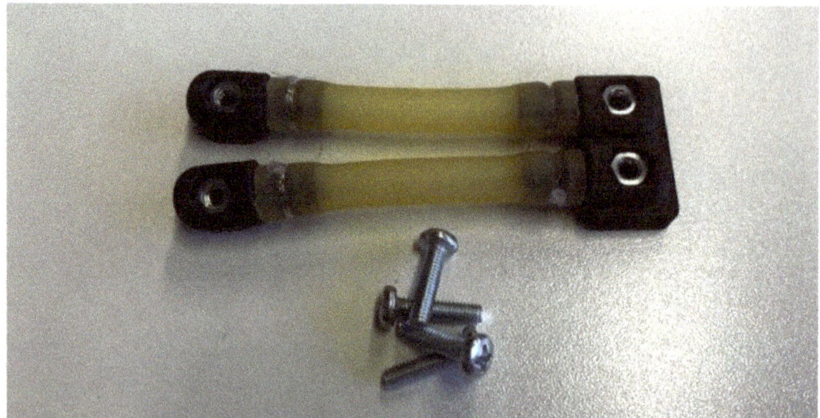

Figure 8. Drop-off system: two pieces are used at left and a single piece is used at right, in both cases standard robustness.

Figure 9. Magnetic closure: single pin closure (**left**) and double pin closure (**right**).

4. Discussion

The device presented aims to resolve several issues identified in the current animal telemetry offers: weight distribution, autonomy, flexibility in design, and price. The device evaluation was carried out using a mixed methods approach [43–45] using quantitative techniques such as laboratory experiments, current offer comparison tables, as well as qualitative (e.g., interviews, focus group, observations) following the Xassess design evaluation method [43]. This evaluation relies on a high number of prototyping iterations of local parts, but also of the entire product (Figure 10).

Several experts from different institutions participated in the evaluation of the product: four veterinarians; three technical personnel who work in continuous contact with animals; four behavioral researchers; and two managers of animal centers. The collars were placed on sheep (rasa aragonesa and roya bilbilitana) and goats (murciano granadina and mestizo de Florida), as seen in (Figure 11), and they have also been placed as part of a horse (hispano-bretón) halter, demonstrating their adaptability.

Figure 10. High fidelity prototypes developed to evaluate the range of options available. The elements that have more than one design proposal are represented in one or the other collar, being totally interchangeable.

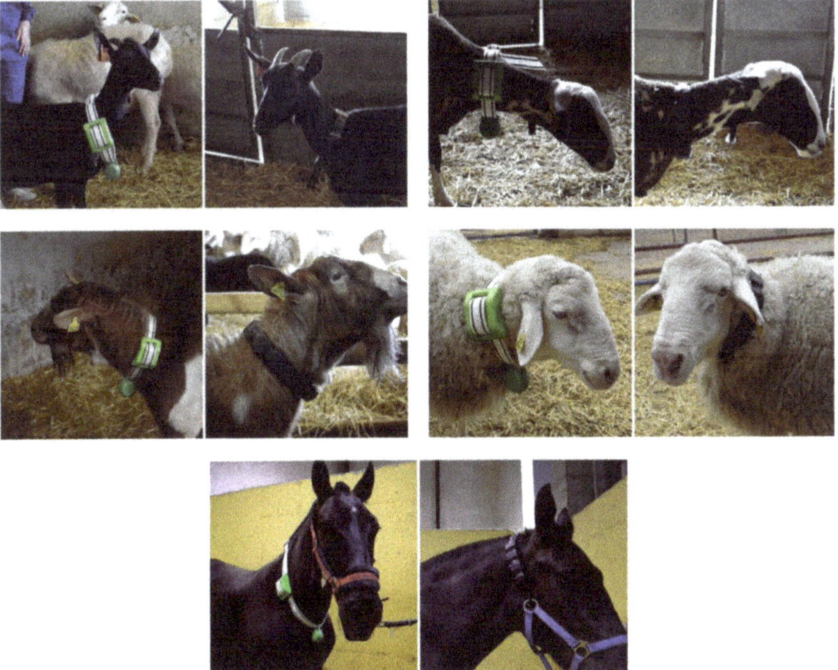

Figure 11. Different animal species with: a commercial collar [36] (left image of each of the pairs, green collar); and with the proposed collar (right image of each of the pairs, black collar).

4.1. Design Flexibility

The recent popularization of telemetry for animal research has led to the emergence of increasingly diverse projects with more specific requirements. This means that, on occasion, the market offers are not adapted to the needs of the project and alternative solutions have to be sought such as individual customization of the devices. Until recently, these modifications involved manufacturing processes whose costs made the project unviable [47], having a negative impact on the size of the animal samples studied [48] and limiting the variability of solutions adapted to different animals or to different technical requirements (batteries of different sizes depending on the needs, adapted communication modules, etc.).

The new concept of modular design of the wearable as well as the particular design of the modules that we propose highlight the fractioning of the electronics, which allow the product to be adapted to the project requirements by customizing the components, and therefore the functions, providing flexibility. This is possible thanks to the modular system of bodies and lids, which adapts the size of the module to the component it houses.

4.2. Weight and Weight Distribution: Comfort and Autonomy

The main weight of the wearables that are currently available on the market is concentrated in the telemetric device itself and this is generally placed at the bottom, so there is no homogeneous distribution of weight (Figure 11). Our proposal distributes the weight along the entire strap, distributing the electronic components in at least six modules (Figure 10). The homogeneous distribution of the elements of the collar also allows the thickness of the collar to be more homogeneous (Table 4). According to the experts, both results are translated into greater comfort for the animal.

Table 4. Current offer comparison table: weight and weight distribution.

Device	Weight	Weight Distribution (L × W × H mm^3)
Our Proposal	210 g (Collar A) 270 g (Collar B)	7 modules: 4 large modules of (30.4 × 40.4 × 11) and 3 small modules of (30.4 × 40.4 × 8)
Personalized Telonics Collar	238 g	1 module (Approx. 55 × 38 × 28)
Telonics, TGW-4570-4	500–880 g	3 modules (73 × 51 × 37)
Telemetry Solutions, Iridium GPS Collar	125–250 g	2 modules (-)
Tellus, Small Personalizable	>600 g	2 modules (76 × 56 × 55)
Advanced Telemetry Systems, G2110E2 Iridium	825 g	2 modules (115 × 80 × 65)
Advanced Telemetry Systems, G5-D Iridium	500 g	2 modules (70 × 50 × 47)
Lotek, Ultimate V6C 176G	278–325 g	1 module (88 × 32 × 30)
Lotek, WILDCELL MG	950 g	2 modules (120 × 86 × 126)
Lotek, PinnaclePro L	630–670 g	3 modules (-)
Ixorigue, GPS Ixotrack	960 g	1 module (83 × 113 × 38)
Open-source collar for terrestrial animals over 8 kg [19]	240 g	1 module (62 × 38 × 32)

The autonomy of the devices is another point of concern. The current autonomy is often not sufficient and therefore the animals cannot be monitored for the desired time [26]. The size and weight of the electronic device of the commercial collars is determined by the battery life; the larger the battery, the larger the electronic element and the heavier the collar. The weight distribution and the modular nature of the proposal show that, although the autonomy of the battery is increased, the resulting increase in weight and size can be distributed along the collar (Table 5).

Table 5. Current offer comparison table: modules on which autonomy depends and operational life.

Device	Autonomy Dependent On	Operational Life
Our Proposal	(at least) 3 modules	Intensive monitoring Smart mode + Activity + Bluetooth Low Energy (BLE) + Sending data—91 days
		Remote monitoring Smart mode (24 gps/day) + Activity + BLE—236 days Smart mode (4 gps/day)—2021 days
Personalized Telonics Collar	1 module	-
Telonics, TGW-4570-4	1 module	4 gps/day, No Very High Frequency (VHF)—6.2 years 4 gps/day, VHF 4 h/day—5.1 years
Telemetry Solutions, Iridium GPS Collar	1 module	-
Tellus, Small Personalizable	1 module	-
Advanced Telemetry Systems, G2110E2 Iridium	1 module	VHF on 8 h/day, 12 locations/day—3 years VHF on 8 h/day, 3 locations/day—4 years
Advanced Telemetry Systems, G5-D Iridium	1 module	VHF on 8 h/day, 6 locations/day, uplinked every 2 days—4 years
Lotek, Ultimate V6C 176G	1 module	60 ppm VHF—776 days
Lotek, WILDCELL MG	1 module	50 min between gps fixes. An SMS message is sent after 7 acquired gps fixes—2 years
Lotek, PinnaclePro L	1 module	VHF beacon is set to operate for 1 h a day at the average. The collar transmits through Iridium after collecting 18 positions, 7 positions/day—4 years
Ixorigue, GPS Ixotrack	1 module	24 gps/day—1 year
Open-source collar for terrestrial animals over 8 kg [19]	1 module	24 gps/day—103 days

4.3. Structure

The strap forms the main structure of the collar, therefore a correct definition of its material (leather or rubber–canvas) and width (the same width as the modules or a little less) is vitally important.

The best material for the strap is leather because it is lighter and adapts better to the movements of the animals and therefore is more comfortable. Rubber–canvas is stronger but leather is sufficient for farm animals: the neck is a protected part of the body and they are herbivorous animals with blunt teeth.

Regarding the width of the strap, a similar reasoning is followed: a narrower width than the modules is less bulky and provides movement to the collar; and therefore, increases comfort by being resistant enough for use with farm animals.

4.4. Unions

The union between the different elements of the design is usually made by rivets and/or nut–screw unions (Table 6), mechanical elements that increase the number of parts of the product and its weight. Our modules are attached on the strap and it is the coating that finishes fixing them. This registration is carried out using double-sided tape, which translates into extra-light joints and with a reduced number of pieces (one piece per module compared to 6–8 on the market).

Table 6. Current offer comparison table: union weight.

Device	Unions Weight (g) [1]
Our Proposal	1.05 g (0.15 × 7 pieces)—7 pieces of double-sided tape
Personalized Telonics Collar	1.2 g (0.15 × 8 pieces)—4 double-sided rivets
Telonics, TGW-4570-4	3.6 g (0.15 × 24 pieces)—12 double-sided rivets
Telemetry Solutions, Iridium GPS Collar	-
Tellus, Small Personalizable	30.4 g (7.6 × 4)—16 pieces, 4 base sets with rods, plate and 2 self-locking nuts
Advanced Telemetry Systems, G2110E2 Iridium	22.8 g (7.6 × 3)—12 pieces, 4 base sets with rods, plate and 2 self-locking nuts
Advanced Telemetry Systems, G5-D Iridium	15.6 g (2.6 × 6) 12 pieces, 6 sets of screw + self-locking nut
Lotek, Ultimate V6C 176G	-
Lotek, WILDCELL MG	15.2 g (7.6 × 2)—8 pieces, 4 base sets with rods, plate and 2 self-locking nuts
Lotek, PinnaclePro L	31.2 g (2.6 × 12) 24 pieces, 12 sets of screw + self-locking nut
Ixorigue, GPS Ixotrack	It does not use mechanical unions to fix the module to the strap, however, it does use a shot on the bottom part of the collar (500 g) to keep it in the correct position.
Open-source collar for terrestrial animals over 8 kg [19]	10.4 g (2.6 × 4) 8 pieces, 4 sets of screw + self-locking nut

[1] The weight of the unions in commercial devices has been estimated from the weights of various commercial mechanical elements. Each piece of a rivet is considered to weigh approximately 0.15 g; each base set with rods, plate, and 2 self-locking nuts is considered to weigh approximately 7.6 grams; each set of screw + self-locking nut is considered to weigh approximately 2.6 grams. The rest of the numerical data were extracted from the characteristics specified by the manufacturers of each device.

4.5. Closing System

The closing of the devices is done mechanically with threaded connections, and is one of the most critical points in the sequence of use as it takes too long (Table 7). However, the proposed closure system allows much faster manipulation of the collar. In addition, it reduces the number of parts of the closure and standardizes the opening tool: a neodymium magnet.

Table 7. Current offer comparison table: ease of use.

Device	Ease of Use
Our Proposal	Magnetic closure
Personalized Telonics Collar	Mechanical nut–screw closure
Telonics, TGW-4570-4	Mechanical nut–screw closure
Telemetry Solutions, Iridium GPS Collar	Mechanical nut–screw closure
Tellus, Small Personalizable	Mechanical nut–screw closure
Advanced Telemetry Systems, G2110E2 Iridium	Mechanical nut–screw closure
Advanced Telemetry Systems, G5-D Iridium	Mechanical nut–screw closure
Lotek, Ultimate V6C 176G	Mechanical nut–screw closure
Lotek, WILDCELL MG	Mechanical nut–screw closure
Lotek, PinnaclePro L	Mechanical nut–screw closure
Ixorigue, GPS Ixotrack	Metal buckle.
Open-source collar for terrestrial animals over 8 kg [19]	Metal buckle.

Regarding the structure of the closures, it is considered that both versions are easy to understand and use and that the force to be applied in the system is correct. However, the double pin version prevents the strap from rotating on itself unlike the single pin version. This limits the movements of the collar and makes the data from the inertial sensors more accurate.

4.6. Integration of the Elements

On many occasions, the elements that make up the wearable are not formally integrated (Table 8), and it is believed that animals that use wearables stand out more among predators [26]. We propose a device that formally adapts to the context and that integrates the elements, thanks to several design decisions: (i) homogeneous distribution of the weights, which favors that the visual mass of the collar is distributed throughout it; (ii) maintain a similar thickness throughout the entire collar; (iii) apply heat shrink tubing as a coating over most of the collar to homogenize the device; (iv) hide shiny elements; (v) use an internal antenna to avoid manipulative elements outside; and (vi) use rounded and smooth shapes that adapt to the morphometry of the animal.

Table 8. Current offer comparison table: integration of elements and formal and aesthetic adaptation.

Device	Integration of Elements and Formal and Aesthetic Adaptation
Our Proposal	Elements with similar thickness (8 mm the minimum and 20.5 mm maximum and the maximum is between a piece of 19 mm and another of 13 mm) distributed along the collar. A single coating. Antenna integrated in PCB, without external elements. Smooth and rounded finishes. Curvature in the body of the module that adapts to the neck of the animal. Hidden shiny elements.
Personalized Telonics Collar	Large main element at the bottom (28 mm). Heat shrinkable in the Drop-Off area. External antenna. Edges at the top and bottom, although rounded at the front. Curvature in the body of the module that adapts to the neck of the animal. Hidden glossy elements except for the closure.
Telonics, TGW-4570-4	Large main element at the bottom (37 mm). It does not use heat shrink, the coating is sandwich type. Internal antenna. Slightly rounded edges. Curvature in the body of the module that adapts to the neck of the animal. Bright elements exposed.
Telemetry Solutions, Iridium GPS Collar	Great main element at the bottom. Heat shrinkable only on modules. Internal antenna. Modules with irregular shapes. No curvature in the body of the module to adapt to the neck of the animal. Bright elements exposed.
Tellus, Small Personalizable	Large main element at the bottom (55 mm). Without cover. Internal antenna. Modules with slightly rounded edges. No curvature in the body of the module to adapt to the neck of the animal. Bright elements exposed.
Advanced Telemetry Systems, G2110E2 Iridium	Large main element at the bottom (65 mm). Without cover. External antenna. Modules with slightly rounded shapes and edges. With curvature in the body of the module to adapt to the neck of the animal. Bright elements exposed.
Advanced Telemetry Systems, G5-D Iridium	Two large main elements at the bottom (47 mm). Without cover. External antenna. Modules with slightly rounded shapes and edges. With curvature in the body of the module to adapt to the neck of the animal. Bright elements exposed.
Lotek, Ultimate V6C 176G	Large main element at the bottom (30 mm). Heat shrinkable coatings in specific locations. External antenna. Modules with slightly rounded edges. With curvature in the body of the module to adapt to the neck of the animal. Hidden glossy elements except for the closure.

Table 8. *Cont.*

Device	Integration of Elements and Formal and Aesthetic Adaptation
Lotek, WILDCELL MG	Large main element at the bottom (126 mm). Without cover. Internal antenna. Module with robust and slightly rounded shapes, lid-body closure not visually integrated. With curvature in the body of the module to adapt to the neck of the animal. Bright elements exposed.
Lotek, PinnaclePro L	Great main element at the bottom. External antenna. Module with robust shapes and sharp edges. With curvature in the body of the module to adapt to the neck of the animal. Bright elements exposed.
Ixorigue, GPS Ixotrack	Large main element at the right side (38 mm). Without cover. Internal antenna. Module with robust and slightly rounded shapes, lid-body closure not visually integrated. No curvature in the body of the module to adapt to the neck of the animal. No shiny elements exposed except the closure.
Open-source collar for terrestrial animals over 8 kg [19]	Large main element at the bottom (32 mm). Without cover. Internal antenna. Edged module. With curvature in the body of the module to adapt to the neck of the animal. Bright elements exposed.

5. Conclusions

Animal telemetry is a topic with a great future within intelligent animal farming, but where serious design-dependent problems are evident. The objective of the project was to cover the niches that have been deduced from the study of scientific literature and the market and to provide solutions from the application of design.

The presented device represents a telemetric option whose design process has put the user at the center, especially the animal user, through an animal-centered design strategy that could be followed in future research. In this way, the concept of the GPS collar has evolved, traditionally chaired by a central module that housed practically all the electronic elements and that did not attend to the premise that wearable devices must be able to collect accurate and reliable data without influencing the behaviors and activities of carrier users [49]. This device solves many of the existing problems in animal telemetry devices and contributes to improving the current offer on the market:

- Homogeneous distribution of weight in at least six modules;
- Three times lighter than devices on the market with the highest number of modules (2–3 modules);
- Design flexibility: modularity and 3D printing;
- Modular electronics on demand of the project with customizable functions;
- Extra-light unions with a reduced number of pieces (one piece per module compared to 6–8 on the market);
- Tightness and resistance to environmental conditions;
- Collar thickness of at least 50% less than that of commercial devices;
- Quick magnetic closure system;
- Wirelessly rechargeable batteries and homogeneous distribution on the collar in case of higher demand; and
- Formal adaptation to the requirements of the context and visually integrated elements.

The results of this research are of interest to designers and manufacturers of animal telemetry, technologists, and professionals in the animal and farm sector, since they contribute to the knowledge about animal monitoring through the design of the device itself and the methodological approach used for its achievement.

Author Contributions: Conceptualization, M.S., T.B. and R.C.; Methodology, M.S., T.B. and R.C.; Hardware R.C.; Validation, M.S., T.B., R.C. and F.R.; Formal analysis M.S., T.B. and R.C.; Investigation, M.S.; Writing, M.S.; Writing—review and editing, T.B., R.C. and F.R.; Supervision, T.B.; Funding acquisition, T.B., R.C. and F.R. All authors have read and agreed to the published version of the manuscript.

Funding: This work was partially supported by the Aragon Regional Government through the program for R&D groups (T27_20R and T59_20R), by the Spanish Government program Torres Quevedo (PTQ2018-010045), and by the Academic Senate of the University of California, San Diego (RG096635-COG506R).

Institutional Review Board Statement: The study was conducted according to the guidelines on the use of animals for experimentation set by the Spanish Royal Decree 53/2013, and has been approved by the Ethical Committee of the University of Zaragoza (PI55/20, 28 October 2020).

Informed Consent Statement: Not applicable.

Data Availability Statement: Not applicable.

Conflicts of Interest: The authors declare no conflict of interest.

References

1. Talcott, M.R.; Akers, W.; Marini, R.P. Chapter 25—Techniques of Experimentation. In *Laboratory Animal Medicine*, 3rd ed.; Fox, J.G., Anderson, L.C., Otto, G.M., Pritchett-Corning, K.R., Whary, M.T., Eds.; American College of Laboratory Animal Medicine; Academic Press: Boston, MA, USA, 2015; pp. 1201–1262. ISBN 978-0-12-409527-4.
2. Cochran, W.W.; Lord, R.D. A Radio-Tracking System for Wild Animals. *J. Wildl. Manag.* **1963**, *27*, 9. [CrossRef]
3. Habib, B.; Shrotriya, S.; Sivakumar, K.; Sinha, P.R.; Mathur, V.B. Three Decades of Wildlife Radio Telemetry in India: A Review. *Anim. Biotelemetry* **2014**, *2*, 4. [CrossRef]
4. Wilmers, C.C.; Nickel, B.; Bryce, C.M.; Smith, J.A.; Wheat, R.E.; Yovovich, V. The Golden Age of Bio-Logging: How Animal-Borne Sensors Are Advancing the Frontiers of Ecology. *Ecology* **2015**, *96*, 1741–1753. [CrossRef]
5. Duran-Lopez, L.; Gutierrez-Galan, D.; Dominguez-Morales, J.P.; Rios-Navarro, A.; Tapiador-Morales, R.; Jimenez-Fernandez, A.; Cascado-Caballero, D.; Linares-Barranco, A. A Low-Power, Reachable, Wearable and Intelligent IoT Device for Animal Activity Monitoring. In Proceedings of the IJCCI 2019—Proceedings of the 11th International Joint Conference on Computational Intelligence, Vienna, Austria, 4–6 September 2019; pp. 516–521. [CrossRef]
6. Jukan, A.; Masip-Bruin, X.; Amla, N. Smart Computing and Sensing Technologies for Animal Welfare: A Systematic Review. *ACM Comput. Surv.* **2017**, *50*, 1–27. [CrossRef]
7. Muminov, A.; Sattarov, O.; Lee, C.W.; Kang, H.K.; Ko, M.C.; Oh, R.; Ahn, J.; Oh, H.J.; Jeon, H.S. Reducing GPS Error for Smart Collars Based on Animal's Behavior. *Appl. Sci.* **2019**, *9*, 3408. [CrossRef]
8. Nakagawa, K.; Kobayashi, H.H. Optimal Arrangement of Wearable Devices Based on Lifespan of Animals as Device Transporter Materials for Long-Term Monitoring of Wildlife Animal Sensor Network. *Sens. Mater.* **2020**, *32*, 13. [CrossRef]
9. Neethirajan, S. Recent Advances in Wearable Sensors for Animal Health Management. *Sens. Bio-Sens. Res.* **2017**, *12*, 15–29. [CrossRef]
10. Zhang, B.; Zhuang, L.; Qin, Z.; Wei, X.; Yuan, Q.; Qin, C.; Wang, P. A Wearable System for Olfactory Electrophysiological Recording and Animal Motion Control. *J. Neurosci. Methods* **2018**, *307*, 221–229. [CrossRef]
11. Lahoz-Monfort, J.J.; Magrath, M.J.L. A Comprehensive Overview of Technologies for Species and Habitat Monitoring and Conservation. *BioScience* **2021**, *71*, 1038–1062. [CrossRef]
12. Sugai, L.S.M. Pandemics and the Need for Automated Systems for Biodiversity Monitoring. *J. Wildl. Manag.* **2020**, *84*, 1424–1426. [CrossRef]
13. Neethirajan, S. Transforming the Adaptation Physiology of Farm Animals through Sensors. *Animals* **2020**, *10*, 1512. [CrossRef]
14. Helwatkar, A.; Riordan, D.; Walsh, J. Sensor Technology for Animal Health Monitoring. *Int. J. Smart Sens. Intell. Syst.* **2014**, *7*. [CrossRef]
15. Neethirajan, S.; Kemp, B. Digital Livestock Farming. *Sens. Bio-Sens. Res.* **2021**, *32*, 100408. [CrossRef]
16. Brandt, S.; Vassant, J.; Baubet, E. Adaptation d'un Collier Émetteur Extensible Pour Sanglier. *Faune Sauvage* **2004**, *263*, 13–18.
17. Cid, B.; da Costa, R.d.C.; Balthazar, D.d.A.; Augusto, A.M.; Pires, A.S.; Fernandez, F.A.S. Preventing Injuries Caused by Radiotelemetry Collars in Reintroduced Red-Rumped Agoutis, Dasyprocta Leporina (Rodentia: Dasyproctidae), in Atlantic Forest, Southeastern Brazil. *Zoologia* **2013**, *30*, 115–118. [CrossRef]
18. Dick, B.L.; Findholt, S.L.; Johnson, B.K. A Self-Adjusting Expandable GPS Collar for Male Elk. *Wildl. Soc. Bull.* **2013**, *37*, 887–892. [CrossRef]
19. Foley, C.J.; Sillero-Zubiri, C. Open-Source, Low-Cost Modular GPS Collars for Monitoring and Tracking Wildlife. *Methods Ecol. Evol.* **2020**, *11*, 553–558. [CrossRef]

20. Haramis, G.M.; White, T.S. A Beaded Collar for Dual Micro GPS/VHF Transmitter Attachment to Nutria. *Mammalia* **2011**, *75*, 79–82. [CrossRef]
21. Holzenbein, S. Expandable PVC Collar for Marking and Transmitter Support. *J. Wildl. Manag.* **1992**, *56*, 473. [CrossRef]
22. Isbell, L.A.; Bidner, L.R.; Omondi, G.; Mutinda, M.; Matsumoto-Oda, A. Capture, Immobilization, and Global Positioning System Collaring of Olive Baboons (*Papio anubis*) and Vervets (*Chlorocebus pygerythrus*): Lessons Learned and Suggested Best Practices. *Am. J. Primatol.* **2019**, *81*, 22997. [CrossRef]
23. Kolz, A.L.; Johnson, R.E. Self-Adjusting Collars for Wild Mammals Equipped with Transmitters. *J. Wildl. Manag.* **1980**, *44*, 273. [CrossRef]
24. Smith, B.L.; Burger, W.P.; Singer, F.J. An Expandable Radiocollar for Elk Calves. *Wildl. Soc. Bull.* **1998**, *26*, 113–117.
25. Dore, K.M.; Hansen, M.F.; Klegarth, A.R.; Fichtel, C.; Koch, F.; Springer, A.; Kappeler, P.; Parga, J.A.; Humle, T.; Colin, C.; et al. Review of GPS Collar Deployments and Performance on Nonhuman Primates. *Primates* **2020**, *61*, 373–387. [CrossRef]
26. Trayford, H.R.; Farmer, K.H. An Assessment of the Use of Telemetry for Primate Reintroductions. *J. Nat. Conserv.* **2012**, *20*, 311–325. [CrossRef]
27. Weaver, S.J.; Westphal, M.F.; Taylor, E.N. Technology Wish Lists and the Significance of Temperature-Sensing Wildlife Telemetry. *Anim. Biotelemetry* **2021**, *9*, 29. [CrossRef]
28. Advanced Telemetry Systems G5-D Iridium/GPS Collar. Available online: https://atstrack.com/tracking-products/transmitters/G5D-Iridium-GPS-Collar.aspx (accessed on 19 October 2020).
29. Advanced Telemetry Systems G2110E2 Iridium/GPS Collar. Available online: https://atstrack.com/tracking-products/transmitters/G2110E2-Iridium-GPS-Collar.aspx (accessed on 19 October 2020).
30. Lotek Ultimate V6C Series. Available online: https://www.lotek.com/wp-content/uploads/2018/12/Ultimate-V6C-Series-Spec-Sheet.pdf (accessed on 30 October 2020).
31. Lotek WildCell Series. Available online: https://www.lotek.com/wp-content/uploads/2018/05/WildCell-Series-Spec-Sheet.pdf (accessed on 30 October 2020).
32. Lotek PinnaclePro Series. Available online: https://www.lotek.com/wp-content/uploads/2018/08/PinnaclePro-Series-Spec-Sheet.pdf (accessed on 30 October 2020).
33. Telemetry Solutions Iridium GPS Collars. Available online: https://www.telemetrysolutions.com/wildlife-tracking-devices/gps-collars/iridium-gps-collars/ (accessed on 30 October 2020).
34. Tellus Tellus GPS Collars. Available online: https://www.followit.se/files/folder/magazines/tellus/tellus_product_sheet.pdf (accessed on 30 October 2020).
35. Telonics GPS/Iridium Terrestrial Systems. Available online: https://www.telonics.com/products/gps4/gps-iridium.php (accessed on 30 October 2020).
36. Ixorigue GPS Ixotrack. Available online: https://ixorigue.com (accessed on 30 October 2020).
37. Blanco, T.; Casas, R.; Manchado-Pérez, E.; Asensio, Á.; López-Pérez, J.M. From the Islands of Knowledge to a Shared Understanding: Interdisciplinarity and Technology Literacy for Innovation in Smart Electronic Product Design. *Int. J. Technol. Des. Educ.* **2017**, *27*, 329–362. [CrossRef]
38. What Is the Framework for Innovation? Design Council's Evolved Double Diamond. Available online: https://www.designcouncil.org.uk/news-opinion/what-framework-innovation-design-councils-evolved-double-diamond (accessed on 12 November 2020).
39. IDEO.org IDEO Tools—Human Centered Design. Available online: https://www.ideo.org/tools (accessed on 12 November 2020).
40. Insitute of Design at Stanford. An Introduction to Design Thinking PROCESS GUIDE. Available online: https://dschool-old.stanford.edu/sandbox/groups/designresources/wiki/36873/attachments/74b3d/ModeGuideBOOTCAMP2010L.pdf (accessed on 12 November 2020).
41. Blanco, T. (Universidad de Zaragoza, Zaragoza, Spain). Cosica_Guía Didáctica Para El Diseño. 2020; Unpublished work.
42. Blanco, T.; Casas, R.; Marín, J.; Marco, Á. (Universidad de Zaragoza, Zaragoza, Spain). Designing in the Internet of Things. A Multidisciplinary Instructional Methodology. 2020; Unpublished work.
43. Blanco, T.; Berbegal, A.; Blasco, R.; Casas, R. Xassess: Crossdisciplinary Framework in User-Centred Design of Assistive Products. *J. Eng. Des.* **2016**, *27*, 636–664. [CrossRef]
44. Brannen, J. Mixing Methods: The Entry of Qualitative and Quantitative Approaches into the Research Process. *Int. J. Soc. Res. Methodol.* **2005**, *8*, 173–184. [CrossRef]
45. Lund, T. Combining Qualitative and Quantitative Approaches: Some Arguments for Mixed Methods Research. *Scand. J. Educ. Res.* **2012**, *56*, 155–165. [CrossRef]
46. Telonics Telonics Breakways Collars. Available online: https://www.telonics.com/products/expansionBreakawayCollars/ (accessed on 30 October 2020).
47. Frankfurter, G.; Beltran, R.S.; Hoard, M.; Burns, J.M. Rapid Prototyping and 3D Printing of Antarctic Seal Flipper Tags. *Wildl. Soc. Bull.* **2019**, *43*, 313–316. [CrossRef]
48. Fischer, M.; Parkins, K.; Maizels, K.; Sutherland, D.R.; Allan, B.M.; Coulson, G.; Di Stefano, J. Biotelemetry Marches on: A Cost-Effective GPS Device for Monitoring Terrestrial Wildlife. *PLoS ONE* **2018**, *13*, e0199617. [CrossRef] [PubMed]
49. Paci, P.; Mancini, C.; Price, B.A. Designing for Wearability. In Proceedings of the Sixth International Conference on Animal-Computer Interaction, Haifa, Israel, 12–14 November 2019; ACM: New York, NY, USA, 2019; pp. 1–12.

 *sensors*

Article

Livestock Informatics Toolkit: A Case Study in Visually Characterizing Complex Behavioral Patterns across Multiple Sensor Platforms, Using Novel Unsupervised Machine Learning and Information Theoretic Approaches

Catherine McVey [1,*], Fushing Hsieh [2], Diego Manriquez [3], Pablo Pinedo [3] and Kristina Horback [1]

1 Department of Animal Science, University of California Davis, Davis, CA 95616, USA; kmhorback@ucdavis.edu
2 Department of Statistics, University of California Davis, Davis, CA 95616, USA; fhsieh@ucdavis.edu
3 Department of Animal Science, Colorado State University, Fort Collins, CO 80523, USA; diego.manriquez_alvarez@colostate.edu (D.M.); pablo.pinedo@colostate.edu (P.P.)
* Correspondence: cgmcvey@ucdavis.edu

Abstract: Large and densely sampled sensor datasets can contain a range of complex stochastic structures that are difficult to accommodate in conventional linear models. This can confound attempts to build a more complete picture of an animal's behavior by aggregating information across multiple asynchronous sensor platforms. The Livestock Informatics Toolkit (LIT) has been developed in R to better facilitate knowledge discovery of complex behavioral patterns across Precision Livestock Farming (PLF) data streams using novel unsupervised machine learning and information theoretic approaches. The utility of this analytical pipeline is demonstrated using data from a 6-month feed trial conducted on a closed herd of 185 mix-parity organic dairy cows. Insights into the tradeoffs between behaviors in time budgets acquired from ear tag accelerometer records were improved by augmenting conventional hierarchical clustering techniques with a novel simulation-based approach designed to mimic the complex error structures of sensor data. These simulations were then repurposed to compress the information in this data stream into robust empirically-determined encodings using a novel pruning algorithm. Nonparametric and semiparametric tests using mutual and pointwise information subsequently revealed complex nonlinear associations between encodings of overall time budgets and the order that cows entered the parlor to be milked.

Keywords: dairy welfare; hierarchical clustering; mutual information; precision livestock farming; time budgets; unsupervised machine learning

Citation: McVey, C.; Hsieh, F.; Manriquez, D.; Pinedo, P.; Horback, K. Livestock Informatics Toolkit: A Case Study in Visually Characterizing Complex Behavioral Patterns across Multiple Sensor Platforms, Using Novel Unsupervised Machine Learning and Information Theoretic Approaches. *Sensors* **2022**, *22*, 1. https://doi.org/10.3390/s22010001

Academic Editor: Yongliang Qiao

Received: 30 October 2021
Accepted: 17 December 2021
Published: 21 December 2021

1. Introduction

Precision livestock farming (PLF) technologies produce prodigious amounts of data [1]. Although the behaviors encoded by such sensors are often much simpler than those that can be quantified by a human observer, the measurement granularity and perseverance provided by these technologies creates new opportunities to study complex behavioral patterns across time and in a wider range of contexts. Observations collected on a single animal over extended observation windows at high sampling frequencies can, however, contain a range of complex temporal patterns, such as cyclicity, non-stationarity, autocorrelation, etc. [2]. Furthermore, when sensors are applied to large heterogenous groups of animals housed socially in spatially restricted environments, recorded behaviors may also contain complex interdependencies between animals at the dyadic, triadic, clique, and herd levels [3–5]. Failing to accommodate all these complex structural and stochastic features in a conventional model-based approach to statistical inference risks returning spurious insights into the underlying behavioral dynamics. Developing such a model with a single PLF data stream can be challenging. Provided multiple data streams, however, the logistical challenges presented by model-based analytical frameworks can rapidly compound, creating significant barriers to cross-sensor inferences, and thereby impeding researchers from extracting more holistic behavioral inferences from increasingly data-rich farm environments.

Unsupervised Machine Learning (UML) tools may provide a more flexible and forgiving approach to knowledge discovery in the context of large sensor datasets [6,7]. Such algorithms excel at identifying and characterizing complex non-random behavioral patterns lying beneath the stochastic surface of a dataset, while often employing relatively few structural assumptions about the data [8–10]. Hierarchical clustering-based techniques offer an intuitive and highly adaptable approach to visualizing high dimensional datasets that is particularly well-suited to exploratory data analysis [4,9]. Indeed, by reducing the complex behavioral signals present in a sensor dataset into a series of discrete clusters, such algorithms may be viewed as an empirical extension of classical ethological techniques. Discrete data, however, can be challenging to work with in most frequentist and even many Bayesian frameworks. Estimators based on information entropy, on the other hand, are purpose-made to quantify uncertainty in discretely encoded data without knowledge of the underlying distribution, and thus naturally complement hierarchical clustering-based algorithms [7,11,12].

Clustering algorithms, by virtue of their incredible flexibility, have successfully been applied to a range of PLF data streams [7,13–18]. In our own previous work, we have highlighted the utility of hierarchical clustering-based approaches in leveraging the behavioral co-dependencies of cows housed socially in large groups, in a production environment, in order to recover complex temporal patterns in behavior [7]. In these analyses, data mechanics algorithms were able to recover complex nonstationarity in the order in which cows entered the milking parlor. Some of these changes in queuing patterns could be attributed to the shift to spring pasture access, but other transient and persistent shifts in entry order recovered in these encodings were driven by environmental factors not experimentally recorded [7,19,20]. Entropy-based nonparametric permutation tests were also successful in recovering preliminary evidence of significant nonlinear associations between encodings of entry-order patterns and activity patterns recorded using ear-tag accelerometers. In this paper we will explore how novel ensemble simulation techniques [11] that emulate and adjust for the complex sources of error in PLF data streams may be used to produced more balanced encodings of multi-dimensional behavioral data. We also introduce a new dendrogram pruning algorithm that is able to efficiently repurpose these same ensemble simulations, to ensure that that the power of hierarchical clustering tools do not exceed the resolution of the sensor. Finally, we demonstrate the utility of information decomposition techniques within our existing nonparametric mutual information testing framework, to better facilitate the visual characterization of complex behavioral patterns across sensor data sets that might be overlooked in more conventional model-based analyses.

2. Materials and Methods

2.1. Description of Data

To demonstrate the efficacy of our analytical approach, data was repurposed from a feed trial assessing the impact of an organic fat supplement on cow health and productivity, through the first 150 days of lactation. All animal handling and experimental protocols were approved by the Colorado State University Institution of Animal Care and Use Committee (Protocol ID: 16-6704AA). The study ran from January through July in 2017, on a USDA Certified Organic dairy in Northern Colorado, enrolling a total of 200 cows over a 1.5-month period into a mixed-parity herd of animals, with predominantly Holstein genetics. Cows were maintained in a closed herd in an open-sided free-stall barn, stocked at roughly half capacity with respect to both feed bunk spaces and stalls. Cows had free access to an adjacent outdoor dry lot while in their home pen, and beginning in April were moved onto pasture at night, to comply with organic grazing standards. Cows were milked three times a day, with free access to TMR between milkings, and were head locked each morning to facilitate data collection and daily health checks. For more details on feed trial protocols, see Manriquez et al. (2018) and Manriquez et al. (2019) [21,22].

In addition to standard production and health assessments, behavioral data was also obtained from several PLF data streams [19]. Milking order, or the sequence in which cows enter the parlor to be milked, is automatically recorded as metadata in all modern RFID-equipped milking systems. Our study cows were milked in a DelPro™ rotary parlor (DeLaval, Tumba, Sweden). At each morning milking, raw milking logs were exported from the parlor software, and the data were processed in order to extract the single-file order that cows entered the rotary [23]. A total of 80 milk order records—26 recorded while cows remained overnight in a free-stall barn, and 54 following the transition to overnight access to spring pasture—were used to create discrete encodings for parlor entry patterns via data mechanics clustering (see McVey et al. for further analytical details) [7]. The dendrograms summarizing the distribution of cow entry-order patterns and subsequent heatmap visualizations will be subjected to further analysis, without modifications to the previously reported encodings.

Animals enrolled in this feed trail were also fitted with a CowManager ear tag accelerometer (Agis Automatisering BV, Harmelen, The Netherlands). This commercial sensor platform, while designed and optimized for disease and heat detection, also provides hourly time budget estimates for total time (min) engaged in five mutually exclusive discrete behaviors—eating, rumination, non-activity, activity, and high activity [24,25]. Time budget data was collected on all animals for a contiguous period of 65 days (1560 h). The observation window began on 17 February, shortly after trial enrollment was completed, and ended on 23 April, when the grazing season commenced and cows were moved overnight beyond the range of the receiver antennae. After eliminating the data of cows that were removed prematurely from the observation herd due to acute clinical illness, as well as several cows with persistent receiver failure, complete sensor records were available for 179 animals. In order to focus fully on the logistical challenges of encoding and characterizing the complex multivariate dynamics of this system, we have chosen to compress this data over the time axis to consider only the overall time budgets of these cows, and will leave explorations of the longitudinal and cyclical complexity of this dataset for future work.

2.2. Improving Empirical Encodings of Overall Time Budget through Simulation

Regardless of its original distribution, data can always be coarsened into a discrete variable [26]. For complex or poorly defined systems, where appropriate cutoffs (binning rules) cannot be inferred *a priori*, an empirically-determined encoding may provide a more flexible and comprehensive approach to discretizing the underlying behavioral signals. One algorithm that provides a model-free approach to pattern encoding within the larger cannon of UML tools is hierarchical clustering. This approach employs a bottom-up agglomerative strategy to group observational units into discrete clusters of variable sizes, progressively building a coherent picture of herd-level global structures from the similarities in behavioral patterns observed between pairs of individuals [9,10]. This series of progressive pairings can be expressed graphically in the form of a dendrogram, which serves as a 2D representation of the data's geometric distribution in its higher dimensional measurement space, and can subsequently be used in data visualizations to highlight the most prominent structural features of a dataset [19].

The efficacy of any hierarchical clustering scheme, however, is largely contingent on the adequacy of the estimator used to quantitatively express the pairwise dissimilarity between observational units [10]. The Euclidean distance (L2 norm) is the default estimator used in most applications of this algorithm [9,10,27], including much of the previous work in precision livestock applications [13,14,18]. The L2 norm is appropriate for many measurement systems where variance is reasonably uniform across a continuous domain of support. Time budget data, however, is distributed multinomially, and as such has significant domain constraints [26]. Put more simply, we know that the minutes logged for each behavior must sum to an hour. So, if a cow has ruminated for 60 min, then there can be no uncertainty in the remaining axes, because we know these values must be zero. These domain constraints impose co-dependencies between the behavioral axes that become stronger as observations shift towards the boundaries of the distribution's support, which in turn warp the intrinsic variability of each axis contingent upon their location within the domain.

This statistical tedium also has some intuitive behavioral implications. Suppose we have two cows, Betty and Bessy, who spend 13 and 14 h a day ruminating, respectively. How "different" are these values? Since both cows are exceeding rumination rates needed to sustain a healthy metabolism, we would not anticipate that this difference would have a significant biological impact on these animals, and may ultimately be explained by relatively trivial behavioral fluctuations. Now, suppose instead that we have two other cows, Daisy and Delilah, who spend only 3 and 4 h a day ruminating, respectively. Given that both these cows are now well below the normal threshold for this behavior, this one-hour difference may have significant biological impacts. With a simple L2 norm, however, these two pairs would be given equivalent dissimilarity estimates for this behavioral axis, and so clearly a better estimator is needed.

Relative entropy, also referred to as the Kullback–Leibler divergence, is a classic information theoretic metric specifically designed to contrast discrete probability distributions, and thus a natural candidate for analysis of time budget data [12]. For any two distributions that utilize the same alphabet of k = 1 . . . K categorical features (i.e.,—use the same ethogram), relative entropy can be calculated using Equation (1), and converted to a symmetric distance measure using Equation (2). By utilizing the proportion of time that an animal invests in each behavior as both a nominal and relative value, this estimator is able to adjust the relative difference between cows by the absolute position of each observation relative to the boundary of the domain.

$$D_{KL}(P||Q) = \sum_{k} P(k)\ log \frac{P(k)}{Q(k)} \tag{1}$$

$$P = normalized\ time\ budget\ vector\ for\ Cow\ A$$

$$Q = normalized\ time\ budget\ vector\ for\ Cow\ B$$

$$D_{KL}(P, Q) = D_{KL}(P||Q) + D_{KL}(Q||P) \tag{2}$$

Domain constraints are not, however, the only stochastic feature that need be accommodated when working with time budget data. There is also the measurement error attributable to the sensor itself. Returning to the previous example, suppose that we also know that our rumination records are only accurate to ± 1 h. Is it then still appropriate to give more weight to the one-hour difference between Daisy and Delilah, than between Betty and Betsy? Since both observations are within the bounds of error, attempting to enhance the underlying biological signal may only succeed in amplifying measurement noise. A closed-form estimator, however, may not be readily generalizable to the wide range of measurement error models encountered with PLF sensors. We therefore propose that a simulation-based approach may offer a more flexible means of accounting for measurement errors in dissimilarity estimates [11].

The LIT package provides a built-in simulation utility for time budget data that seeks to mimic the stochastic error structure of the original data while still preserving the underlying behavioral signal [11]. Data is provided as a tensor, with cow indexed on the first axis, time indexed on the second, and the component behaviors on the final axis. The count data at each cow-by-time index is then used to redraw a simulated datapoint from one of three optional distributions [26]. In the first, the user may sample directly from a multinomial distribution centered around the normalized observed count vector. This model assumes that measurement error should shrink as a cow dedicates larger proportions of an observation window to specific behaviors, and intrinsically prevents estimates from being generated outside the domain of support. Variance can be underestimated at the extremes of the domain, however, if the probability for a behavior is non-negligible, but the observed count is zero due to under-sampling. This issue may be addressed in sampling option two, where samples are redrawn from a multivariate beta distribution (MBD), also known as a Dirichlet distribution, again parameterized using the normalized observed count. While this sampling strategy slightly biases the simulation towards the center of the distribution, it prevents under-sampling at the extremes of the domain. Finally, users may combine these sampling strategies in sampling option three, wherein the probability vector used to parameterize the multinomial is drawn first from the Dirichlet, in order to further increase the uncertainty in the simulated data. After simulation has been completed by redrawing samples at the finest level of temporal granularity supported by the sensor, the data can then be conditionally or fully aggregated along temporal axis as required for downstream analysis as a time budget.

This simulation routine was used to create an ensemble of B = 500 simulated overall time budget matrices that mimicked the stochasticity attributable to a reasonable approximation of the measurement error of the sensor. Stored as a tensor with replication on the last axis, the variance of the ensemble of simulations could then be easily calculated for each combination of cow index and behavioral axis. If the underlying simulation strategy is a reasonable representation of the noise in the sensor, then these variance terms will then serve as a sufficient approximation of the relative uncertainty in each data point. We propose that that this information can then be incorporated into the calculation of dissimilarity estimates by serving as penalty terms in the calculation of an ensemble-weighted distance estimator defined in Equation (3).

$$D_{EW}(P,Q) = \sum_k \frac{(P_{obs}(k) - Q_{obs}(k))^2}{\sigma^2_{P^*(k)} + \sigma^2_{Q^*(k)}} \tag{3}$$

$$\sigma^2_{P^*(k)} = Variance\ of\ ensemble\ of\ simulated\ values\ for\ Cow\ A\ for\ behavior\ k$$

$$\sigma^2_{Q^*(k)} = Variance\ of\ ensemble\ of\ simulated\ values\ for\ Cow\ Q\ for\ behavior\ k$$

The rescaling strategy employed in our proposed dissimilarity estimator is strongly inspired by traditional analysis of variance (ANOVA) techniques, thereby providing several insights into its anticipated behavior. First, because the simulations were generated using the multinomial or one of its analogs, we can infer that these penalty terms will not be homogenous across the domain of support, but should shrink as observations approach the boundary. This will allow the ensemble-weighted distance estimator to emulate the rescaling dynamic achieved with the KL distance; however, rescaling at the extremes of the domain will ultimately be bounded by our simulated measurement error, so as not to exceed the precision of the sensor. Second, because we have here emulated measurement error in our simulation using sampling uncertainty, the central limit theorem will apply [9]. Thus, we can anticipate that as the number of observations per animal increases, the impact of measurement error on our inferences will shrink, allowing progressively more subtle differences between animals to come into resolution. Taking this property to its limit, however, can it be said that with enough observation minutes the differences between cows can be inferred with near certainty? That intuition, of course, is at odds with our characterization of a dairy herd as a complex system, and highlights an additional stochastic element that must be accommodated—the behavioral plasticity of the cows themselves in response to changes in the production environment [4].

Given the extended observation window of this particular data set, it would be possible to recalculate time budget conditional on the day of observation, and then use the variance in daily time budget along each behavioral axis as a penalty term. Such estimates would collectively reflect heterogeneity in variance attributed to domain constraints, measurement error, and behavioral plasticity. Such an approach would not, however, be feasible for datasets collected over shorter time intervals with fewer replications, or in applications with behavioral responses where there is no clear hierarchy in the temporal structure of the same. We therefore propose that our stochastic simulation model can be extended to also provide a generalizable means to approximate the uncertainty of the underlying behavioral signal.

As before, the measurement error was simulated by redrawing samples at the finest temporal granularity provided by the sensor. Prior to compression along the temporal axis, however, a random subsample of observations days was selected across all cows, and only these values were used to calculate the simulated overall time budget. If all cows demonstrated comparable levels of consistency in their daily time budgets, then reducing the effective sample size of our simulated data sets through a subsampling routine would increase the ensemble variance estimates. This, in turn, would make our approximation of measurement error hyper-conservative, but this increase would be uniform across all cows. If, instead, some cows were less consistent in their time budgets across days, then the sampling error imposed by the subsampling routine would be greater, resulting in a larger ensemble variance estimate. Thus, we would expect a stronger penalty to be applied to cows who demonstrated greater plasticity in their behavioral responses to both transient and persistent changes in the production environment. For small datasets with a limited number of replications, the number of subsamples could be set quite close to the size of the complete sample, and would thus emulate a jackknife approach to variance estimation [9,10,28]. For larger datasets, however, the subsample size could be set smaller, to make the resulting ensemble variance estimates progressively more sensitive to the uncertainty in the underlying behavioral signal.

To evaluate the empirical performance of these dissimilarity estimators, distance matrices were calculated for the 177 cows with complete CowManager time budget records. Euclidean distance and KL Distance were calculated using base R utilities, with speed up options utilizing the *Rfast* package [23,29]. An ensemble-weighted dissimilarity matrix was first calculated using simulated values accounting only for measurement error using the most conservative joint Dirichlet-multinomial sampling scheme, hereafter referred to as noise-penalized distance. A second ensemble-weighted dissimilarity matrix was then calculated using the same sampling scheme for measurement noise but aggregated over a 14-day subsample to account for behavioral plasticity in daily time budgets, hereafter referred to as plasticity-penalized distance. The LIT package provides users a clustering visualization utility, which converts dissimilarity matrices into a dendrogram using the *hclust* utility in base R with default Ward D2 linkage [23], and the generates heatmap visualizations of the resulting clustering results using the *pheatmap* package [30]. Heatmaps were generated on a grid of cluster values from k = 1 . . . 10 for each of the four dissimilarity estimators, with complete results provided in Supplementary Materials, the results for k = 10 clusters are provided. The LIT package also provides users with a plotting utility to visually contrast the broader patterns between behavioral encodings. Outputs from the clustering utility are passed in to create a contingency matrix generated using *ggplot2* with cells colored by their corresponding cell count [31]. The heatmap visualizations for each encoding are then added to the row and column margins of the contingency matrix using the *ggpubr* package [32], and arranged such that each row cluster in either heatmap matches the order of the contingency matrix reading either up-down or left-to-right, allowing for direct and detailed visual comparison of the discretized behavioral patterns. Comparisons between the noise-penalized and plasticity-penalized encodings are provided.

2.3. *Improving Tree Pruning Decisions through Simulation*

An optimal encoding strategy seeks to minimize the loss of relevant information by retaining as much of the underlying deterministic signal as possible, while hemorrhaging only noise [26]. In a hierarchical clustering framework, this is achieved by pruning the dendrogram built from the dissimilarity matrix at the point where the branches cease to represent differences in the underlying signal. Standard pruning strategies allow users to either: (1) provide a dissimilarity cutoff, below which value all further branches are grouped into the same bin, or (2) extract the first K branches of the tree [9,10]. As with the default Euclidean distance dissimilarity estimator, this approach may be appropriate for datasets with relatively homogenous variance structures. For data drawn from intrinsically heterogenous distributions, however, the branch lengths cannot be directly compared across the domain of support, making globally-defined pruning rules a suboptimal strategy for analysis of time budget data.

More fundamentally, a homogenous pruning strategy may be too simplistic for many PLF sensor datasets, for which the underlying signal often represents a complex composite of behavioral mechanisms that operate at multiple scales. Although some environmental factors might be expected to have an impact on cattle behaviors that are uniform across the herd, other factors might elicit responses that differ in magnitude for different subgroups within the larger population, or even become isolated within smaller social cliques. For example, we might expect the number of times cows are moved each day for milking will place similar constraints on the time left to lie down across all animals, but overstocking with respect to stall spaces might have a much larger magnitude of impact on the lying patterns of subordinate heifers than the more dominant older cows [33]. In such a complex system, we would expect the heterogeneity imposed by the underlying biological signal to differ in scale across the dataset. Subsequently, in attempting to employ a global cutoff decision to encode information for such a dataset, we would always be faced with the difficult decision to either ignore the subtler behavioral patterns present in some branches of the tree, or else allow noise to contaminate our encoding of other branches with intrinsically coarser behavioral patterns.

Although all the components that contribute to the signal in a complex livestock system might be difficult to anticipate *a priori*, we propose that a more dynamic pruning algorithm might still be achieved, by again employing flexible simulation-based approaches to emulate the comparably simpler sources of uncertainty. If each branch of the dendrogram is viewed as a pairwise contrast between two groups of animals, then we need only to determine whether the bifurcation under inspection represents a difference in the underlying signal that can be reliably distinguished from noise. If it can, then the two groups should be split in the final encoding to capture this feature of the data's distribution. If a branch falls below the intrinsic resolution of the data, however, then the branch may be pruned so that all animals are placed into the same cluster, with no loss of meaningful information. By implementing such a branch-level test recursively, we can gradually work our way down the tree with adaptive locally-defined pruning decisions.

To evaluate the reliability of the behavioral signal encoded at each bifurcation of the tree [34], our branch test utility utilizes two mimicries. The first set of simulations are generated under the alternative hypothesis that assumes a branch contains an underlying deterministic signal that is only partially obscured by stochastic noise. Thus, we can simply repurpose the ensemble of simulated data sets used previously to calculate the ensemble weighted dissimilarity metrics by mimicking the uncertainty in the observed data. The second set of simulations are generated under the null hypothesis that a given branch contains only noise. As the null implies that animals demonstrate equivalent patterns of behavior within the resolution of the sample, this mimicry can be generated quite efficiently using a standard bootstrapping routine [28], wherein time budgets simulated under the alternative are unconditionally resampled from amongst all animals in a given branch. HClustering is then performed independently on each data mimicry in either ensemble, and the first k branches are extracted to create an ensemble of discrete encodings.

Under the alternative hypothesis, a strong signal should produce a robust tree structure such that, even after the addition of simulated noise, the resulting encoding would still closely mirror that of the original observed data. As the stochastic component of a dataset becomes stronger relative to the signal, these bifurcation points will become progressively less stable, and the subsequent encodings less reliably aligned with the original data. When the signal falls below the resolution of the data, the tree structures of the simulated data would then seldom match that of the original data, and so would become poorly distinguished from encodings generated under the null, with no signal component. We propose that mutual information, which can be calculated without any additional distributional assumptions, can be used to quantify the similarity between the observed data and each mimicked dataset, and subsequently used to determine if simulations under the alternative are distinguishable from the null [12]. In our study, a bifurcation was determined to be significant if less than 5% of the MI values calculated for data simulated under the alternative hypothesis fell below the 95th quantile of MI values calculated for data simulated under the null. If a bifurcation was instead deemed insignificant, the branch was pruned and all cows within it assigned to the same cluster in subsequent encodings.

In evaluating the significance of a bifurcation, it seems intuitive that a k = 2 binary encoding should be utilized. For complex systems subject to the influence of multiple competing drivers of behavioral responses, however, a false negative result can occur with this parameterization if the addition of stochastic noise perturbs the order in which two significant mechanisms with similar magnitudes of impact are bifurcated. Such trivial destabilizations of the tree structures can be readily identified in visualizations of the distributions of MI values calculated against simulations under the alternative, as the "flip flopping" between bifurcation points produces clear evidence of multimodality (see Figure 1). To circumvent this issue, the LIT package provides users the option to re-test any bifurcations deemed insignificant, using a binary encoding with a more granular discretization (k > 2). This effectively allows the algorithm to "look down the branch" to absorb any irrelevant flip-flopping between competing signals, thereby preventing spurious over-pruning that would hemorrhage information on significant behavioral patterns from the final encoding.

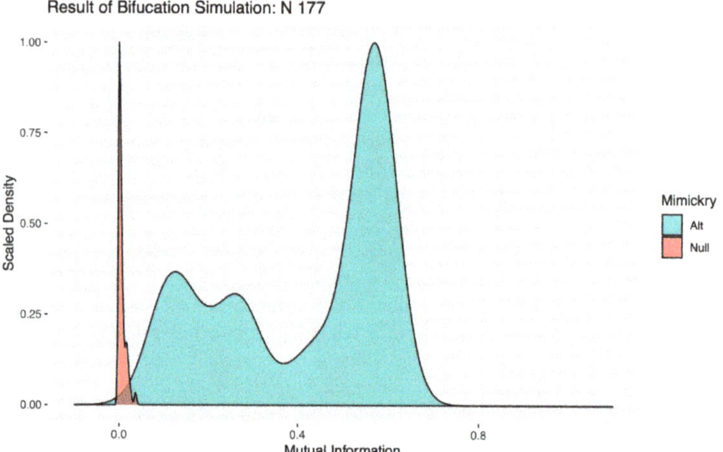

Figure 1. Visualization of the test–branch results for the first bifurcation of the Euclidean distance time budget dendrogram, cut using the noise-penalized ensemble of data mimicries. In simulations under the alternative hypothesis, the addition of noise intended to mimic measurement error has destabilized the tree, causing it to "flip-flop" between first isolating cows with more moderate time budgets, and animals at the two extremes of the tradeoff between eating and ruminating. Although both branches are distinguishable from measurement errors, this ambiguity in bifurcation order has produced bimodality in the distribution of mutual information estimates against the encoding for the observed data. Retesting with more clusters allows the algorithm to "look down the branch" to produce better separation between encodings under the null and the alternative, and thereby avoid spurious over-pruning.

Full results for the application of our ensemble-cut algorithm to dendrograms generated using each of the four dissimilarity estimators discussed in the previous section, and using both the noise-penalized and plasticity-penalized encodings, are provided in the Supplementary Materials. A summary of results for the application of the ensemble-cut algorithm applied to dendrograms generated from the noise- and plasticity-penalized ensemble-weighted distance metrics, the noise-penalized and plasticity-penalized encodings, respectively, are provided.

2.4. An Information Theoretic Framework for Cross-Sensor Inferences

Equipped with an appropriate encoding to discretely represent the heterogeneity in overall time budgets within this herd, and provided with the encoding of longitudinal patterns in parlor entry position from previous work with this data set, a potential question to ask would be: how does a cow's time budget, which is largely determined by her behaviors in the home pen, relate to her behavior in the milking queue? There are a number of nonparametric and parametric techniques available to evaluate the overall strength of association between two discrete variables, by evaluating the distribution of animals in the joint encoding [26]. There is, however, perhaps greater practical utility in characterizing low and high points within the joint encodings, which would provide more detailed insights into the tradeoffs between specific behavioral patterns recovered from the data streams in these distinct farm contexts. Towards this end, information theory offers a more comprehensive approach to decomposing the stochasticity within discretely encoded variables, and thus may provide a more holistic approach to evaluating both the global and local features of a joint encoding, while employing few structural assumptions [12].

First, to evaluate the strength of the overall relationship between two discretized behavioral responses, the LIT package provides users a permutation-based bivariate testing utility that uses the mutual information estimator to quantify the amount of information entropy that is redundant between the two encodings [7,12]. We can anticipate, however, that the efficacy of this test in recovering significant relationships between the underlying biological signals will be affected by the resolutions of the encodings. Suppose that a single latent biological factor impacts the behavioral responses collected by both PLF data streams, creating informational redundancy between the two encodings. If we cut the trees above the intrinsic magnitude of its impact on a given behavior, its influence may be overlooked and mutual information underestimated. On the other hand, if we prune the tree far below the magnitude of its impact, our inferences can lose power, as bin sizes in the joint encoding become progressively smaller, weakening the empirical estimation of the joint probability distribution and thereby increasing estimation error in the MI estimator. The resolution of our encodings must, therefore, be optimized to match the dynamics of the system, or a false negative result may be returned. To further complicate matters, however, we cannot necessarily assume that the magnitude of impact of a given latent factor will be uniform across behaviors, nor should we expect in a complex farm environment that behaviors will be influenced by a single latent factor.

To overcome this logistical challenge without falling back on dubious *a priori* assumptions, the LIT package implements mutual information-based permutation tests on a grid, varying the cluster resolutions across both behavioral axes [7]. Under the null hypothesis that no significant bivariate relationship exists between data streams, cow ID labels are randomly permuted within each tree, preserving the marginal distribution of the data along each axis, but destroying any latent bivariate relationships. These permuted trees are then cut, and the mutual information of the joint encoding estimated for each combination of cluster counts on the grid. A p-value is then generated by comparing the observed MI value of the joint encoding at each grid point against the corresponding distribution of MI values simulated under the null. Just as a scientist varies the focus of a microscope to bring microbes of different size into resolution, we can expect that geometric features of the joint probability distribution imposed by latent deterministic variables, that vary in scale of impact, will come into and fall out resolution as these meta-parameters are varied across the grid of cluster counts. To help the user visually identify where such features have come into resolution, the LIT package also returns a heatmap visualization of the observed MI value for each grid point that is centered and scaled, relative to the distribution of MI values under the null. For behavioral measurements subject to the influence of multiple biological and environmental factors operating simultaneously, this exhaustive approach to parameterization enables users not only to build a more complete picture of a complex behavioral system, but may also provide insight into the hierarchy of these behavioral responses.

Unfortunately, as the resolution of the encodings is increased, MI estimates not only become less precise, but they may also become less accurate. Bias is introduced when empirical estimates of the joint probability distribution become so granular (i.e.,a high number of bins relative to the total sample size) that regions with low but nonzero probabilities go unsampled. These zero-count bins cause the total entropy calculated from the empirical joint probability distribution to be underestimated which, in turn, causes the relative amount of redundant information to be overestimated. Although the magnitude of this bias is partially dependent on the total sample size, it is also contingent on the structure of the joint probability distribution itself, namely the number of low-probability cells. Given that the joint probability distribution under the null, which is randomly permuted to intentionally remove any nonrandom features in the sample, can be expected to have a more uniform distribution of probability than the observed dataset, we can anticipate that the magnitude of the bias may differ between these two distributions as the sample becomes more granular, preventing MI estimates from being directly comparable. To overcome this issue, the LIT package by default provides entropy estimators based on the Maximum Likelihood frequency estimates, but allows users to select from a range of bias-corrected frequency estimates available in the *entropy* package [35]. Based on the simulation work by Hausser and Strimmer (2009), the JS "shrink" estimator was used in our study to conduct bias-corrected mutual information permutation tests [36].

Not only can the impact of latent factors on behavioral measures differ in magnitude, we can also anticipate that responses may differ in both strength and direction for different subgroups within the herd. Such nonlinear dynamics are easily captured in a model-free MI test, but further inspection of the contingency table is needed to fully characterize such complex bivariate relationships between sensor outputs. If either marginal encoding has roughly the same number of observations in each bin, then the cell counts in the joint contingency table can be directly compared, as under the null we would expect each cells to be equiprobable. For empirically defined encodings, however, bin sizes can vary significantly to better capture the underlying geometry of the univariate data distribution. Such differences in marginal probabilities prevent the raw cell counts from being directly compared. To better identify which cells in an empirically defined joint encoding are driving a significant overall relationship between two data streams, mutual information can be decomposed into pointwise mutual information (PMI) values [37]. The LIT package provides users the option in the *compareEncodings* plotting utility to color cells in the joint contingency table by PMI estimate, to better facilitate direct visual comparisons of the encodings. To further enhance visualizations of the joint probability distribution that significantly differs from expected cell counts under the null, users may also specify a probability threshold above which PMI values should not be displayed, which was determined here by simulating PMI estimates under the null by redrawing from a multinomial distribution using the outer product of the marginal distributions.

Bivariate tree tests were applied to the time budget encodings, using both the noise- and plasticity-penalized dissimilarity metrics, and pruned using the more conservative plasticity-penalized mimicry, against the encoding of parlor entry order data produced using data mechanics clustering from our previous work [7]. A 2:10 $\times$ 2:10 grid was used to determine the optimal resolution for the bivariate relationship, with the optimal meta-parameters used to create visualizations of the joint encoding, wherein pointwise mutual information values were used to color cell counts that were significant at the alpha = 0.05 significance level. To further explore latent factors that might explain significant associations between entry position and time budgets, bivariate tree tests and pointwise mutual information tests were also applied separately to the encodings of both PLF data streams and health records.

3. Results and Discussion

3.1. Improving Empirical Encodings of Overall Time Budget through Simulation

Figure 2 provides a visual comparison of the time budget encodings for the four candidate dissimilarity metrics. In each heatmap visualization, individual cows are arranged along the row axis, and the mutually exclusive behaviors that comprise the overall time budget are ordered along the columns. Each cell within the heatmap is subsequently colored to reflect the proportion of time that a given cow is recorded by the accelerometer system to engage in a specific behavior over the observation window. Few cows dedicated more than half of their time to any one behavioral axis, which is not surprising, given that total lying time in this system is split between the nonactive and rumination axes [33]. Time recorded as eating and time recorded as ruminating were the highest magnitude behavioral axes, but time spent eating demonstrated far greater range and heterogeneity. Time spent nonactive was lower in overall magnitude, but still showed a fair amount of heterogeneity across cows. The active and highly active axes, however, were both quite low in magnitude and generally demonstrated less systematic heterogeneity across the herd. The order of cows along the row axis in each heatmap is determined by the dendrogram calculated for each dissimilarity matrix. The dendrogram can be interpreted as an approximate 2D representation of the distribution of the cows with the 5D multinomial space of the time budget, and thus serves to bring out in the heatmap systematic differences in time budget across the herd. Gaps were added between rows to indicate branches that have been pruned, such that all cows within a given branch received the same discrete value in the final time budget encoding.

A cursory appraisal of all four encodings summarized in Figure 2 reveals that, regardless of the dissimilarity metric utilized, there was a considerable amount of heterogeneity in the distribution of overall time budgets across this herd. Looking more closely at the clustering tree produced from the unweighted Euclidean dissimilarity metric in Figure 2A, we can see that the higher magnitude eating and rumination axis entirely dominated the first handful of bifurcations of the dendrogram. Even for users not accustomed to reading dendrograms, this dynamic is clearly animated by parsing through the grid of heatmap visualizations provided by the *encodePlot* utility (see Supplementary Materials). Heterogeneity in the moderate-magnitude nonactivity appears to have been largely ignored in the first half-dozen bifurcations, with the first 10 clusters extracted from this dendrogram being ultimately quite variable in the nonactivity response. Nor is there clear evidence that either activity axes influenced the first 10 bifurcations of this tree. This dynamic is almost certainly attributable to the lack of intrinsic scaling with this estimator. While a behavioral axis that represents a larger proportion of a cow's time investments may warrant additional consideration, these results clearly demonstrate that the Euclidean norm does so to nearly the complete exclusion of lower-magnitude behavioral axes that might still convey important ethological information. The Euclidean distance heatmap is also annotated on the row axis with a number of auxiliary data fields for each cow, which included: age (birth date); calving date; an estimate of peak lactation; nutrition supplementation treatment, and health status during the observation window (see Supplementary Materials for details on the encoding of these auxiliary cow attribute variables). A cursory visual inspection reveals that most clusters appear to be fairly homogenous with respect to cow age, tenure in the pen, and feed supplementation status. Sick cows, however, appear to be slightly overrepresented in some groups, namely the smaller branches representing the more extreme time budget tradeoffs.

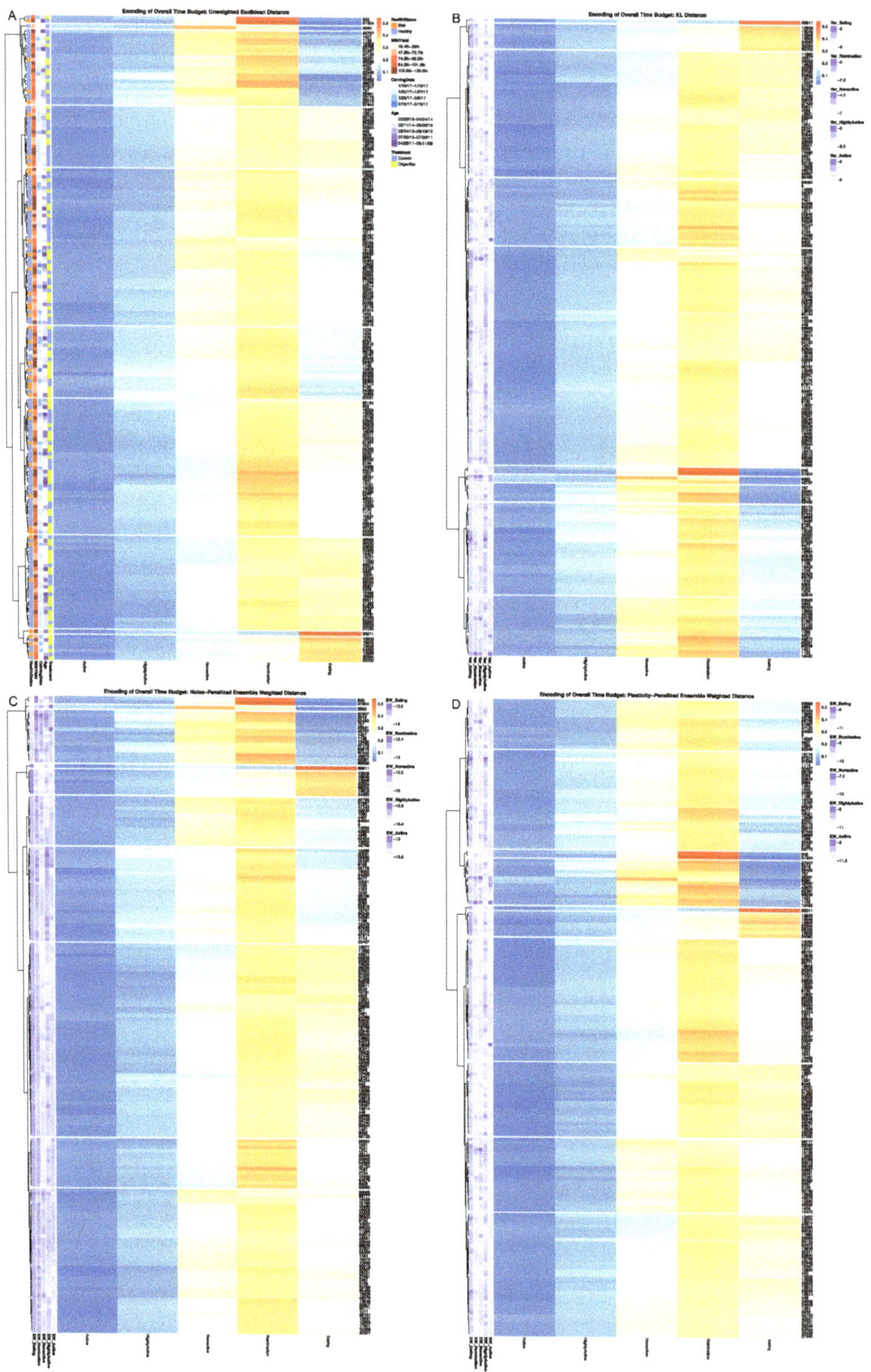

Figure 2. Comparison of overall time budget encodings derived from different dissimilarity metrics.

In each heatmap cows are arranged along the row axis, and the mutually exclusive behaviors along the column axis, such that each cell is colored to represent the proportion of time that a given cow is recorded engaging in a specific behavior. Row gaps have been added within each heatmap to reflect the first 10 branches of the corresponding dendrogram, which here are numerically indexed reading from top to bottom (**A**) Euclidean norm encoding with row annotations representing cow-level attributes. (**B**) KL Divergence encoding with row annotations representing log-scaled variance in observed daily time budgets. (**C**) Noise-penalized ensemble-weighted Euclidean distance encoding with row annotations representing the log-scaled ensemble variances. (**D**) Plasticity-penalized ensemble-weighted Euclidean distance encoding with row annotations representing the log-scaled ensemble variances. See Supplemental Materials for full-scale versions of these images.

Looking next at the hierarchical clustering results visualized in Figure 2B, the KL distance seems to have provided a slightly more holistic encoding of the data that better balances the input across the five behavioral axes. Again, extremes in eating and rumination drive the first few bifurcations of the tree structure, but tradeoffs between time spent eating and nonactivity are considered much earlier in the bifurcation decisions within this tree. Some systematic heterogeneity was also revealed across the herd in the high activity axis, despite its lower magnitude. Unfortunately, the KL distance also appears to have over-stratified cows whose time budgets lie at the extremes. In particular, the cows with extremely low time spent eating (clusters 6–8) were divided into clusters that are likely too small and narrowly defined to facilitate cross-sensor inferences in downstream analyses, and thus may obscure important behavioral dynamics in this dataset. The KL distance heatmap is also annotated on the row axis with the variance in observed daily time budgets for each behavioral axis. Given that time budgets have been normalized here and expressed as proportions, the resulting variance terms were quite small in magnitude (less than zero), and so have been re-expressed on a log-scale, where an increasingly negative value represents a smaller relative magnitude of variation. The fact that all five axes ranged over several orders of magnitude in these variance estimates reveals that there was an appreciable amount of variability in the time budgets across days. Visual appraisal revealed very little systematic patterns in this heteroskedasticity across clusters, however, suggesting that differences in relative plasticity in daily time budget observations may be attributed more to the individual than to any specific pattern in overall time budget.

The noise-penalized ensemble-weighted distance, visualized in Figure 2C, displays clustering dynamics that fall somewhere in between the two extremes of Figure 2A,B. Time spent eating and ruminating still dominate bifurcations nearer the root of the tree, as with the unweighted Euclidean distance, but the most extreme tradeoffs between these axes were here pulled off without over-cutting the tree, as with the KL distance. In the later branches of the tree, however, cows with more moderate time budgets are divided with greater input from the nonactive and highly active axes. Although the ensemble-rescaled estimator does appear to have succeeded in curbing the rescaling of dissimilarity estimates at the extremes of the distribution, the noise-penalized ensemble distance did still bifurcate several cows with anomalously high values in the eating, ruminating, and nonactive axes into their own clusters of size $n = 1$. Although isolating these animals into their own branches will effectively exclude them from cross-sensor inferences in downstream analysis, this encoding may still be appropriate if these datapoints represent authentic outliers that cannot be explained by typical variation in the sensor system. The heatmap was also annotated on the row axis with the ensemble variance terms used to penalize the squared distance estimates. We see that, as anticipated, the magnitude of error in the noise-penalized ensemble variance terms is substantially smaller than the observed variance in observed daily time budgets, confirming that, with so many samples over an extended observation window, measurement error was not contributing substantially to the overall uncertainty in observed time budgets. Closer appraisal of the clear systematic differences in these ensemble variance terms observed across clusters, however, confirms that these penalty terms appear to be effectively mimicking the intrinsic heteroskedasticity in this multinomial sampling space.

Ensemble variances calculated for each cow via the plasticity-penalized simulation routine closely matched ($R \geq 0.99$) the variances in observed daily time budget estimates for all five time budget axes, thereby validating the efficacy of the jackknifing routine. Figure 3 directly contrasts the first 10 clusters extracted from the dendrograms generated by the noise- and plasticity-penalized ensemble-weighted distance measures. In this visualization, clusters are numbered in each heatmap from top to bottom, and so directly align with the row and column indices of the contingency table. For example, we can easily confirm from this graphic that the first three cows constituting the first two clusters in the noise-penalized heatmap were the same cows isolated into the third and fourth clusters in the plasticity-penalized heatmaps—a determination that can be easily confirmed by zooming in on this high-definition rendering to compare Cow ID values. Further comparisons revealed that cluster designations for cows with extremely high time spent eating, extremely low time spent ruminating, and relatively low time spent nonactive (clusters 5 and 6 in the noise- and plasticity-penalized encodings respectively) were virtually identical. In the plasticity-penalized dendrogram, the extremely low eating time cluster (cluster 3) shrunk by just a few animals, compared with the noise-penalized encoding (cluster 4). Additionally, after penalizing for behavioral consistency, the cow with the highest time spent nonactive in the sample (cow 6580) was not isolated as an outlier. This bifurcation was instead shifted to the cows with more moderate time budgets (clusters 7–10), serving to better distinguish between cows with relatively high and only moderate times spent eating. The plasticity-penalized dissimilarity estimator was also notably more generous in assigning cows to the cluster characterized by slightly higher rates of rumination, while all other axes remained relatively low (cluster 7), and appeared to place greater emphasis on the nonactive axis to determine the remaining clusters. Despite these differences, both ensemble-weighted dissimilarity metrics succeeded in producing encodings that provide a more holistic and balanced description of this dataset, and ultimately serve to better visualize heterogeneity in the tradeoffs between all five behavioral axes.

3.2. Improving Tree Pruning Decisions through Simulation

For all dendrograms pruned using the ensemble of simulations that accounted only for measurement noise, an extremely fine-grained encoding was returned. A total of 39 clusters were returned for the unweighted Euclidean distance, 31 for the KL distance, and 38 clusters for the noise-penalized dissimilarity metric. In Figure 4A, the heatmap visualization of the noise-penalized encodings helps to illustrate just how far down each branch the pruning algorithm was able to penetrate before the signal was lost to simulated measurement error. In fact, amongst the first dozen bifurcations in this dendrogram, the only branch not validated was that which would have isolated the cow with the highest observed time spent eating (cow 63911) into her own branch. This result is not necessarily surprising, given the extended observational period over which sensor records were recorded. With over 1500 min of observation for each cow, even in using a relatively conservative simulation strategy that very likely overestimated the noise intrinsic to this sensor, we should expect by the CLT that the standard error attributable to measurement error would ultimately be quite small after averaging over so many sampled timepoints. Subsequently, these results reinforce that the sensors themselves should impose few limitations on downstream inferences for this dataset, and that inconsistencies in the environment and the animals themselves should be the true limiting factor for the resolution of this encoding.

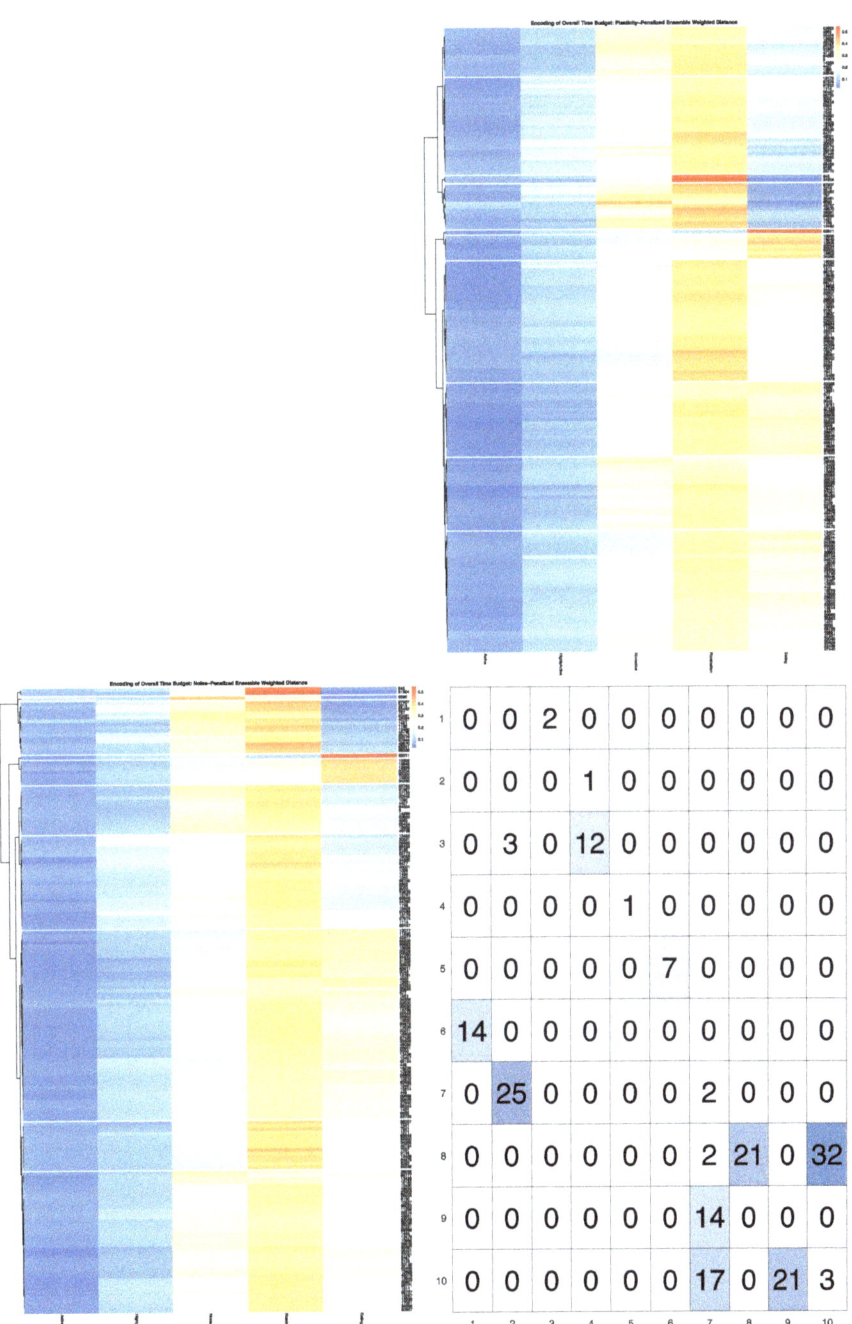

Figure 3. Visualization produced using the *compareEncoding* utility. The noise-penalized encoding is represented on the row axis and the plasticity-penalized encoding is represented on the column axis. Clusters in either heatmap are numbered from top to bottom, and so align directly with the corresponding row and column margins of the contingency table reading up-down and left-right

respectively. Cell counts show that these two encodings are quite similar at the extremes of the time budget distribution, but differ slightly in cutoffs amongst the more moderate time budget clusters. See Supplemental Materials for larger versions of these images.

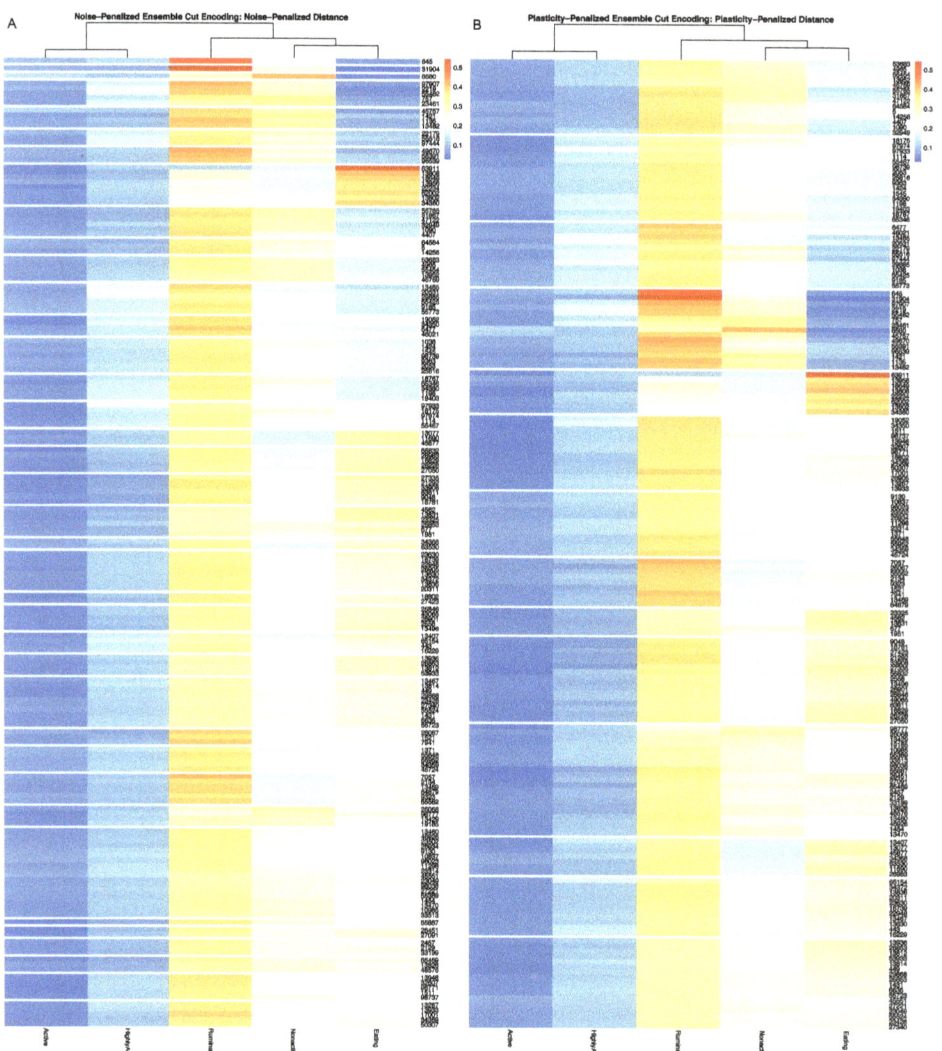

Figure 4. Encodings produced by the cutreeEnsemble algorithm. (**A**) Dendrogram produced by the noise-penalized ensemble weighted dissimilarity metric cut using the noise-penalized data mimicry. The extremely fine encoding with 38 stochastically validated clusters demonstrates that, with so many recorded observations over this extended observation window, the accuracy of the sensor itself should impose few constraints on our behavioral inferences. (**B**) Dendrogram produced by the plasticity-penalized ensemble weighted dissimilarity metric cut using the plasticity-penalized data mimicry. A courser encoding is returned when uncertainty in time budget observations attributable to the behavioral plasticity of the animal itself is taken into consideration.

As expected, the dendrograms pruned with the ensemble of simulations that accounted for both measurement error and longitudinal consistency of the underlying behavioral pattern produced encodings that were far more granular. A total of 13 clusters were returned for the unweighted Euclidean metric, 17 for KL distance, and 14 for both the noise-penalized and the plasticity-penalized dissimilarity metrics. In Figure 4B, the heatmap visualization of these pruning results for the plasticity-penalized dissimilarity metric reveal an encoding that is coarser but ultimately quite well balanced, with the pruning heights modulated to produce cluster sizes that were reasonably uniform across the domain of support. Closer inspection revealed that this final encoding largely matched the order of bifurcations in the original tree, except that this pruning strategy left no animals isolated in anomalous clusters. It should be noted, however, that the granularity of this encoding is not entirely intrinsic to this system, but was dependent on the size of the subsample used to calculate the overall time budget in each simulation. While we can expect cows that were more inconsistent in their daily time budgets to be subjected to a stronger penalty with this estimator due to relatively higher rates of sampling error imposed by the subsampling routine, we can also anticipate that the overall scale of the sampling error imposed on all cows should grow as the size of the subsample is reduced. This would in turn modulate how quickly the underlying behavioral signals would be drowned out by simulated noise within the tree. This suggest that, for larger samples where a greater range of subsample sizes can be utilized, this simulation value can also be treated as a meta-parameter to tune the granularity of the final encoding. Given that the plasticity-penalized mimicry was created for this data set by subsampling only 14 out of 65 observation days, the resolution achieved in the pruned encodings for all four dissimilarity metrics reinforces that this herd was overall fairly consistent in their daily time budgets, and that this data set will support fairly detailed inferences against a strong underlying behavioral pattern.

3.3. An Information Theoretic Framework for Cross-Sensor Inferences

Encodings of the overall time budgets produced using both the noise and plasticity-penalized dissimilarity estimators, wherein both were pruned using the more conservative plasticity-penalized ensemble, produced similar behavioral insights when compared against longitudinal patterns in parlor entry positions across the herd. For the bivariate analyses run with encodings for all 177 cows with complete records, highly significant associations with entry order were recovered for both the noise-penalized ($p = 0.006$) and plasticity-penalized ($p = 0.005$) time budget encodings. The bivariate relationship was optimized for both time budget encodings with a five-cluster encodings of entry-order patterns. The noise-penalized encoding produced the strongest associations with entry order, with seven time budget clusters, whereas the plasticity-penalized encoding performed better with a finer encoding of nine clusters, the key difference being the degree of stratification among animals with the most moderate time budgets.

Visualization of the contingency tables for the optimized encodings colored by their PMI estimates revealed that the significant overall association between the two data streams was driven predominantly by animals in the latter half of the milking queue. Figure 5 displays the results for the noise-penalized encoding. We see first that cows that entered consistently at the very rear of the queue (cluster 1) were significantly overrepresented in the time budget cluster, characterized by moderate time spent eating, low time nonactive, and high rates of rumination (cluster 4). Cows that entered nearer the back of the queue (cluster 2), just ahead of the cows that consistently brought up the rear, were also overrepresented in the same time budget cluster—a trend that was statistically significant for the plasticity-penalized encoding, but only marginally significant for the noise-penalized encodings. In fact, very few animals that entered in the front half of the queue were found to have this time budget pattern, with cows entering just behind the leaders being significantly underrepresented in this time budget cluster. One potential interpretation of this pattern might be that, if these cows were prioritizing time investments in rumination, then this strategy may include hanging back towards the later part of the queue, where they may be able to chew their cud while avoiding the more serious contention for parlor entry position. Further analysis that could facilitate visualization of the cyclical patterns in this time budget data would be needed, however, to confirm this suspicion, and will be left for future work.

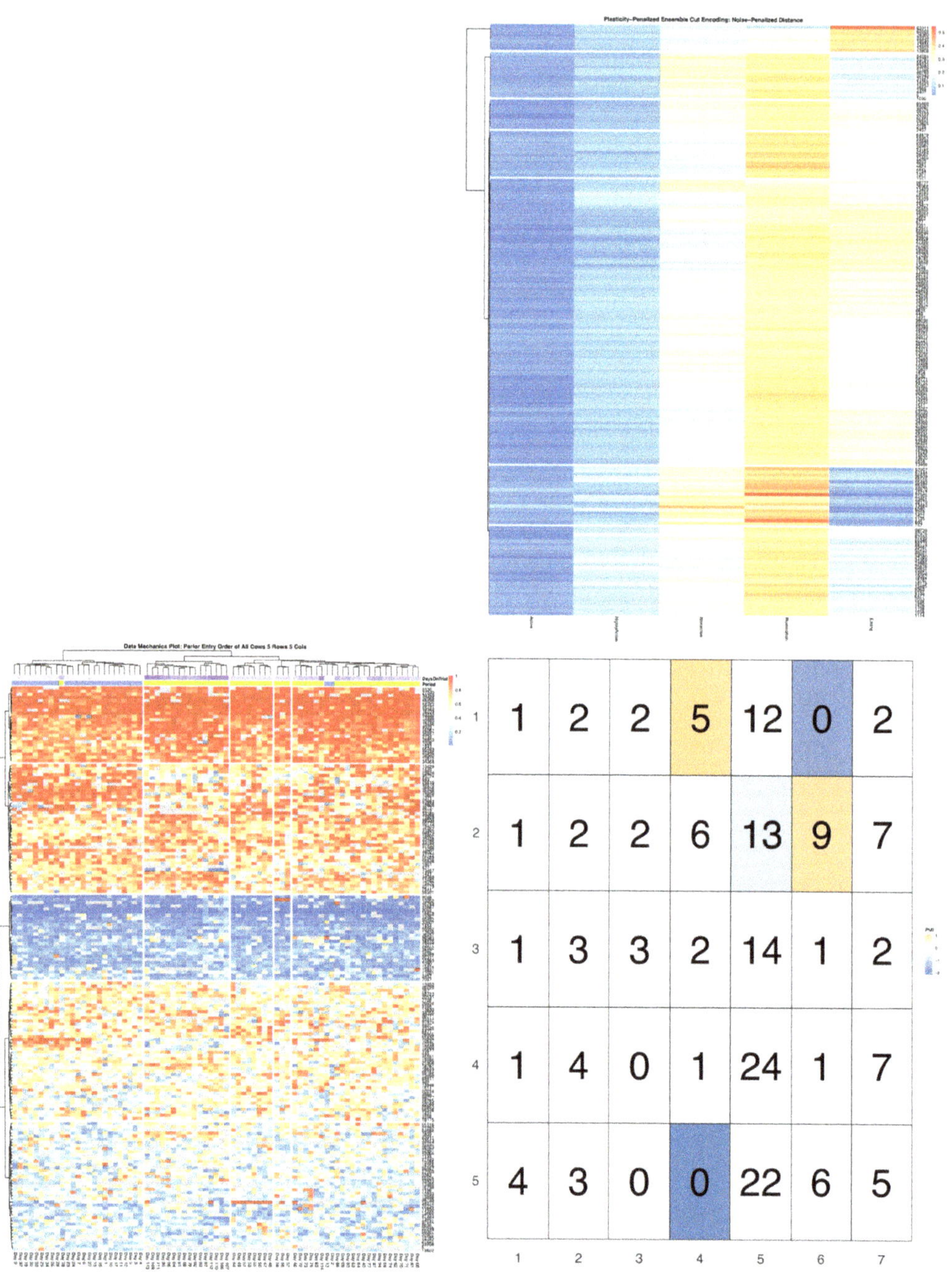

Figure 5. Visualization produced using the *compareEncoding* utility with cells colored by pointwise

mutual information estimates significant at the alpha = 0.05 significance level after simulations using multinomial resampling. Data mechanics encoding of parlor entry position is presented to the row margin of the contingency table, wherein the heatmap contains row annotations representing days on trial and the observation period, such that the pen period corresponds with the observation window of the overall time budget. The noise-penalized encoding of overall time budget is represented on the column axis of the contingency table. Pointwise mutual information values reveal that the significant MI test between these two encodings is driven predominantly by behavioral patterns amongst cows in the latter half of the milking queue.

While this more moderate tradeoff between rumination and nonactivity demonstrated a fairly straightforward and progressive trend across the milking queue, which might readily have been captured by a linear model, more complex dynamics were found for the time budget cluster characterized by extremely low time spent eating and high time spent ruminating and nonactive (cluster 6). Cows that consistently entered at the very end of the queue were significantly underrepresented in this extreme time budget, while the cows that entered just ahead of them were significantly overrepresented. While an extreme tradeoff in eating and ruminating might be explained by issues with sensor placement, that such cows were not evenly dispersed across the herd may instead indicate a biological driver. Health status naturally comes to mind with such an extreme time budget, and indeed several previous studies have reported higher rates of health complications amongst animals in the latter part of the milking queue [38–40]. However, health status alone would not necessarily explain the inversion in association pattern between these two adjacent queue groups.

Previous analyses of milk order records have also revealed that, although cows in general tend to be more consistent in their parlor entry order than would be expected in a purely random system, cows at both the front and rear of the herd tend to be particularly persistent their queuing position [38–42]. It remains unclear in analyses of milk order alone, however, to what degree this pattern is attributable to the cows themselves, and any broader behavioral strategies (syndromes) that they may have adopted, and how much is driven by the natural domain constraints intrinsic to this measurement system [7,41]. In older observational studies of movement patterns in cattle, it has been noted that cow herds appear to be "led" from both the front and the rear of the queue [43,44]. One interpretation may then be that sick cows, who cannot maintain a normal time budget, may also be pushed back by competition for entry position in the milking queue, but they cannot be pushed behind this small group of cows that may be "leading from the rear", effectively serving as a "caboose" for the longer train of animals, as they move between locations to ensure that stragglers are not left behind. Although this behavioral pattern has not been reported in previous experimental studies, it would also not be surprising that such a nonlinear dynamic might be overlooked in analyses relying on linear modeling methods. Indeed, competition between these two behavioral mechanisms, as part of a more complex behavioral system, may explain why relationships between parlor entry position, home pen behaviors, and health status have proven particularly difficult to reliably establish in previous work [5,40,42].

Follow-up bivariate tests with health records confirmed that cows with health complications were indeed overrepresented in the later third of the milking queue (see Supplementary Materials). Mutual information and PMI values did not, however, reveal a significant link between health status and a finer stratification of these late-entering cows. This result, which considers only confirmed cases of acute illness, may however be under-powered, if this behavioral relationship is also influenced by subclinical illnesses not reflected in these health records. More perplexingly, in bivariate analyses with either encoding of overall time budget, cows recorded with acute illness were not found to be significantly overrepresented among the time budget cluster, with extremely low rates of eating. In Figure 6 we can see that sick cows were, in fact, only significantly overrepresented in the time budget cluster characterized by relatively low time spent eating, moderate nonactivity, and elevated rates of high activity—an association that was significant for both the noise and plasticity-penalized time budget encodings. Interestingly, in repeating the bivariate analyses using an encoding of entry-order patterns fit only to animals with no recorded health events, the same time budget cluster with elevated rates of illness was also found to be overrepresented amongst cows entering near the end of the queue, just in front of the "caboose cows" (see Figure 7). Conversely, in this analysis, absent animals with clinical disease, and animals entering at the very rear of the herd were shown to be overrepresented in the time budget cluster that was perhaps best-characterized as demonstrating the most balanced time investments across all five behavioral axes, whereas cows entering just ahead of the caboose cows were underrepresented in this moderate time budget cluster. These results may add weight to the suspicion that ambiguities between clinical and subclinical illness may be obscuring the role of latent health status as at least one key biological link between home pen and milking queue behavior.

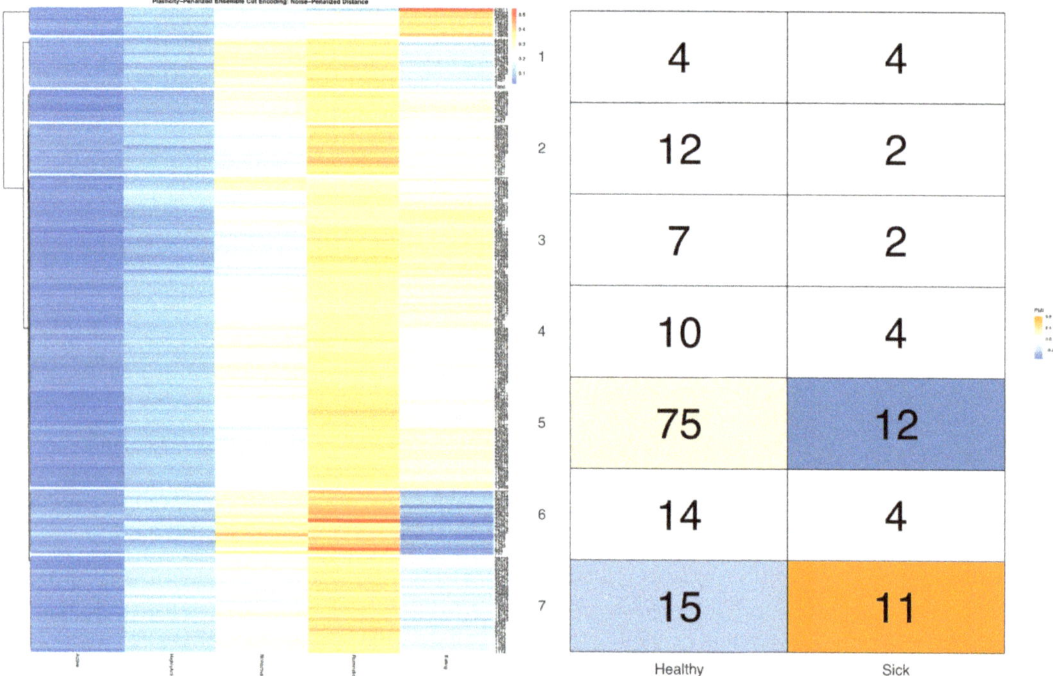

Figure 6. Visualization produced using the *compareEncoding* utility with cells colored by pointwise mutual information estimates significant at the alpha = 0.05 significance level after simulations using multinomial resampling. Cows with the most moderate time budgets were overrepresented among animals with no recorded health events, while sick cows were overrepresented in the overall time budget cluster characterized by relatively low time spend eating and low-to-moderate amounts of time spent nonactive.

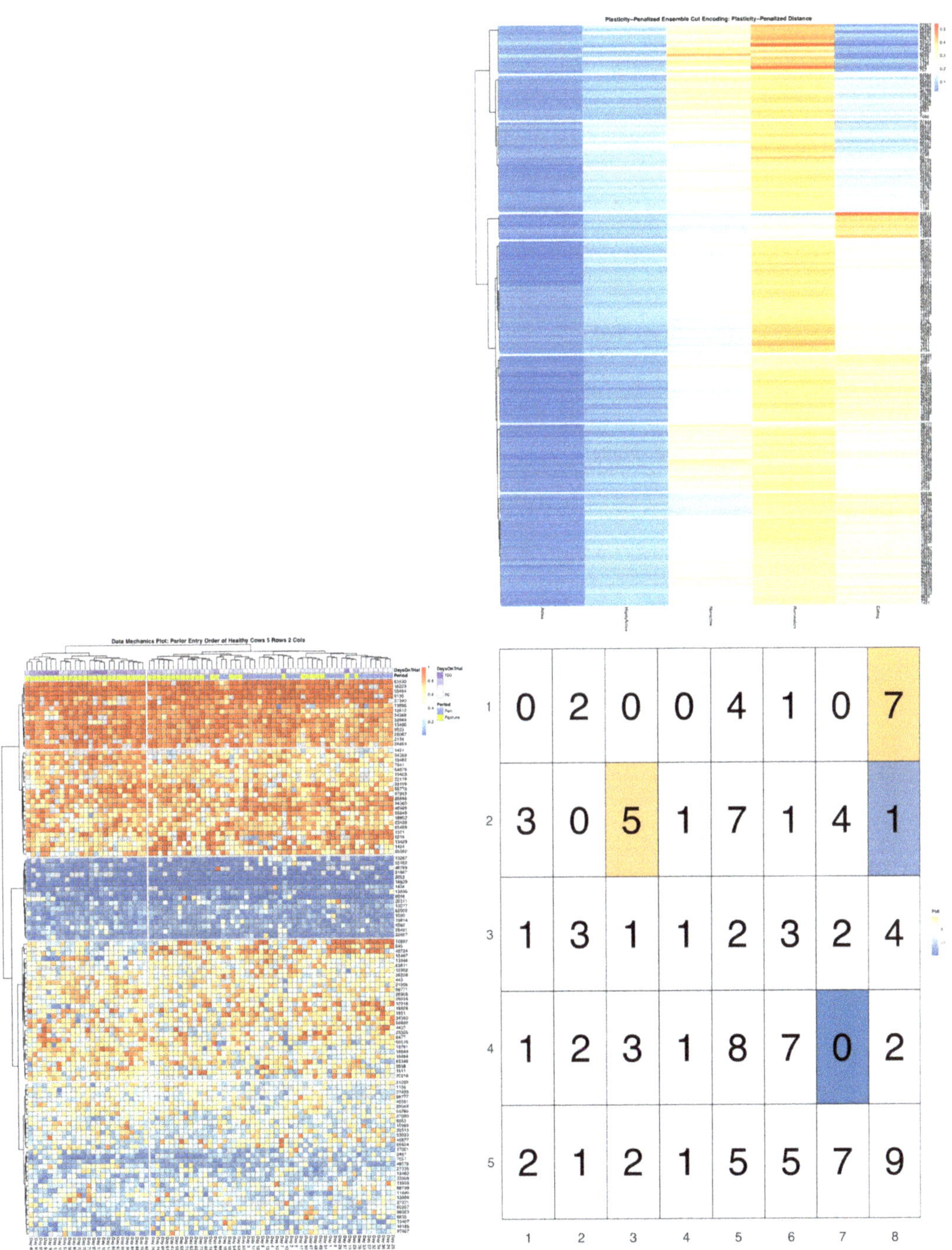

Figure 7. Visualization produced using the *compareEncoding* utility with cells colored by pointwise

mutual information estimates significant at the alpha = 0.05 significance level after simulations using multinomial resampling. Data mechanics encoding of time budget data using only cows with no recorded health events is represented on the row axis, and the plasticity-penalized encoding of overall time budget is represented on the column axis. Among cows with no acute illness, cows at the very end of the queue are now overrepresented in the time budget cluster characterized with fairly high time spent eating (cluster 8). Cows entering just ahead of them are not only underrepresented in this high eating time cluster, but are also overrepresented in the cluster with relatively low eating time cluster with low-to-moderate nonactivity (cluster 3) that was independently associated with higher rates of clinical illness.

4. Conclusions

Time budgets provide a convenient and intuitive means of quantitatively summarizing the behavioral tradeoffs of animals, but multinomial-distributed data present a number of analytical challenges. The results of this analytical case study have highlighted how a novel simulation-based approach may be employed to simultaneously accommodate both the codependency structures fundamental to multivariate-distributed data formats and the complex multi-faceted sources of measurement uncertainty that may be encountered across a broader range of PLF data streams. While such simulations may be more computationally expensive than closed-form estimators, we have demonstrated that an ensemble of data mimicries can be efficiently repurposed throughout the analytical pipeline to improve not only the visualization of these behavioral tradeoffs, but also the compression of such information into robust empirically-defined discrete encodings. It should be noted, however, that the utility of these novel clustering techniques is not restricted to time budget data. The ensemble-penalized dissimilarity estimator and ensemble-cut algorithm that we have introduced in this case study are both fundamentally nonparametric. This means that their implementation is in no way intrinsically restricted to any particular class of data. Subsequently, the choices that a user makes in constructing an appropriate error simulation model are restricted only by their own creativity, allowing this analytical framework to be easily generalized to a much wider array of PLF data streams, and the wider array of complex error structures that they have to offer.

Additionally, while discrete data is typically seen as an impediment to statistical analysis in most model-based approaches, we hope that this analytical case study has served to demonstrate the comparable ease with which insights may be extracted from encoded data when an information theoretic approach is employed. For large, structurally complex, and often informationally redundant PLF data streams, an efficient encoding may be far easier to achieve than a comprehensive model that can fully accommodate the temporal dynamics of behavioral responses in complex farm environments. This may be especially true for data sets where all the factors driving such behavioral responses are not measurable. By avoiding entirely any form of least-squared optimization utilized in most model-based approaches, we have shown that an entirely model-free approach is able to recover nonlinear dynamics between entry order and overall time budget, which likely would have been overlooked if an assumption of linearity had been employed. Although more formal model-based inferences may be warranted for further analysis of the underlying causes of this relationship, the exploratory data analysis tools provided by the LIT pipeline have undoubtably served to create a more comprehensive picture of the complex behavioral dynamics hiding within these two under-utilized data streams.

Supplementary Materials: The following are available online at https://www.mdpi.com/article/10.3390/s22010001/s1: A zip file containing an RMarkdown document with full documentation of all code development and files containing all rendered data visualizations. The most current version of the LIT package can also be downloaded at https://github.com/cgmcvey/LIT, accessed on 16 December 2021.

Author Contributions: Conceptualization C.M., F.H. and K.H. Methodology, All Authors. Software, C.M. Validation, C.M. and F.H. Formal Analysis, C.M. and F.H. Investigation C.M., F.H. and K.H.

Resources P.P., D.M. and K.H. Data curation C.M., D.M. and P.P. Writing—Original draft preparation, C.M. and F.H. Writing—review and editing, All Authors. Visualization C.M. and F.H. Supervision K.H. Project Administration K.H and P.P. Funding acquisition C.M and K.H. All authors have read and agreed to the published version of the manuscript.

Funding: Research support was provided to C.M. through the National Science Foundation Graduate Research Fellowship and the Deans Distinguished Graduate Fellowship from the UC Davis College of Agriculture and Environmental Science.

Institutional Review Board Statement: All animal handling and experimental protocols were approved by the Colorado State University Institution of Animal Care and Use Committee (Protocol ID: 16-6704AA).

Informed Consent Statement: Not applicable.

Data Availability Statement: The datasets generated for this study are available on request to the corresponding author.

Acknowledgments: The authors thank Aurora Organic Dairy for facilitating the use of animals and facilities, as well as all the assistance from the farm's personal in the create of this dataset. The authors also thank Samuel Loomis of the UC Davis Complexity Sciences group for his advice in the development of the mutual information test framework.

Conflicts of Interest: The authors declare no conflict of interest. The funders had no role in the design of the study; in the collection, analyses, or interpretation of data; in the writing of the manuscript, or in the decision to publish the results.

References

1. Stygar, A.H.; Gómez, Y.; Berteselli, G.V.; Costa, E.D.; Canali, E.; Niemi, J.K.; Llonch, P.; Pastell, M. A Systematic Review on Commercially Available and Validated Sensor Technologies for Welfare Assessment of Dairy Cattle. *Front. Veter. Sci.* **2021**, *8*, 634338. [CrossRef] [PubMed]
2. Pinheiro, J.; Bates, D. *Mixed-Effects Models in S and S-PLUS*; Springer Science & Business Media: Berlin, Germany, 2006.
3. Farine, D.R. A guide to null models for animal social network analysis. *Methods Ecol. Evol.* **2017**, *8*, 1309–1320. [CrossRef] [PubMed]
4. McCowan, B.; Beisner, B.; Bliss-Moreau, E.; Vandeleest, J.; Jin, J.; Hannibal, D.; Hsieh, F. Connections Matter: Social Networks and Lifespan Health in Primate Translational Models. *Front. Psychol.* **2016**, *7*, 433. [CrossRef] [PubMed]
5. Cooper, M.D.; Arney, D.R.; Webb, C.R.; Phillips, C.J. Interactions between housed dairy cows during feeding, lying, and standing. *J. Vet. Behav.* **2008**, *3*, 218–227. [CrossRef]
6. Valletta, J.J.; Torney, C.; Kings, M.; Thornton, A.; Madden, J. Applications of machine learning in animal behaviour studies. *Anim. Behav.* **2017**, *124*, 203–220. [CrossRef]
7. McVey, C.; Hsieh, F.; Manriquez, D.; Pinedo, P.; Horback, K. Mind the Queue: A Case Study in Visualizing Heterogeneous Behavioral Patterns in Livestock Sensor Data Using Unsupervised Machine Learning Techniques. *Front. Vet. Sci.* **2020**, *7*, 523. [CrossRef]
8. Kirby, M. *Geometric Data Analysis: An Empirical Approach to Dimensionality Reduction and the Study of Patterns*; John Wiley & Sons: New York, NY, USA, 2001.
9. James, G.; Witten, D.; Hastie, T.; Tibshirani, R. *An Introduction to Statistical Learning*; Springer: New York, NY, USA, 2013; Volume 103.
10. Hastie, T.; Tibshirani, R.; Friedman, J. *The Elements of Statistical Learning: Data Mining, Inference, and Prediction*; Springer Science & Business Media: New York, NY, USA, 2009.
11. Hsieh, F.; Chou, E.; Chen, T.-L. Mimicking Complexity of Structured Data Matrix's Information Content: Categorical Exploratory Data Analysis. *Entropy* **2021**, *23*, 594. [CrossRef]
12. MacKay, D.J.C. *Information Theory, Inference, and Learning Algorithms*; Cambridge University Press: Cambridge, UK, 2003.
13. Adamczyk, K.; Cywicka, D.; Herbut, P.; Trześniowska, E. The application of cluster analysis methods in assessment of daily physical activity of dairy cows milked in the Voluntary Milking System. *Comput. Electron. Agric.* **2017**, *141*, 65–72. [CrossRef]
14. Schwager, M.; Anderson, D.M.; Butler, Z.; Rus, D. Robust classification of animal tracking data. *Comput. Electron. Agric.* **2007**, *56*, 46–59. [CrossRef]
15. Dutta, R.; Smith, D.; Rawnsley, R.; Bishop-Hurley, G.; Hills, J.; Timms, G.; Henry, D. Dynamic cattle behavioural classification using supervised ensemble classifiers. *Comput. Electron. Agric.* **2015**, *111*, 18–28. [CrossRef]
16. Xu, H.; Li, S.; Lee, C.; Ni, W.; Abbott, D.; Johnson, M.; Lea, J.M.; Yuan, J.; Campbell, D.L.M. Analysis of Cattle Social Transitional Behaviour: Attraction and Repulsion. *Sensors* **2020**, *20*, 5340. [CrossRef] [PubMed]

17. Brenninkmeyer, C.; Dippel, S.; Brinkmann, J.; March, S.; Winckler, C.; Knierim, U. Investigating integument alterations in cubicle housed dairy cows: Which types and locations can be combined? *Animal* **2016**, *10*, 342–348. [CrossRef] [PubMed]
18. Lee, M.; Lee, S.; Park, J.; Seo, S. Clustering and Characterization of the Lactation Curves of Dairy Cows Using *K*-Medoids Clustering Algorithm. *Animal* **2020**, *10*, 1348. [CrossRef]
19. Fushing, H.; Liu, S.-Y.; Hsieh, Y.-C.; McCowan, B. From patterned response dependency to structured covariate dependency: Entropy based categorical-pattern-matching. *PLoS ONE* **2018**, *13*, e0198253. [CrossRef] [PubMed]
20. Guan, J.; Fushing, H. Coupling Geometry on Binary Bipartite Networks: Hypotheses Testing on Pattern Geometry and Nestedness. *Front. Appl. Math. Stat.* **2018**, *4*, 38. [CrossRef]
21. Manriquez, D.; Chen, L.; Albornoz, G.; Velez, J.; Pinedo, P. Case Study: Assessment of human-conditioned sorting behavior in dairy cows in farm research trials. *Prof. Anim. Sci.* **2018**, *34*, 664–670. [CrossRef]
22. Manriquez, D.; Chen, L.; Melendez, P.; Pinedo, P. The effect of an organic rumen-protected fat supplement on performance, metabolic status, and health of dairy cows. *BMC Vet. Res.* **2019**, *15*, 1–14. [CrossRef]
23. R Core Team. R: A Language and Environment for Statistical Computing. R Foundation for Statistical Computing. 2018. Available online: https://www.R-project.org/ (accessed on 17 August 2018).
24. Bikker, J.; Van Laar, H.; Rump, P.; Doorenbos, J.; Van Meurs, K.; Griffioen, G.; Dijkstra, J. Technical note: Evaluation of an earattached movement sensor to record cow feeding behavior and activity. *J. Dairy Sci.* **2014**, *97*, 2974–2979. [CrossRef]
25. Pereira, G.; Heins, B.; Endres, M. Technical note: Validation of an ear-tag accelerometer sensor to determine rumination, eating, and activity behaviors of grazing dairy cattle. *J. Dairy Sci.* **2018**, *101*, 2492–2495. [CrossRef]
26. Agresti, A. *Categorical Data Analysis*, 3rd ed.; John Wiley & Sons, Inc.: Hoboken, NJ, USA, 2013.
27. Shirkhorshidi, A.S.; Aghabozorgi, S.; Wah, T.Y. A Comparison Study on Similarity and Dissimilarity Measures in Clustering Continuous Data. *PLoS ONE* **2015**, *10*, e0144059. [CrossRef]
28. Efron, B.; Tibshirani, R.J. *An Introduction to the Bootstrap*; Springer: New York, NY, USA, 1993.
29. Papadakis, M.; Tsagris, M.; Dimitriadis, M.; Fafalios, S.; Tsamardinos, I.; Fasiolo, M.; Borboudakis, G.; Burkardt, J.; Zou, C.; Lakiotaki, K.; et al. Rfast: A Collection of Efficient and Extremely Fast R Functions (1.9.9). 2020. Available online: https://CRAN.R-project.org/package=Rfast (accessed on 10 March 2020).
30. Kolde, R. pheatmap: Pretty Heatmaps. 2019. Available online: https://CRAN.R-project.org/package=pheatmap (accessed on 4 January 2019).
31. Wickham, H. *ggplot2: Elegant Graphics for Data Analysis*; Springer: New York, NY, USA, 2016; Available online: http://ggplot2.org (accessed on 6 July 2020).
32. Kassambara, A. ggpubr: "ggplot2" Based Publication Ready Plots (0.4.0). 2020. Available online: https://CRAN.R-project.org/package=ggpubr (accessed on 6 July 2020).
33. Tucker, C.B.; Jensen, M.B.; De Passillé, A.M.; Hänninen, L.; Rushen, J. Invited review: Lying time and the welfare of dairy cows. *J. Dairy Sci.* **2021**, *104*, 20–46. [CrossRef]
34. Rand, W.M. Objective Criteria for the Evaluation of Clustering Methods. *J. Am. Stat. Assoc.* **1971**, *66*, 846–850. [CrossRef]
35. Hausser, J.; Strimmer, K. entropy: Estimation of Entropy, Mutual Information and Related Quantities (1.3.0). 2021. Available online: https://CRAN.R-project.org/package=entropy (accessed on 9 August 2021).
36. Hausser, J.; Strimmer, K. Entropy Inference and the James-Stein Estimator, with Application to Nonlinear Gene Association Networks. *J. Mach. Learn. Res.* **2009**, *10*, 1469–1484.
37. Johnson, M. Confidence Intervals on Likelihood Estimates for Estimating Association Strength. Technical Report. 1999. Available online: http://web.science.mq.edu.au/~mjohnson/papers/sigdiff.pdf (accessed on 10 September 2021).
38. Rathore, A. Order of cow entry at milking and its relationships with milk yield and consistency of the order. *Appl. Anim. Ethol.* **1982**, *8*, 45–52. [CrossRef]
39. Gadbury, J.C. Some preliminary field observations on the order of entry of cows into herringbone parlours. *Appl. Anim. Ethol.* **1975**, *1*, 275–281. [CrossRef]
40. Berry, D.; McCarthy, J. Genetic and non-genetic factors associated with milking order in lactating dairy cows. *Appl. Anim. Behav. Sci.* **2012**, *136*, 15–19. [CrossRef]
41. Beggs, D.; Jongman, E.; Hemsworth, P.; Fisher, A. Short communication: Milking order consistency of dairy cows in large Australian herds. *J. Dairy Sci.* **2018**, *101*, 603–608. [CrossRef]
42. Soffié, M.; Thinès, G.; De Marneffe, G. Relation between milking order and dominance value in a group of dairy cows. *Appl. Anim. Ethol.* **1976**, *2*, 271–276. [CrossRef]
43. Kilgour, R.; Scott, T.H. Leadership in a Herd of Dairy Cows. *Proc. N. Z. Soc. Anim. Prod.* **1959**, *19*, 36–43.
44. Reinhardt, V. Movement Orders and Leadership in a Semi-Wild Cattle Herd. *Behaviour* **1983**, *83*, 251–264. [CrossRef]

 MDPI

Article

A Non-Invasive Millimetre-Wave Radar Sensor for Automated Behavioural Tracking in Precision Farming—Application to Sheep Husbandry

Alexandre Dore [1,2,*], Cristian Pasquaretta [2], Dominique Henry [1,2], Edmond Ricard [3], Jean-François Bompa [3], Mathieu Bonneau [4], Alain Boissy [5], Dominique Hazard [3], Mathieu Lihoreau [2,†] and Hervé Aubert [1,†]

1 Laboratory for Analysis and Architecture of Systems, Toulouse University, CNRS, INPT, 31400 Toulouse, France; dhenry@laas.fr (D.H.); herve.aubert@toulouse-inp.fr (H.A.)
2 Research Center on Animal Cognition (CRCA), Center for Integrative Biology (CBI), CNRS, University Paul Sabatier-Toulouse III, 31400 Toulouse, France; cristian.pasquaretta@univ-tlse3.fr (C.P.); mathieu.lihoreau@univ-tlse3.fr (M.L.)
3 GenPhySE, Toulouse University, INRAE, ENVT, 31326 Castanet Tolosan, France; edmond.ricard@inrae.fr (E.R.); jean-francois.bompa@inrae.fr (J.-F.B.); dominique.hazard@inrae.fr (D.H.)
4 URZ, INRAE, Petit-Bourg, 97170 Guadeloupe, France; mathieu.bonneau@inrae.fr
5 UMR Herbivores, Clermont University, INRAE, VetAgro Sup, 63122 Saint-Genès Champanelle, France; alain.boissy@inrae.fr
* Correspondence: alexandre.dore@univ-tlse3.fr
† These authors contributed equally to the work.

Citation: Dore, A.; Pasquaretta, C.; Henry, D.; Ricard, E.; Bompa, J.-F.; Bonneau, M.; Boissy, A.; Hazard, D.; Lihoreau, M.; Aubert, H. A Non-Invasive Millimetre-Wave Radar Sensor for Automated Behavioural Tracking in Precision Farming—Application to Sheep Husbandry. *Sensors* **2021**, *21*, 8140. https://doi.org/10.3390/s21238140

Academic Editors: Yongliang Qiao, Lilong Chai, Dongjian He and Daobilige Su

Received: 2 November 2021
Accepted: 20 November 2021
Published: 6 December 2021

Abstract: The automated quantification of the behaviour of freely moving animals is increasingly needed in applied ethology. State-of-the-art approaches often require tags to identify animals, high computational power for data collection and processing, and are sensitive to environmental conditions, which limits their large-scale utilization, for instance in genetic selection programs of animal breeding. Here we introduce a new automated tracking system based on millimetre-wave radars for real time robust and high precision monitoring of untagged animals. In contrast to conventional video tracking systems, radar tracking requires low processing power, is independent on light variations and has more accurate estimations of animal positions due to a lower misdetection rate. To validate our approach, we monitored the movements of 58 sheep in a standard indoor behavioural test used for assessing social motivation. We derived new estimators from the radar data that can be used to improve the behavioural phenotyping of the sheep. We then showed how radars can be used for movement tracking at larger spatial scales, in the field, by adjusting operating frequency and radiated electromagnetic power. Millimetre-wave radars thus hold considerable promises precision farming through high-throughput recording of the behaviour of untagged animals in different types of environments.

Keywords: radar sensors; radar signal processing; animal farming; computational ethology; signal classification; wavelet analysis

1. Introduction

Behavioural research increasingly requires automated recording and analyses of animal movements [1]. This is exemplified by emerging methods for high-throughput monitoring and statistical analyses of movements that enable the quantitative characterisation of behaviour on large numbers of individuals, the discovery of new behaviours, but also the objective comparison of behavioural data across studies and species [2,3]. These quantitative approaches are particularly powerful to study inter-individual behavioural variability or personalities in animal populations [4]. In livestock, for instance, large-scale genetic selection programmes are based on the measurements of several hundreds (if not thousands) of farm animals [5]. Many behavioural tests have been developed to assess personality traits in these animals [6], with some applications in breeding programmes, for instance to discard the more aggressive individuals [7]. However, in these studies behavioural measures are frequently obtained from direct observations by the experimenters or farmers [8], which considerably limits the possibility to quantify behavioural traits at the experimental or commercial farm level.

Animal tracking methods involving on-board devices, such as Global Positioning Systems (GPS) [9], radio telemetry [10], radio frequency identification (RFID) [11] or harmonic radar [12], are hardly suitable for detailed high throughput behavioural phenotyping due to the limited accuracy and duration of measurements. Best available approaches therefore involve image-based analyses [13]. So far however, these techniques often require large computational resources to fit the classification model and to process images [14], and are sensitive to light variation [15]. Moreover, video processing using machine learning is typically limited to the detection of one type of target (e.g., the focal animal species), which means that other potentially important information in the signal (e.g., the presence of a farmer) is ignored.

Recently, Frequency-Modulated Continuous-Wave (FMCW) radars operating in the millimetre-wave frequency band have been proposed for the automated tracking of the behaviour of a large diversity of animals (sow: [16], bees: [17]; sheep: [18]). In this approach, it was shown it is possible to record one-dimensional movements (distance to radar) of individual sheep in an arena test [18]. Tracking animals with FMCW radars has the great advantage of being non-invasive (does not require a tag), insensitive to light intensity variations, and fast (does not require large memory resource). FMCW radars therefore provide considerable advantages for the development of automated high-throughput analyses of behaviour in comparison to more conventional approaches like video and infrared cells. The radar signal processing does not require fitting a model to detect targets, which relaxes the need to collect thousands of data before application. In addition, it offers the possibility to detect targets placed behind a non-transparent wall, which can be used to hide the tracking device, or to study the effect of physical obstacles on an animal's behaviour.

Here we report a millimetre-wave FMCW radar system for the automated tracking and analysis of the 2D trajectories of freely moving animals. We illustrate our approach with the analysis of the movements of 58 sheep in an experimental farm. The measurements were performed during a behavioural test commonly used to estimate the sociability of individual sheep in genetic selection [8,19]. First, we compared the estimate of the sheep position with the radar and standard video tracking and infrared cells. Second, using the radar data we identified new behavioural estimators that could be used for large-scale behavioural phenotyping. Third we showed that the radar system can also operate for long-distance tracking, in the field, by adjusting radar emission frequency and radiated electromagnetic power.

2. Material and Methods

2.1. Sheep

We ran the experiments in July 2019 at the experimental farm la Fage of the French National Research Institute for Agriculture, Food, and Environment (INRAE), France (43.918304, 3.094309). We tested 58 lambs (29 males, 29 females) *Ovis aries* with known weight (range: 12–31.3 kg) and age (range: 59–88 days). Ewes and their lambs were reared outdoor on rangelands. After weaning, lambs were reared together outside and tested for behaviour 10 days later. This delay enabled the development of social preferences for conspecifics instead of preference for mother.

All the lambs were previously tested in a "corridor test" to estimate their docility towards humans. Briefly, the test pen consisted of a closed, wide rectangular circuit (4.5×7.5 m) with opaque walls [8]. A non-familiar human entered the testing pen and walked at constant speed through the corridor until two complete tours had been achieved. The corridor was divided into 6 virtual areas. Every 5 s, the areas in which the human and the animal were located were recorded and the mean distance separating the human and the lamb was calculated. The walking human also recorded with a stopwatch the total duration when he could see the head of the lamb to discriminate between fleeing and following lambs. The reactivity criteria towards an approaching human was constructed by combining both distance and duration measurements (for more details see [20]). The higher the resulting variable (i.e., "docility" variable in the present study), the more docile the animal.

2.2. Arena Test

We measured sheep behaviour in a standard protocol ("the arena test") used to assess the sociability of sheep through measures of inter-individual variability in social motivation in the absence or presence of a shepherd [8,19]. A sheep (focal sheep) was introduced in the pen (2 m × 7 m) (Figure 1A) (for more details see [21]). Three other sheep from the same cohort (social stimuli) were placed behind a grid barrier, on the opposite side of the arena entrance. The test involved three phases (Figure 1B):

- In phase 1, the focal sheep could explore the arena for 15 s and see its conspecifics through a grid barrier;
- In phase 2, visual contact between the focal sheep and the social stimuli was disrupted using an opaque panel pulled down from the outside of the pen for 60 s. This phase was used to assess the sociability of the sheep towards its conspecifics;
- In phase 3, visual contact between the focal sheep and its conspecifics was re-established and a human was standing still in front of grid barrier for 60 s. This phase was used to assess the sociability of the focal sheep towards conspecifics in presence of a immobile human.

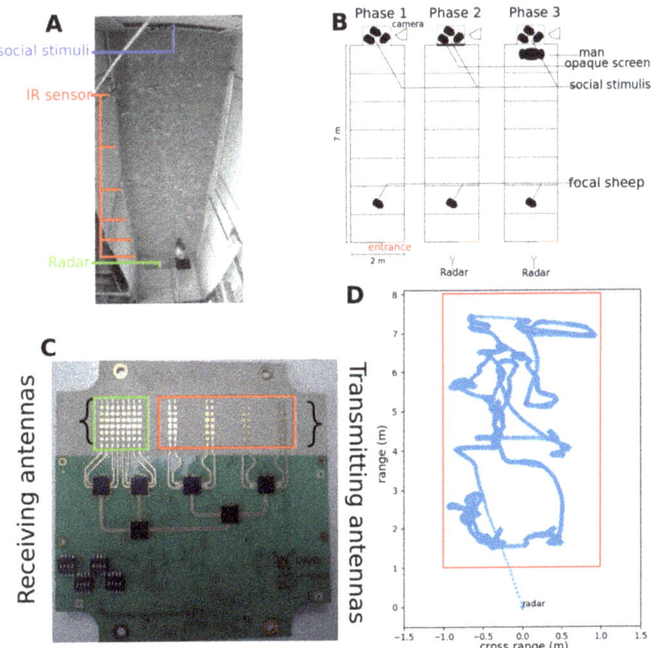

Figure 1. Corridor test. (**A**) Top view of the focal sheep and the social stimuli in the corridor (example image extracted from video data). (**B**) Schematic representation of experimental phases 1, 2 and 3. (**C**) Image of the FMCW radar frontend (phot credit AD). Each rectangle corresponds to patch [22]. (**D**) Example of a trajectory of a sheep obtained with radar tracking after removing the clutter and normalizing the estimated value. The red rectangle represent the pen walls.

Sets of 2 infrared cells were placed at the height of the sheep's body and every meter along the arena test to define 7 virtual areas of 1 m. Analyses of the data resulting from the activation of the infrared cells by the sheep were performed with Fortran algorithms to compute longitudinal displacements of the sheep in the device.

2.3. Data Collection

We measured the displacement of the focal sheep in phases 2 and 3 of the arena test (phase 1 is the initiation phase) using three automated tracking systems: (1) infrared sensors, (2) a video camera, and (3) a millimetre-wave FMCW radar. During the measurement, an experimenter also recorded the number of high-pitched bleats by the focal sheep, a proxy of sociability [8]. A proximity score was computed as the time spent in each virtual area weighted according to the virtual area delimited by the infrared receptor in such a way that a high score indicated high proximity to conspecifics [20]. Crossing rate measured the number of virtual areas crossed during arena test phases 2 and 3.

2.4. Video and Radar Tracking

We compared the efficiency of the radar system and standard video tracking for monitoring the 2D movements of the sheep. For the video tracking, we placed a camera on one end of the arena (opposite to entrance side, Figure 1B). The camera was elevated 2 m above ground in order to film the entire arena, producing black and white images of size 720 p × 576 pixels every 25 ms. Sheep movements were tracked in 2D. For image processing, we applied a detection algorithm using the state-of-the-art image object detector tiny-YOLO V3 (You Only Look Once) network, which is a version of the YOLO model adapted for faster processing allowing 244 images of 0.17 mega pixels (416 × 416 pixels) per second (on a TITAN X graphics card) [23]. This Convolutional Neural Network (CNN) was pre-trained on the PASCAL Visual Object Classes Challenge dataset [24]. YOLO detected all the objects on the image, including the focal sheep, possibly some parts of the background and the human when entering inside the arena. To differentiate between the sheep and non-sheep detected objects, we used another CNN, Alexnet, that we parameterized using transfer learning [25]. A set of 40 sheep and 40 non-sheep images were used to re-train the network. Finally, for some images the focal sheep was not detected, especially when it was located at the opposite of the camera. In these cases, the location of the sheep was extrapolated by continuing the trajectories with a constant speed between the two known locations.

For the radar tracking, we placed a millimetre-wave FMCW radar (Figure 1C, see technical characteristics in Table 1) at one end of the arena test (i.e., entrance side, Figure 1B). The radar was setup outside of the test pen behind a Styrofoam wall transparent to millimetre-waves [26]. The transmitting antenna array radiated a repetition over time of a so-called chirp (i.e., saw-tooth frequency-modulated signal [27]). The chirp was backscattered by the targeted focal sheep, but also by the surrounding scene which provides undesirable radar echoes called the electromagnetic clutter. The total backscattered signal was then collected by the receiving antenna array and processed to mitigate the clutter and to derive the sheep 2D trajectory from radar data. In the millimetre-wave frequency range, the detectability of the sheep depends mainly on the bandwidth of the frequency modulation, the beamwidth of the radar antennas, and the radiated electromagnetic power [27].

Processing of radar data included two main steps. First, we extracted the position of the animal. Next, we computed behavioural parameters to characterize the movement of the animal. We extracted the distance of the focal sheep to the radar and its direction in the horizontal plane of the scene. To mitigate the electromagnetic clutter, we estimated the mean value and standard deviation of the radar signal in absence of the sheep and we derived the signal, denoted by D, from the signal S delivered by the radar in presence of the animal, as follows:

$$D(t,r,\theta) = \frac{S(t,r,\theta) - mean(r,\theta)}{std(r,\theta)}$$

where r is the radar-to-sheep separation distance, *mean* is the time-averaged radar signal at the range r and angular position θ, *std* is the time-standard deviation of the radar signal. Figure 1D shows an example of position estimations of a sheep over time after removing the electromagnetic clutter.

2.5. Extraction of New Behavioural Parameters Form the Radar Data

We used the radar estimated 2D trajectories to extract new behavioural parameters characterizing sheep movements using three approaches.

- 1: Behavioural classes;

We statistically identified broad classes of behaviour using Gaussian Mixture Models (GMM). First, we divided the trajectories into time windows of 1 s for each sheep and for each experimental phase. Next, we extracted movement parameters from each window: average speed, sinuosity (total displacement over distance between the first position and the last) and total displacement distance. Then, because the social stimuli (i.e., the three conspecifics) were located at one end of the corridor, we split the speed vector into two components: along the two lateral walls of the corridor and across the two longitudinal walls. Finally, to derive behavioural classes we performed a GMM on the extracted movement parameters for each lamb [28]. The number of classes (i.e., the number of Gaussians to be used) was determined by comparing models using 1 to 15 classes. We selected the model with the lowest Akaike score, which represents the model with the features best explaining the parameter under consideration [29]. The GMM was performed using the Python package scikit-learn [30]. We estimated the rate of time spent in each movement classes for the two phases.

Table 1. Technical characteristics of the FMCW radar used for indoor tracking [22] and outdoor tracking [31].

Name	Indoor Tracking	Outdoor Tracking	Note
Operating frequency	77 GHz	24 GHz	This frequency is also called the carrier frequency of the frequency-modulated signal transmitted by the radar
Modulation Bandwidth	3 GHz	800 MHz	Frequency interval, centred at the operating frequency, used for the saw-tooth frequency modulation of the transmitted signal
Ramp time	256 μs	1 ms	Up-ramp duration of the saw-tooth frequency-modulated signal (or chirp duration)
Repetition time	50 ms	30 ms	Period of the transmitted frequency-modulated signal (or chirp repetition interval)
Number of linear arrays of the transmitting antenna array	4	1	One linear array composed of 8×2 rectangular patches radiating elements
Number of linear arrays of the receiving antenna array	8	2	Eight linear arrays composed of 8 rectangular patches radiating elements
Main lobe beamwidth of the transmitting antenna array in the horizontal plane	50°	58°	Angular range (or field of view) of the radar illumination in the horizontal plane
Transmitted power	100 mW	100 mW	Power delivered at the input terminals of the transmitting array antenna (the radiated power is defined as the product of the transmitted power by the efficiency of the antenna)

- 2: Behavioural transitions;

We determined behavioural changes over time using Ricker wavelet processing [32]. Wavelet processing consists in filtering the sheep position signal using a wavelet as a filter [33]. The use of wavelet analysis to describe animal behaviour was previously used in [34]. This type of filtering is applied to several time scales, thus allowing the detection of a change in the direction and speed of the sheep, depending on when the changes occur or the duration of the change. Our aim was to determine the precise moments when the focal sheep changed its way of moving, which was estimated using the spectrum described by each scale of the used wavelet. We observed that the number of local maxima in the wavelet transform coefficients is sensitive to the number of changes in the way of moving and the size of the wavelet will determine if the change is global or punctual. This estimation of changes was done on the lateral and longitudinal movement and for the two last phases of the experiment.

- 3: Space coverage;

We investigated the space occupied across time by the focal sheep using heatmaps representing the areas the sheep spent time in during the measurements. The use of heatmaps to describe animal behaviour was previously used in [35]. We partitioned the arena into a grid of 80 virtual zones of 44 × 40 cm^2 each (i.e., 16 partitions along the arena length and 5 partitions along the arena width). We chose this grid dimension because it is the width of a small lamb [36]. We counted the number of zones (i.e., the heatmap score) the focal sheep remained in for more than 200 ms. This count was used to extract behavioural features for the two last phases of the experiment.

2.6. Outdoor Radar Tracking

We ran outdoor experiments in order to demonstrate the applicability of our radar system for the tracking of sheep in field conditions. These measurements were done in an open space with no obstacles (60 m × 15 m asphalt place). A human experimenter moved within the radar catching area in order to induce animal movements. We tested one female sheep. To enhance detection range to 40 m, we used a FMCW radar with the lower operating frequency of 24 GHz. At fixed transmitted power, lower frequencies enable reduction of the free-space attenuation of the radiated electromagnetic power [27]. The gain due to the free-space attenuation is 10.13 dB.

2.7. Statistical Analyses

We ran all analyses using the programming environment R [37]. Raw trajectory data extracted from radar and video measures are available in Dataset S1.

- Analysis of new movement features

We tested the influence of sheep characteristics (docility, and sociability) in interaction with the two test phases on the proportion of time spent in the behavioural classes using a generalized Linear Mixed Model with binomial family error distribution. We tested all possible dual interactions of each variable with the test phase. Three-way Interactions were excluded to avoid over-fitting of the model [38]. Sheep identity was included as a random effect. We ran a model selection on all feature combinations (docility, sociability the phases and their interactions) using the Akaike score. The model with the lowest score was retained as the best model. When the second best model have an AIC score equivalent to the best model (i.e., when the difference is lower than 2) an average model was performed with those that have equivalent AIC. We used a similar procedure to test the influence of the sheep individual characteristics on continuous wavelet transforms estimated on lateral and longitudinal movements (Gaussian family error distributions) and heatmaps (Poisson family error distribution).

- Classification of behavioural types;

To improve the interpretation of the sheep behaviour in the corridor, we reduced our four movement features (proportion of fast movements, changes in longitudinal and transversal movements, space coverage) for phases 2 and 3, using a Principal Component Analysis (PCA). The PCA was performed using the R package FactoMineR [39]. We explored afterward whether our new automated estimators could be used to replace estimators recorded manually using a General Linear Model (GLM, using the R package stats) approach.

3. Results

3.1. Radar Tracking Is Faster and More Accurate Than Video Tracking

To test the efficiency of the radar tracking system, we compared the data obtained from the infrared cells, the video and the radar. This efficiency was estimated by comparing the proximity score estimated using the infrared cell, video and radar detection but also by using the crossing rate estimated by the infrared cell and the mean speed along the longitudinal axis estimated by the video and radar detection. We analysed data from 58 individuals (29 males, 29 females). Both data collected by the radar and the video enabled to capture information given by infrared cells with high fidelity. Proximity scores and crossing rates obtained from infrared cells were positively correlated with data obtained from the radar (Pearson correlation; proximity: $r = 0.77$, $p < 0.001$; crossing rate: $r = 0.87$, $p < 0.001$) and the video (Pearson correlation test; proximity: $r = 0.91$, $p < 0.001$; crossing rate: $r = 0.34$, $p < 0.001$).

Radar tracking had additional advantages over video tracking in terms of data processing (Table 2). The radar produced two times more measures per second. Radar processing was also much faster (50 frames per second for radar and 4 for video processing) and therefore, it may be used for real time analyses. Radar measurement data were of similar size as video measures (ROM), but required approximately seven times less memory (RAM) to process. Finally, radar processing did not require a learning phase with important data collection and a time-consuming training phase that can last several hours just for the adaptation of the model, or several days if the network is not trained beforehand.

Table 2. Comparison of data processing characteristics with radar and video tracking systems.

Tracking Method	Radar	Video
Number of measures per second	50	25
Read Only Memory (ROM) for all measures of a sheep	151 Mo	62 Mo
Random Access Memory (RAM) per measure	524 Kb	3.7 Mb
Processing time per measure	<20 ms	250 ms
Distance to target centre	1.1 m	1.5 m

3.2. New Behavioural Indicators from the Radar Data

The following analyses were made on the 58 sheep. The 2D radar trajectory data offered the opportunity for high resolution analyses of sheep movements.

- Behavioural classes: detection of slow and fast movements

In order to classify the different types of movements exhibited by the sheep, we applied the GMM procedure to statistically identify behavioural classes from the trajectory data. We found four behavioural classes (Figure 2A):

Class 1 (51.3% of the measures) was characterized by null or slow movements ("slow movements");

Class 2 (35.48% of the measures) was characterized by fast movements with low sinuosity ("fast movement");

Class 3 (10.2% of the measures) was characterized by fast movements with high sinuosity ("fast tortuous");

Class 4 (3.01% of the measures) was characterized by slow movements with high sinuosity ("slow tortuous").

Each of the two behavioural classes with strong sinuosity (classes 3 and 4) represented less than 10% of all data. We thus focused our analyses on slow and fast movements only (classes 1 and 2). We tested the effects of the individual characteristics of sheep on the rate of time spent in each in the two main behavioural classes using GLMMs. The best (using Akaike criterion) model (See Table S1) retained the docility, sociability indicators and the phase of the test to explain the two main behavioural classes extracted by the radar, i.e. the rate of slow movement and fast movement. In phase 3, all the sheep tended to move less than in phase 2 (estimate = −1.24, std. = 0.008, $p < 0.001$). In phase 2, highly sociable sheep moved less than little sociable sheep (estimate = −0.11, std. = 0.015, $p < 0.001$). This trend was reduced in phase 3 for both sociable and docile sheep (sociability: estimate = −0.12, std. = 0.039, $p < 0.001$ docility: estimate = 0.16, std. = 0.0074, $p < 0.001$) (Table S1 and Figure 2).

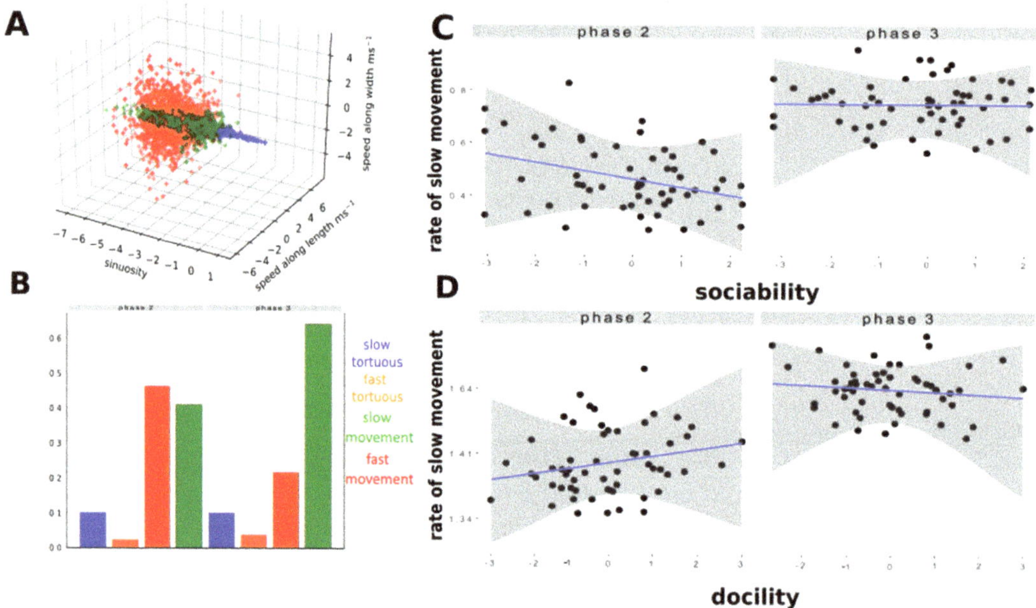

Figure 2. Analyses of behavioural classes. (**A**) Distribution of the four behavioural classes after a Gaussian Mixture Model. (**B**) Frequency of behavioural classes during phase 2 and phase 3 of the corridor test. (**C**) Correlation between the proportion of time spent in slow movements and the sociability score of sheep during phase 2 and 3 (see details of models in Table 3). (**D**) Correlation between the proportion of time spent in slow movements and the docility score of sheep during phase 2 and 3. N = 58 sheep.

Table 3. Analyses of behavioural classes. Results of the best GLMM (binomial family, after model selection—see Table S1). The model tested the effects of phase, docility, sociability, and dual interaction of each variable with phase, on the proportion of time spent in fast movements (behavioural class 2). Lamb identity was included as a random factor. Significant effects ($p < 0.05$).

	Estimate	Std. Error	z Value	Pr (>|z|)
(Intercept)	0.11	0.055	2.08	0.037
Sociability	0.13	0.039	3.47	<0.001
phase 3	−1.24	0.0086	−144.04	<0.001
Docility	−0.11	0.047	−2.43	0.015
sociability:phase 3	−0.12	0.0061	−19.90	<0.001
Docility: phase 3	0.16	0.0074	21.31	<0.001

- Wavelet analysis: detection of erratic behavioural transitions;

Our second approach to describe the sheep behaviour was to quantified changes in movements (i.e., variation in speed, direction, or both) through time. This was done using continuous wavelet analyses (Figure 3). We tested the effects of the individual characteristics of sheep on the frequency of these changes using GLMMs and model selection (Tables S2 and S3). When considering longitudinal displacements (i.e., wavelet Y) along the arena device (Table 4), we found that highly sociable sheep made more changes in the pattern of displacement during both phases of test (estimate = 16.98, std. = 4.68 $p < 0.001$) (Figure 3A,C). In general the movements were less erratic in phase 2 than in phase 3 (estimate = −91.50, std. = 9.07, $p < 0.001$). When considering transversal movements (i.e., wavelet X) across the arena device (Table 4), we found that sheep made more changes in the way of displacement during phase 2 than phase 3 of test (estimate = −53.15, std. = 8.26, $p < 0.001$) (Figure 3B,D). However, this trend was reduced for the docile sheep (estimate = 19.19, std. = 7.11, $p = 0.009$).

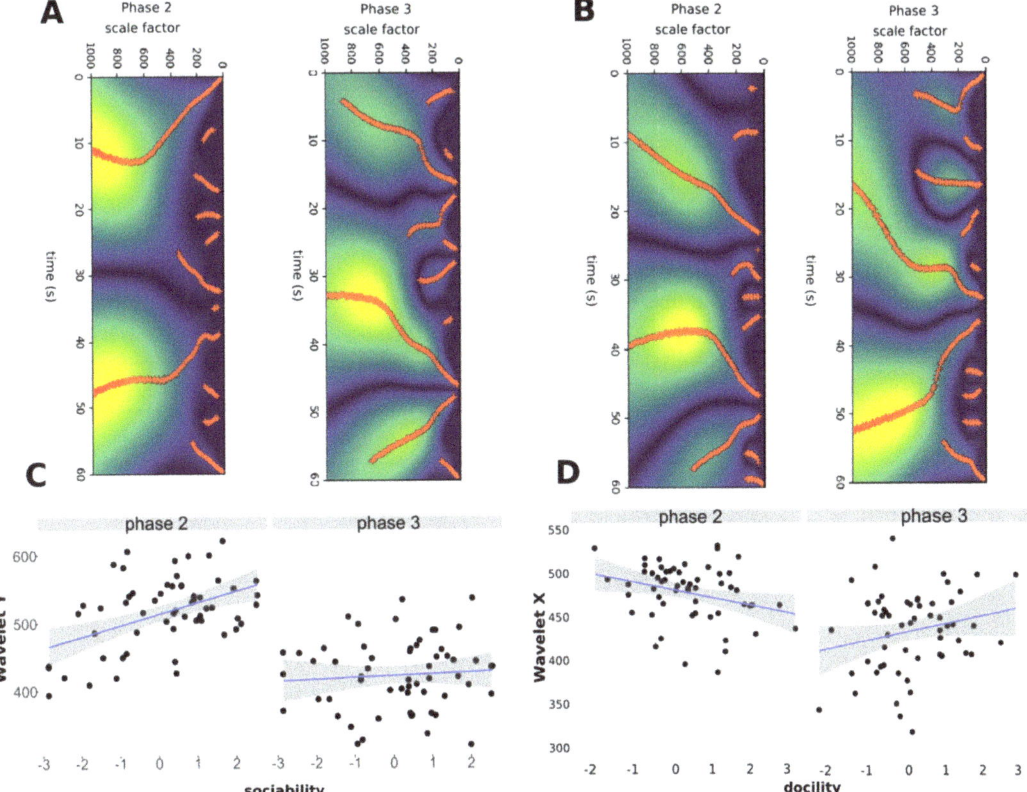

Figure 3. Wavelet analyses. (**A**) Example of wavelet transform for lateral movements (X). Red dots correspond to the detection of a change in the displacement at scale factor and time position (i.e., a local maxima of the wavelet transform of the signal position). (**B**) Example of wavelet transform for longitudinal movements (Y). (**C**) Relationship between the number of local maxima (red dots in (**A**,**B**)) in the wavelet extraction and the degree of sociability of sheep during phases 2 and 3. (**D**) Relationship between the number of wavelets and the degree of docility of sheep during phases 2 and 3. See details of models in Table 4. N = 58 sheep.

Table 4. Wavelet analyses. Results of the best GLMM (Gaussian family, after model selection—see details in Tables S2 and S3). The model tested the effects of phase, docility, sociability, and binary interactions of each variable with phase, on the number of wavelets. Lamb identity was included as a random factor. Significant effects ($p < 0.05$) are shown in bold. Wavelet Y: longitudinal movements. Wavelet X: transversal movements.

Wavelet Y	Estimate	Std. Error	Df	t Value	Pr (>|t|)
(Intercept)	**514**	**6.64**	**110**	**77.3**	**<0.001**
sociability	**17**	**4.68**	**110**	**3.63**	**<0.001**
phase 3	**−91.5**	**9.07**	**55**	**−10.1**	**<0.001**
docility	−3.12	5.72	110	−0.545	0.587
Sociability:phase 3	−14.4	6.4	55	−2.25	0.05
Docility:phase 3	4.7	7.81	55	0.602	0.55
Wavelet X	**Estimate**	**Std. Error**	**df**	**t Value**	**Pr (>|t|)**
(Intercept)	**467**	**6.04**	**110**	**77.3**	**<0.001**
sociability	0.526	4.26	110	0.124	0.902
phase 3	**−53.2**	**8.26**	**55**	**−6.43**	**<0.001**
Docility	−9.61	5.2	110	−1.85	0.0673
Sociability:phase 3	7.36	5.82	55	1.26	0.212
Docility: phase 3	**19.2**	**7.11**	**55**	**2.7**	**<0.05**

- Heatmap analyses: Detection of spatial coverage

Finally we quantified the spatial coverage by individual sheep (number of zones occupied in the arena) using heatmaps (Figure 4). Overall, the sheep used 2.37 (std. 1.03) time less space in phase 3 than in phase 2. We tested the effects of the individual characteristics on the number of zones in which the sheep spent more than 200 ms using GLMMs and model selection. Here we describe the most explanatory model considering AIC, but the three best models gave a similar trend on the sheep behaviour (see Table S4), so that an average model was ultimately performed using the models with n difference of AIC lower than 2 with the best model. Using a spatial resolution of the grid similar to the dimension of a lamb body size (i.e., dimension: 0.44×0.40 m; example Figure 4A) revealed that sheep tended to use less space in phase 3 than in phase 2 (estimate = −0.765, std. = 0.053, $p < 0.001$), and that highly sociable sheep used more space in phase 2 than less sociable sheep (estimate = 0.048, std. = 0.024, $p = 0.043$). It also showed that most docile sheep used less space in phase 2 than less docile sheep (estimate = −0.066, std. = 0.031, $p = 0.0389$) but the phenomenon was reduced in phase 3 (estimate = 0.099, std. = 0.046, $p = 0.032$) (Table 5). Therefore, the influence of sociability on spatial coverage decreased in phase 3.

Table 5. Heatmap analyses. Results of the best GLMM (Gaussian family, after model selection—see details in Table S4). The model tested the effects of phase, docility, sociability, and dual interactions of each variable with phase, on the number of areas where the lamb spent more than 1 s. Lamb identity was included as a random factor. Significant effects ($p < 0.05$) are shown in bold.

Heatmap	Estimate	Std. Error	z Value	Pr (>|z|)
(Intercept)	**2.95**	**0.037**	**79.00**	**$<2 \times 10^{-16}$**
docility	**−0.066**	**0.031**	**2.07**	**0.039**
phase 3	**−0.77**	**0.053**	**14.27**	**$<2 \times 10^{-16}$**
sociability	**0.048**	**0.023**	**2.022**	**0.043**
phase 3: docility	**0.099**	**0.046**	**2.15**	**0.032**
phase 3: sociability	−0.020	0.038	0.52	0.60

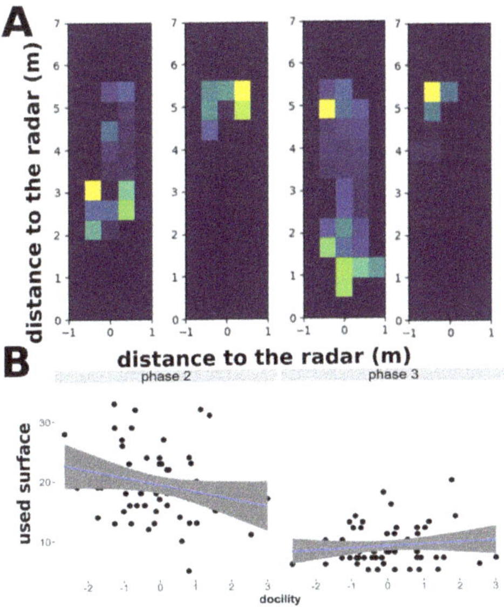

Figure 4. Heatmap analyses. Relationship between the numbers of areas occupied by the lambs and the degree of docility in phase 2 and phase 3. (**A**) Resolution grid (cell dimension: 0.44 × 0.40 m). (**B**) Relationship between the surface used by the sheep the degree of docility of sheep during phases 2 and 3. See details of models in Table 5. N = 58 sheep.

3.3. Sheep Behavioural Phenotype

We explored whether the new movement features extracted from the radar data could capture information from behavioural traits measured manually by the experimenter in the arena test. We focused on docility and sociability. We ran a PCA based on the eight behavioural measures extracted from the radar data in phase 2 and phase 3: proportion of fast movements (class 1) out of all movements (class 1 + class 2), longitudinal movements (wavelets Y), transversal movements (wavelets X) and space coverage (heatmaps). We retained two PCs using the Kaiser–Guttman criterion [40]. PC1 explained 30.65% of the variance and PC2 explained 19.31% of the variance (Table 6). The eigenvalues associated to the 3 first components are: 2.8928914, 1.7375911, 0.9738257. PC1 was positively associated with all behavioural variables (Figure 5A). Sheep with high PC1 values moved more often fast, made more changes in the way of displacement, and used more zones than sheep with low PC1 values. We therefore interpreted PC1 as a "movement" component. PC2 was positively associated with the four behavioural variables of phase 3 and negatively associated with the four behavioural variables of phase 2 (Figure 5A). Sheep with high PC2 values showed a more important increase of time spent moving fast, of the frequency of changes in the way of displacement, and numbers of zones occupied between phase 2 and phase 3 than sheep with low PC2 levels. We interpreted PC2 as a variable of "movement in response to social isolation". Using PC1 and PC2, we investigated contribution of the docility and sociability of the sheep on these components. It showed that the first was linked to the sociability (estimate = 0.2690, std. = 0.1054, $p = 0.0135$) and the second was linked to docility (estimate = 0.28296, std. = 0.1111, $p = 0.0137$). The link between PC1 and docility and PC2 and sociability was not significant.

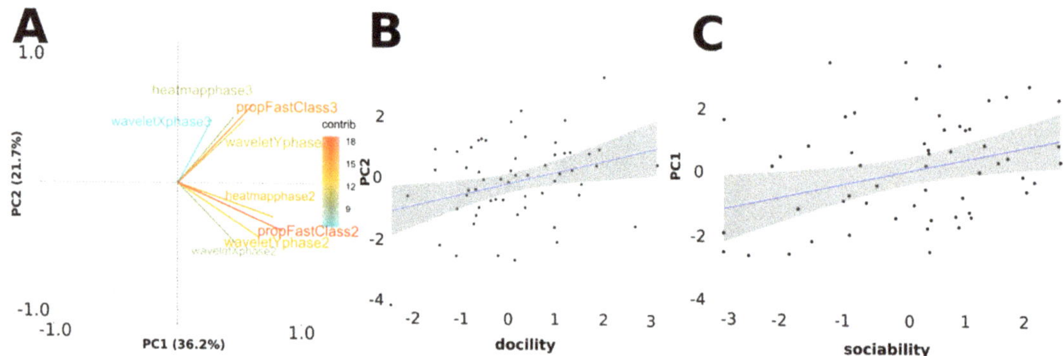

Figure 5. (**A**) Correlations between the two first components (PCs) of the principal component analysis (PCA). Arrows represent the eight behavioural variables on PC1 (movement speed) and PC2 (movement increase between phases). Contribution of variables to the variance explained is color-coded. Each data point represents the PC1 and PC2 scores of a given lamb (N = 58). (**B**) Relationship between PC1 and sociability. (**C**) Relationship between PC2 and docility. Blue lines represent linear models (see main text). N = 58 sheep.

Table 6. Eigenvalue for each component (PC) of the Principal Component Analysis using the eight behavioural features extracted using the radar tracking.

Component	Eigenvalue	Variance Explained
PC 1	2.893	30.65
PC 2	1.738	19.31
PC 3	0.974	13.04
PC 4	0.833	9.27
PC 5	0.564	7.20
PC 6	0.492	6.69

3.4. Outdoor Radar Tracking

To demonstrate that our radar tracking system could be used at larger spatial scales, in the field, we sat up a radar with a lower operating frequency in an outside corridor (10 × 60 m; Figure 6A). We successfully monitored the 2D trajectory of one sheep over a maximum distance of 45 m the backscattering signal was not detectable using one radar measurement (Figure 6B). The presence of a human to induce sheep movement did not deteriorate sheep tracking (Figure 6C).

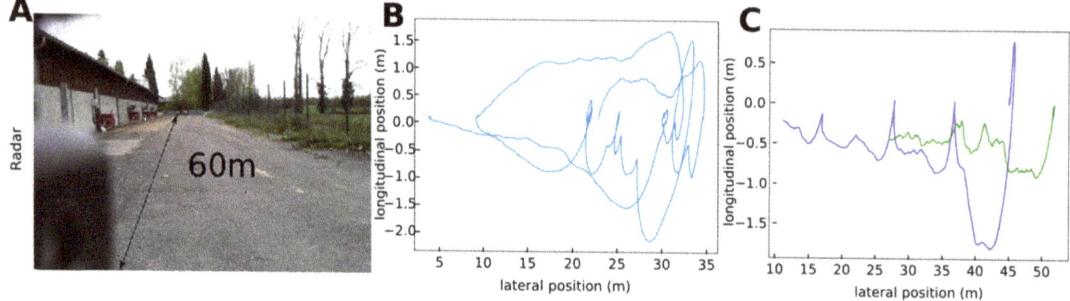

Figure 6. (**A**) Picture of the outside corridor used for radar tracking of a sheep (credit AD). The radar was positioned 60 m from the end of the corridor. (**B**) Example of trajectory of a sheep derived from the radar data. (**C**) Example of trajectory of a sheep (red) and a man (green) derived from the radar data.

4. Discussion

Research in animal behaviour increasingly requires automated monitoring and annotation of animal movements for comparative quantitative analyses [2,3]. Here we introduced a radar tracking system suitable to study the 2D movements of sheep indoor and outdoor, within a range of 45 m. A summary of the method is shown on Figure 7. The system is non-sensitive to light variations, compatible with real time data analyses, transportable, fast processing and adaptable to various species and experimental contexts. Moreover, it does not require tags or transponders to track animals. It is therefore suitable for the collection of large sets of behavioural data in an automated way required in many areas of biological and ecological research, as well as applied ethology for precision farming as illustrated here.

We recently used FMCW radars to track the behaviour of sheep [18], pigs [16] and bees [17]. Here, however, for the first time, we demonstrate the applicability of this approach to monitor 2D trajectories of untagged walking animals within a range of 45 m. Others methods can be used to estimate the sheep position, such as video detection [24] which can detect sheep in 2D up to 20 m but with a precision from 50 cm (at 5 m) to 1 m (at 20 m) and GPS detection [41], but this requires to equip the animals with transponders. We showed that the radar acquisition system has several advantages over these more conventional methods, and in particular video tracking. It collects more data per second (50 measures per second for the radar versus 25 for the video), requires less RAM (524 Kb for one radar measurement versus 3.7 Mb for one video frame). It also requires 10 times less processing time (e.g., does not require to train neural networks) and generates less false detection rates (15% of false detection for video processing and 5.2% for radar processing). Importantly, the radar is not dependent on brightness and can be used for outside tracking over long distances by adjusting operating frequencies. It also enables the tracking of individualized animals without tags, based on the size and shape of the radar echoes of the different targets.

Our application of radar-based tracking to behavioural phenotyping of sheep shows that the radar analysis is consistent with current semi-automated analyses (i.e., infrared sensors and video). Using the radar, we found that sheep tend to have a greater displacement in phase 2 than in phase 3 of the arena test. This agrees with previous studies showing that sheep are more active when socially isolated from conspecifics [20,21]. Higher behavioural activity in a social isolation context, for instance through locomotion and vocalization behaviours, may be interpreted as the way for the isolated animal for searching for social contact with conspecifics as described in the ewe-lamb relationships [42] or between familiar lambs [43].

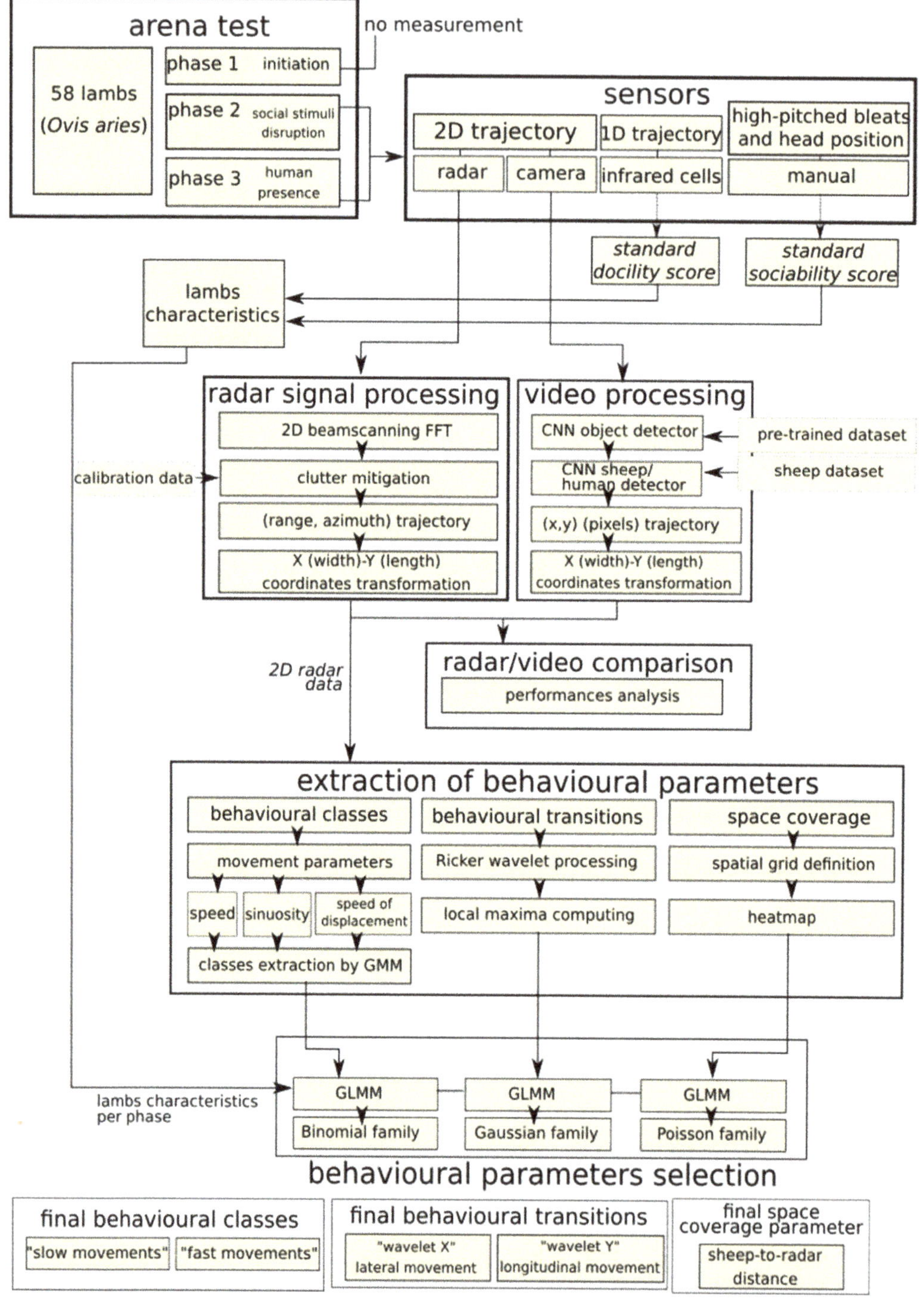

Figure 7. Summary of the method described in the study, from the behavioural test and the acquisition of the data with radar to the extraction of the new behavioural parameters form the trajectory data.

In addition, the high resolution 2D, in theory 5 cm in range and 6° in azimuth, trajectories obtained from the radar enabled identification of new behavioural estimators that could greatly benefit the fast and automated identification of behavioural phenotypes. For example, our application of unsupervised behavioural annotation to identify statistically significant behaviours by sheep in the arena test showed that sheep exhibit less fast movements in phase 3 than in phase 2. The wavelet analysis, considering the way that the sheep moves (i.e., referred to here as "way of displacement") revealed the occurrence of "erratic" displacements. Here low erratic displacements corresponded to displacements showing a constant speed whereas high erratic displacements corresponded to a high level of alternation in slow and fast displacements. These erratic displacements may be linked to the sociability and/or docility of sheep. Finally, space occupation analysis showed that individuals exploit narrower areas in phase 3 than in phase 2 of the arena test. All these results are consistent with previous observations using semi-automated recording methods. Indeed, social isolation from conspecifics (i.e., phase of test 2) resulted in the expression of on average higher behavioural activity (i.e., individual variability exists), including displacements, than in presence of conspecifics and a motionless human (i.e., phase of test 3). The higher displacement activity during social isolation resulted in a higher exploration of the arena whereas, in presence of conspecifics and a motionless human, lambs showed limited displacement. The combination of these new automatically computed estimators appears to be complementary to behavioural traits of interest that were until now measured (i.e., for instance no or slight relationship with sociability or docility) and could be used for more detailed characterization of animal behavioural profiles. Note, however, that this first study is based on relatively low sample sizes (58 individuals) and further measurements are needed to verify the biological trends observed on a much larger number of sheep.

Beyond the case study of the arena test described here, our system could be tuned to suit a large diversity of animal sizes and experimental contexts. Several ways can be considered. For instance, the range and resolution of detection could be improved using different radars. Here, we had to place the radar at 1 m from the arena fences in order to illuminate and monitor the entire arena. Antennas with larger beamwidth may allow placing the radar on the arena fences. Moreover, the detection was limited to a few meters, but it is possible to detect a sheep at tens of meters using a radar operating at a lower frequency (24 GHz) and/or transmitting higher electromagnetic power. It is also possible to improve radar detection by using more antennas. Indeed, by multiplying the number of antennas, we multiply the number of signal estimations and then the noise from the radar can be decreased. The same radar technology could be used to track individuals in groups over longer distances in open fields, for instance to explore the mechanisms underpinning social network structures and collective behaviour [44]. The processing of the radar signal can also be improved for tracking large number of sheep simultaneously by using deep radar processing but this would require the use of a large amount of annotated data to train the neural networks [45]. Individual tracking within groups could also be improved with non-invasive passive tags that depolarize radar signal in specific directions [46]. Note that at the moment, we do not know the long-term effects of the use of millimetre waves on these animals and this should be investigated in further studies.

5. Conclusions

We demonstrated the feasibility of tracking a sheep in a restricted area using a millimetre-wave FMCW radar. This detection is possible even if each wall of the arena backscatters the transmitted electromagnetic signal. This radar tracking system can also be advantageously used to extract features that are correlated to the movement of the sheep and can estimate if it is erratic, fast and the space occupied in the corridor. In contrast to other short-range tracking methods, our radar detection approach does not require pre-annotated data and can be applied in real time. This flexibility holds considerable premises for tracking the behaviour of animals of various sizes and environments in a wide range of contexts and research fields.

Supplementary Materials: The following are available online at https://www.mdpi.com/article/10.3390/s21238140/s1, Table S1: Model selection for behavioural class analyses. Null model, best model, second and third best models are displayed. Table S2: Model selection for X wavelet analyses (latitudinal movements). Null model, best model, second and third best models are displayed. Table S3: Model selection for Y wavelet analyses (longitudinal movements). Null model, best model, second and third best models are displayed. Table S4: Model selection for heatmap analyses (low spatial resolution). Null model, best model, second and third best models are displayed. Dataset S1: list of the sheep trajectory during the behavioural test and list of all behavioral score.

Author Contributions: Conceptualization, A.D., E.R., J.-F.B., D.H. (Dominique Hazard), A.B., M.L. and H.A.; investigation, A.D., E.R. and J.-F.B.; formal analysis, A.D., C.P. and M.B.; writing—original draft preparation, A.D.; writing—review and editing, all authors; funding acquisition, M.L. and H.A. All authors have read and agreed to the published version of the manuscript.

Funding: This research was funded by the CNRS, a grant fom the Région Occitanie to ML and HA (SIDIPAR), a grant from the Agence National de la Recherche to ML and HA (ANR-19-CE37-0024—3DNaviBee), and a ERC Consolidator grant to ML (GA101002644—BEE-MOVE).

Institutional Review Board Statement: The experiments described here fully comply with applicable legislation on research involving animal subjects in accordance with the European Union Council directive (2010/63/UE). The investigators who carried out the experiments were certified by the relevant French governmental authority. All experimental procedures were performed according to the guidelines for the care and use of experimental animals and approved by the local ethics committee (approval number SSA_2018_011).

Informed Consent Statement: Informed consent was obtained from all subjects involved in the study.

Data Availability Statement: The dataset used to compare radar features and behavioural scores are available in Dataset S1.

Acknowledgments: We thank Sara Parisot for the management of the experimental farm La Fage, and Sébastien Douls, Christian Durand and Gaëtan Bonnafe for managing the experimental flock, for animal care and for their active role in collecting experimental data. We are also grateful to Eric Delval for his involvement in the thinking about behavioral criteria.

Conflicts of Interest: The authors declare no conflict of interest.

References

1. Branson, K.; Robie, A.A.; Bender, J.; Perona, P.; Dickinson, M.H. High-Throughput Ethomics in Large Groups of Drosophila. *Nat. Methods* **2009**, *6*, 451–457. [CrossRef]
2. Anderson, D.J.; Perona, P. Toward a Science of Computational Ethology. *Neuron* **2014**, *84*, 18–31. [CrossRef]
3. Brown, A.E.X.; de Bivort, B. Ethology as a Physical Science. *Nat. Phys.* **2018**, *14*, 653–657. [CrossRef]
4. Morand-Ferron, J.; Cole, E.F.; Quinn, J.L. Studying the Evolutionary Ecology of Cognition in the Wild: A Review of Practical and Conceptual Challenges. *Biol. Rev.* **2016**, *91*, 367–389. [CrossRef]
5. Huang, W.; Pilkington, J.G.; Pemberton, J.M. Patterns of MHC-Dependent Sexual Selection in a Free-Living Population of Sheep. *Mol. Ecol.* **2021**. [CrossRef] [PubMed]
6. Canario, L.; Mignon-Grasteau, S.; Dupont-Nivet, M.; Phocas, F. Genetics of Behavioural Adaptation of Livestock to Farming Conditions. *Animal* **2013**, *7*, 357–377. [CrossRef] [PubMed]
7. Phocas, F.; Boivin, X.; Sapa, J.; Trillat, G.; Boissy, A.; Neindre, P.L. Genetic Correlations between Temperament and Breeding Traits in Limousin Heifers. *Anim. Sci.* **2006**, *82*, 805–811. [CrossRef]
8. Boissy, A.; Bouix, J.; Orgeur, P.; Poindron, P.; Bibé, B.; Le Neindre, P. Genetic Analysis of Emotional Reactivity in Sheep: Effects of the Genotypes of the Lambs and of Their Dams. *Genet. Sel. Evol.* **2005**, *37*, 381–401. [CrossRef]
9. Tomkiewicz, S.M.; Fuller, M.R.; Kie, J.G.; Bates, K.K. Global Positioning System and Associated Technologies in Animal Behaviour and Ecological Research. *Philos. Trans. R. Soc. B Biol. Sci.* **2010**, *365*, 2163–2176. [CrossRef]
10. Cadahía, L.; López-López, P.; Urios, V.; Negro, J.J. Satellite Telemetry Reveals Individual Variation in Juvenile Bonelli's Eagle Dispersal Areas. *Eur. J. Wildl. Res.* **2010**, *56*, 923–930. [CrossRef]
11. Voulodimos, A.S.; Patrikakis, C.Z.; Sideridis, A.B.; Ntafis, V.A.; Xylouri, E.M. A Complete Farm Management System Based on Animal Identification Using RFID Technology. *Comput. Electron. Agric.* **2010**, *70*, 380–388. [CrossRef]
12. Riley, J.R.; Smith, A.D.; Reynolds, D.R.; Edwards, A.S.; Osborne, J.L.; Williams, I.H.; Carreck, N.L.; Poppy, G.M. Tracking Bees with Harmonic Radar. *Nature* **1996**, *379*, 29–30. [CrossRef]
13. Pérez-Escudero, A.; Vicente-Page, J.; Hinz, R.C.; Arganda, S.; De Polavieja, G.G. IdTracker: Tracking Individuals in a Group by Automatic Identification of Unmarked Animals. *Nat. Methods* **2014**, *11*, 743–748. [CrossRef]

14. García-Martín, E.; Rodrigues, C.F.; Riley, G.; Grahn, H. Estimation of Energy Consumption in Machine Learning. *J. Parallel Distrib. Comput.* **2019**, *134*, 75–88. [CrossRef]
15. Dell, A.I.; Bender, J.A.; Branson, K.; Couzin, I.D.; de Polavieja, G.G.; Noldus, L.P.; Pérez-Escudero, A.; Perona, P.; Straw, A.D.; Wikelski, M. Automated Image-Based Tracking and Its Application in Ecology. *Trends Ecol. Evol.* **2014**, *29*, 417–428. [CrossRef] [PubMed]
16. Dore, A.; Lihoreau, M.; Billon, Y.; Ravon, L.; Bailly, J.; Bompa, J.-F.; Ricard, E.; Aubert, H.; Henry, D.; Canario, L. Millimetre-Wave Radars for the Automatic Recording of Sow Postural Activity. In Proceedings of the 71. Annual Meeting of the European Association for Animal Production (EAAP), Porto, Portugal, 17 December 2020.
17. Dore, A.; Henry, D.; Lihoreau, M.; Aubert, H. 3D Trajectories of Multiple Untagged Flying Insects from Millimetre-Wave Beamscanning Radar. In Proceedings of the 2020 IEEE International Symposium on Antennas and Propagation and North American Radio Science Meeting, Montréal, QC, Canada, 5–10 July 2020; IEEE: Piscataway, NJ, USA, 2020; pp. 1209–1210.
18. Henry, D.; Aubert, H.; Ricard, E.; Hazard, D.; Lihoreau, M. Automated Monitoring of Livestock Behavior Using Frequency-Modulated Continuous-Wave Radars. *Prog. Electromagn. Res.* **2018**, *69*, 151–160. [CrossRef]
19. Hazard, D.; Moreno, C.; Foulquié, D.; Delval, E.; François, D.; Bouix, J.; Sallé, G.; Boissy, A. Identification of QTLs for Behavioral Reactivity to Social Separation and Humans in Sheep Using the OvineSNP50 BeadChip. *BMC Genom.* **2014**, *15*, 778. [CrossRef]
20. Hazard, D.; Bouix, J.; Chassier, M.; Delval, E.; Foulquie, D.; Fassier, T.; Bourdillon, Y.; François, D.; Boissy, A. Genotype by Environment Interactions for Behavioral Reactivity in Sheep. *J. Anim. Sci.* **2016**, *94*, 1459–1471. [CrossRef]
21. Ligout, S.; Foulquié, D.; Sèbe, F.; Bouix, J.; Boissy, A. Assessment of Sociability in Farm Animals: The Use of Arena Test in Lambs. *Appl. Anim. Behav. Sci.* **2011**, *135*, 57–62. [CrossRef]
22. Haderer, A.; Wagner, C.; Feger, R.; Stelzer, A. A 77-GHz FMCW Front-End with FPGA and DSP Support. In Proceedings of the 2008 International Radar Symposium, Wroclaw, Poland, 21–23 May 2008; IEEE: Piscataway, NJ, USA, 2008; pp. 1–6.
23. Redmon, J.; Farhadi, A. Yolov3: An Incremental Improvement. *arXiv* **2018**, arXiv:1804.02767.
24. Everingham, M.; Eslami, S.A.; Van Gool, L.; Williams, C.K.; Winn, J.; Zisserman, A. The Pascal Visual Object Classes Challenge: A Retrospective. *Int. J. Comput. Vis.* **2015**, *111*, 98–136. [CrossRef]
25. Bonneau, M.; Vayssade, J.-A.; Troupe, W.; Arquet, R. Outdoor Animal Tracking Combining Neural Network and Time-Lapse Cameras. *Comput. Electron. Agric.* **2020**, *168*, 105150. [CrossRef]
26. Dietlein, C.R.; Bjarnason, J.E.; Grossman, E.N.; Popović, Z. Absorption, Transmission, and Scattering of Expanded Polystyrene at Millimeter-Wave and Terahertz Frequencies. In Proceedings of the Passive Millimeter-Wave Imaging Technology XI, Orlando, FL, USA, 16–20 March 2008; International Society for Optics and Photonics: Bellingham, WA, USA, 2008; Volume 6948, p. 69480E.
27. Balanis, C.A. *Modern Antenna Handbook*; John Wiley & Sons: Hoboken, NJ, USA, 2011.
28. Reynolds, D.A. Gaussian Mixture Models. *Encycl. Biom.* **2009**, *741*, 659–663.
29. Burnham, K.P. Model Selection and Multimodel Inference. In *A Practical Information-Theoretic Approach*; Springer: New York, NY, USA, 1998.
30. Pedregosa, F.; Varoquaux, G.; Gramfort, A.; Michel, V.; Thirion, B.; Grisel, O.; Blondel, M.; Prettenhofer, P.; Weiss, R.; Dubourg, V. Scikit-Learn: Machine Learning in Python. *J. Mach. Learn. Res.* **2011**, *12*, 2825–2830.
31. Simon, W.; Klein, T.; Litschke, O. Small and Light 24 GHz Multi-Channel Radar. In Proceedings of the 2014 IEEE Antennas and Propagation Society International Symposium (APSURSI), Memphis, TN, USA, 6–11 July 2014; IEEE: Piscataway, NJ, USA, 2014; pp. 987–988.
32. Ryan, H. *Ricker, Ormsby; Klander, Bntterwo-A Choice of Wavelets*; CSEG Recorder: Calgary, AB, Canada, 1994.
33. Poirier, J.-R.; Aubert, H.; Jaggard, D.L. Lacunarity of Rough Surfaces from the Wavelet Analysis of Scattering Data. *IEEE Trans. Antennas Propag.* **2009**, *57*, 2130–2136. [CrossRef]
34. Gaucherel, C. Wavelet Analysis to Detect Regime Shifts in Animal Movement. *Comput. Ecol. Softw.* **2011**, *1*, 69.
35. Bains, R.S.; Cater, H.L.; Sillito, R.R.; Chartsias, A.; Sneddon, D.; Concas, D.; Keskivali-Bond, P.; Lukins, T.C.; Wells, S.; Acevedo Arozena, A. Analysis of Individual Mouse Activity in Group Housed Animals of Different Inbred Strains Using a Novel Automated Home Cage Analysis System. *Front. Behav. Neurosci.* **2016**, *10*, 106. [CrossRef]
36. Idris, A.; Moors, E.; Budnick, C.; Herrmann, A.; Erhardt, G.; Gauly, M. Is the Establishment Rate and Fecundity of Haemonchus Contortus Related to Body or Abomasal Measurements in Sheep? *Animal* **2011**, *5*, 1276–1282. [CrossRef]
37. Team, R.C. *R: A Language and Environment for Statistical Computing*; R Foundation for Statistical Computing: Vienna, Austria, 2014.
38. Stroup, W.W. *Generalized Linear Mixed Models: Modern Concepts, Methods and Applications*; CRC Press: Boca Raton, FL, USA, 2012.
39. Lê, S.; Josse, J.; Husson, F. FactoMineR: An R Package for Multivariate Analysis. *J. Stat. Softw.* **2008**, *25*, 1–18. [CrossRef]
40. Kaiser, H.F. Coefficient Alpha for a Principal Component and the Kaiser-Guttman Rule. *Psychol. Rep.* **1991**, *68*, 855–858. [CrossRef]
41. Roberts, G.; Williams, A.; Last, J.D.; Penning, P.D.; Rutter, S.M. A Low-Power Postprocessed DGPS System for Logging the Locations of Sheep on Hill Pastures. *Navigation* **1995**, *42*, 327–336. [CrossRef]
42. Sebe, F.; Nowak, R.; Poindron, P.; Aubin, T. Establishment of Vocal Communication and Discrimination between Ewes and Their Lamb in the First Two Days after Parturition. *Dev. Psychobiol. J. Int. Soc. Dev. Psychobiol.* **2007**, *49*, 375–386. [CrossRef] [PubMed]
43. Sebe, F.; Ligout, S.; Porter, R. Vocal Discrimination of Kin and Non-Kin Agemates among Lambs. *Behaviour* **2004**, *141*, 355–369. [CrossRef]
44. Ginelli, F.; Peruani, F.; Pillot, M.-H.; Chaté, H.; Theraulaz, G.; Bon, R. Intermittent Collective Dynamics Emerge from Conflicting Imperatives in Sheep Herds. *Proc. Natl. Acad. Sci. USA* **2015**, *112*, 12729–12734. [CrossRef] [PubMed]

45. Huang, H.; Gui, G.; Sari, H.; Adachi, F. Deep Learning for Super-Resolution DOA Estimation in Massive MIMO Systems. In Proceedings of the 2018 IEEE 88th Vehicular Technology Conference (VTC-Fall), Chicago, IL, USA, 27–30 August 2018; pp. 1–5.
46. Lui, H.-S.; Shuley, N. Resonance Based Radar Target Identification with Multiple Polarizations. In Proceedings of the 2006 IEEE Antennas and Propagation Society International Symposium, Albuquerque, NM, USA, 9–14 July 2006; IEEE: Piscataway, NJ, USA, 2006; pp. 3259–3262.

 sensors

Article

Prediction of Cow Calving in Extensive Livestock Using a New Neck-Mounted Sensorized Wearable Device: A Pilot Study

Carlos González-Sánchez , Guillermo Sánchez-Brizuela , Ana Cisnal, Juan-Carlos Fraile *, Javier Pérez-Turiel and Eusebio de la Fuente-López

ITAP (Instituto de las Tecnologías Avanzadas de la Producción), Universidad de Valladolid, Paseo del Cauce 59, 47011 Valladolid, Spain; cgonzalezs90@gmail.com (C.G.-S.); guillermo.sanchez.brizuela@uva.es (G.S.-B.); ana.cisnal@uva.es (A.C.); turiel@eii.uva.es (J.P.-T.); efuente@eii.uva.es (E.d.l.F.-L.)
* Correspondence: jcfraile@eii.uva.es

Abstract: In this study, new low-cost neck-mounted sensorized wearable device is presented to help farmers detect the onset of calving in extensive livestock farming by continuously monitoring cow data. The device incorporates three sensors: an inertial measurement unit (IMU), a global navigation satellite system (GNSS) receiver, and a thermometer. The hypothesis of this study was that onset calving is detectable through the analyses of the number of transitions between lying and standing of the animal (lying bouts). A new algorithm was developed to detect calving, analysing the frequency and duration of lying and standing postures. An important novelty is that the proposed algorithm has been designed with the aim of being executed in the embedded microcontroller housed in the cow's collar and, therefore, it requires minimal computational resources while allowing for real time data processing. In this preliminary study, six cows were monitored during different stages of gestation (before, during, and after calving), both with the sensorized wearable device and by human observers. It was carried out on an extensive livestock farm in Salamanca (Spain), during the period from August 2020 to July 2021. The preliminary results obtained indicate that lying-standing animal states and transitions may be useful to predict calving. Further research, with data obtained in future calving of cows, is required to refine the algorithm.

Keywords: cow; extensive livestock; sensorized wearable device; monitoring; parturition prediction

Citation: González-Sánchez, C.; Sánchez-Brizuela, G.; Cisnal, A.; Fraile, J.-C.; Pérez-Turiel, J.; Fuente-López, E.d.l. Prediction of Cow Calving in Extensive Livestock Using a New Neck-Mounted Sensorized Wearable Device: A Pilot Study. *Sensors* **2021**, *21*, 8060. https://doi.org/10.3390/s21238060

Received: 19 October 2021
Accepted: 30 November 2021
Published: 2 December 2021

1. Introduction

The study and monitoring of livestock has always been a subject of great interest. Indeed, quantitative measurement of animal behaviour is an important tool for understanding their reproduction, survival, welfare, and interaction with other animals [1]. Animal activity is one of the most important indicators associated with animal health and welfare [2], and animal behaviour is an indicator of the well-being and health of cows [3]. Detecting changes in the behaviour and activity of cows is a good preventive tool to determine the animal's health status [4].

Every year an average of 8.5% of perinatal calves are lost due to natural abortions, stillbirths, and complications during parturition (calving) [5], which translates into higher economical costs and reduced animal wellbeing. Ideally the calving process should be carefully overviewed by experts to avoid or correct any problems that may arise (e.g., dystocia). Even today, the analysis of cow behaviour and calving detection is mainly carried out by experienced workers through unaided monitoring. These approaches are however expensive and time consuming, and not effective for extensive livestock farming, where many animals are kept under grazing in the open air on large areas of surface.

An automated solution based on cow data collection from sensors could provide better calving predictions. This will allow the farmer to better identify those cows that require intensive supervision and to focus on caring for cows with upcoming calving, reducing possible risks and improving the health and wellbeing of the animals.

Internet of things (IoT), an already mature and effective technology, can help improve the efficiency and productivity in agriculture and livestock production systems [6]. IoT has initially spread into the agriculture and farming industry, and mainly aims to supervise the well-being of animals, thus enhancing the profitability of farms by increasing productivity [7]. Most connected livestock solutions are developed for cattle, especially cows, due to the valuable price of such animals [8].

Precision livestock farming (PLF) aims to manage individual animals by continuously monitoring their health, welfare, production/reproduction and/or environmental impact in real time [9]. This is achieved through real-time image [10,11] and/or sound analyses or by wearable devices with sensors that monitor physical (position, direction of movement, speed ...) [12–15], and physiological variables (heart rate, breathing rate, temperature ...) [16–18] of each animal.

Different physiological and behavioural parameters associated with calving can be monitored through sensors. The analysis of the internal temperature and its evolution, usually measured in the vulva, the rectum, or the rumen of the animal, is one of the most accurate calving predictors [19]. It has been demonstrated that a decrease in vaginal temperature equal or greater than 0.3 °C in cows bearing singletons can predict calving within the next 36 h in 83.3% of cases and up to 100% within 60 h [20]). However, the core temperature of the animal is difficult to be measured in a non-intrusive way, and the available commercial solutions require intensive veterinary care to install and to check the correct location of the measuring device. This approach is therefore not preferred for use in PLF.

An exhaustive meta-analysis of the different publications related to calving detection in cows is showed in [21]. It concludes that automated monitoring and detection of calving, as well as of dystocia incidents, is possible. However, behavioural changes associated with calving vary between individual animals. Behaviour associated with feeding and rumination descent gradually in the two weeks leading up to calving and is drastically reduced during calving [22]. The duration of rumination descends up to 33% the day when calving takes place in comparison with the previous day. This behaviour could be successfully measured using ear or neck-mounted devices [23–26].

Another indicator of calving is the increase in lying bouts (LB). This behaviour is associated with the restlessness that the animal feels due to the imminency of the calving. The frequency of lying bouts and their mean duration increase greatly as the calving event approaches, starting already 48 h before and being maximized on the day of calving [26]. This increase [6] (from 9.3 $\pm$ 1.31 LB/day four days before calving to 13.0 $\pm$ 1.02 LB/day the day of calving) is especially important in heifers, but also multiparous dam show more activity prior to calving, and can be observed on average 6 h before calf birth [27].

Lying bouts can be easily identified using a leg-mounted accelerometer [28], but detection is much more challenging using ear or neck-based sensors due to the similarities in the signals from the accelerometer when the animal is standing and lying. A recent study [29] reported the use of a neck-based accelerometer to distinguish between those states by detecting the characteristic movement associated with the transition between states. After calving, the number of steps per hour stays elevated, whereas lying bouts tend to gradually decrease as the animal transitions between pre and postpartum states.

Tail-raising patterns have been observed to change in the 24 h prior to calving [30,31] and can be monitored using tail-mounted accelerometers. This new approach is however not viable for long-term monitoring due to the weight limitation of these devices, overall reduced stability, and possible damage to the skin of the animal [32].

In calving prediction by traditional methods, the farmer makes a visual inspection of the cow to know its status. This is an error-prone task in which even experts may fail to provide an accurate prediction of calving date. In extensive farming, animals move freely in a wide area, which makes it more difficult for the farmer to properly monitor and manage pregnant cows. An automated solution based on cow data collected from sensors and processed by algorithms, can provide better delivery predictions than visual observation. This will allow the farmer to have a more accurate estimation of the expected calving date of the cow and to identify those cows that require intensive supervision due to the proximity of calving. It will make possible both reducing the workload of farmers, who can focus on caring for cows with upcoming calving and improving the health of the cow.

In this paper a low-cost neck-mounted sensorized wearable device was designed and an algorithm was developed to detect the onset of calving of cows in extensive livestock farming. This work was divided in three main phases: (1) development of a wearable solution for data collection based on different sensors; (2) data collection in extensive livestock farming by using the aforementioned solution and human-based observations; and (3) development of algorithms to detect the onset of calving and creation of a decision function based on the frequency and duration of lying and standing behaviour (lying bouts).

This paper is organized as follows. Section 2 presents the sensorized wearable device, the software developed to collect data from different cows by human observers, gives an overview of the collected data, as well as the methodology followed during this study and introduces the proposed algorithm for parturition detection. Results and discussion of this algorithm are presented in Section 3. Finally, Section 4 shows the conclusions and future work.

2. Materials and Methods

The goal of the first phase of the study is the development of a solution for recording large quantities of behavioural information, which will later be combined with human observation of the animals. The captured information will be used for the development, validation, and quantification of algorithms for calving prediction.

A new low-cost sensorized wearable device was developed and integrated into a collar, which can be placed around the cow's neck. The developed device incorporates three sensors: an Inertial Measurement Unit (IMU), a GNSS and a thermometer. Human observers helped with monitoring, labelling, and recording the animals' state using our own development PC software. The designed collars were tested on cows from an extensive livestock farm in Salamanca (Spain), during the period August 2020–July 2021.

2.1. Sensorised Wearable Device

The sensorized wearable device is a collar which is placed on the animal's neck. The collar houses a nRF52840-dongle (Nordic Semiconductor, Trondheim, Norway) microcontroller, three sensors (thermometer, 9-axis IMU—3-axis accelerometer, 3-axis gyroscope and 3-axis magnetometer—and GNSS), a microSD card breakout board (AdaFruit, NY, EEUU) for data storage and lithium batteries. Figure 1 shows the overall architecture of the collar.

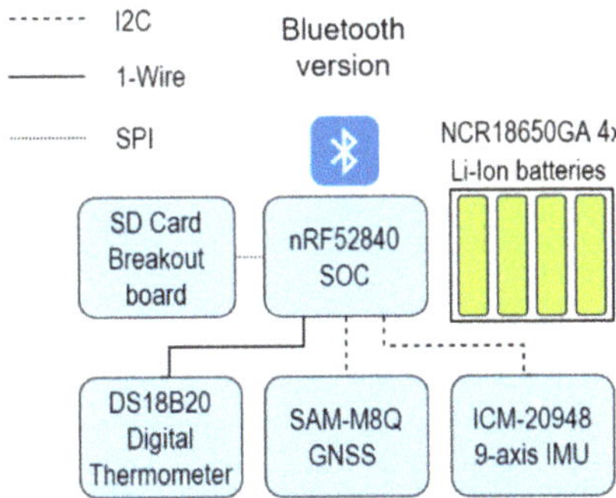

Figure 1. Overall architecture of sensorized collar.

The microcontroller nRF52840 is a small, low-cost device built around the 32-bit ARM® Cortex™-M4 CPU which supports short-range wireless standards including BLE. For measuring the cow's body temperature, a DS18B20 digital thermometer with a temperature range of −55 to 125 °C (accuracy ± 0.5 °C) is placed in such a way that contact between the skin of the cow and the sensor occurs. The microcontroller acquires the 9–12 bits configurable Celsius temperature measurements using a unique 1-wire interface, which only requires one port pin for the communication. The IMU integrated in the collar is the ICM-20948 (InvenSense, Berkeley, CA, USA, EEUU), which is a low-power 9-axis motion tracking device embedding a 3-axis gyroscope, a 3-axis accelerometer, and a 3-axis compass. A SAM-M8Q (U-blox, Thalwil, Switzerland) receiver is used for precise geographic positioning of the animal and is configured to work in PSM (Power Save Management) mode to minimize power consumption. Both the IMU and the GNSS communicate with the nRF52840 microcontroller using I^2C at 400 KHz. The microSD card, which allows the storage of data from the sensors, communicates with the microcontroller using SPI. Four lithium ion NCR18650GA (Sanyo, Osaka, Japan) batteries of 3.7 V and 3350 mA were used to power the device.

Figure 2 shows the developed collar. To avoid damage to the electronic components and the batteries, a protective box was designed and produced using additive manufacturing in a 3D printer (Ultimaker® 2+). The polymer selected was ABS, a low-cost plastic material with good mechanical properties. The cover box has two areas (Figure 2a,b). The lower part contains the electronics, whereas the upper area houses the batteries. This design facilitates the replacement of the batteries in the collar without affecting the electronic components. To waterproof the container, a 2 mm diameter nitrile rubber O-ring was fitted in the junction between the two covers. The box was attached to the neck of the cows using an adjustable leather belt (Figure 2c), which allows for a good fit on animals of different sizes. The material of the belt is soft, which maximises the animal's comfort.

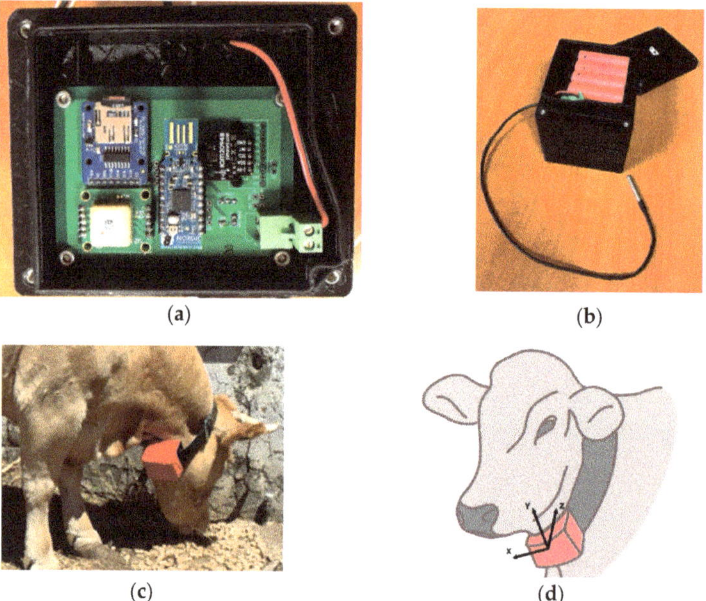

Figure 2. Collar developed for data collection. (**a**) PCB with electronic components; (**b**) collar batteries and temperature sensor; (**c**) collar with belt placed on the cow's neck; (**d**) IMU axis orientation in the collar.

All electronic components of the collar were configured to work in low power mode, to minimize power consumption, and to extend battery life. The data from the accelerometer and the gyroscope embedded in the IMU were sampled at a frequency of 17.6 Hz. The temperature and geolocation information (longitude, latitude, altitude, and speed) were collected at 1 Hz. The orientation of the IMU axis is shown in Figure 2d.

Initial tests were performed using BLE 4.2 for the communication between the board and a computer where data was registered. Due to high power consumption, however, it was decided to store the information onboard instead, using a microSD card. This approach increased the average collar battery life to up to 15 days.

Each datapoint includes the information from the sensors and a timestamp, using the format described in Table 1. The data are then saved to the microSD card using binary format. To minimize data loss if the collar runs out of batteries or fails, a new file is created every hour.

Table 1. Raw data collected by the sensorized collar and saved in the microSD card.

Variable	Datatype	Units
Timestamp	int32	ms since Unix Epoch
Temperature IMU	float32	degrees C
Temperature DB18B20	float32	degrees C
Longitude (Lon)	int32	degrees ($\times 10^{-7}$)
Latitude (Lat)	int32	degrees ($\times 10^{-7}$)
Altitude above sea level (Alt)	int32	m ($\times 10^{-3}$)
Speed	int32	m s^{-1} ($\times 10^{-3}$)
Acceleration axis X (ax)	float32	$\times$g
Acceleration axis Y (ay)	float32	$\times$g
Acceleration axis Z (az)	float32	$\times$g
Rotation X axis (gx)	float32	degrees s^{-1}
Rotation Y axis (gy)	float32	degrees s^{-1}
Rotation Z axis (gz)	float32	degrees s^{-1}

Although the data transmission is no longer performed wirelessly, the collar still makes use of BLE: during collar initialization a timestamp will be exchanged between the collar and the laptop, which allows for later data synchronization. A keepalive message is also sent periodically (every minute), from collar to laptop, using BLE, to allow the human observer monitoring the cow to detect if the collar is still operative or another action is required (e.g., replacement of batteries).

2.2. Data Annotation with Computer Software

Direct visual observations were used to collect states and actions of cows in their natural environment at the cattle farm. A PC-based software (Figure 3a) was designed and developed to help the user register the observed state of multiple cows. The program includes the following functionalities:

Collar initialization

During collar initialization, a timestamp is sent to the collar using BLE. This timestamp will be used for the synchronization of the sensor data collected by the collar and the states and actions of the cows recorded by the human observer.

Collar management (Figure 3b)

According to the European Commission, individual identification and registration of bovine livestock is mandatory to ensure full traceability and, consequently, enhance food safety and better safeguard animal health. The application uses an alpha-numeric code as the cowl's ID. Similarly, each device is identified by a unique 64-bit collar ID, which corresponds to the serial number of the microcontroller housed in the collar.

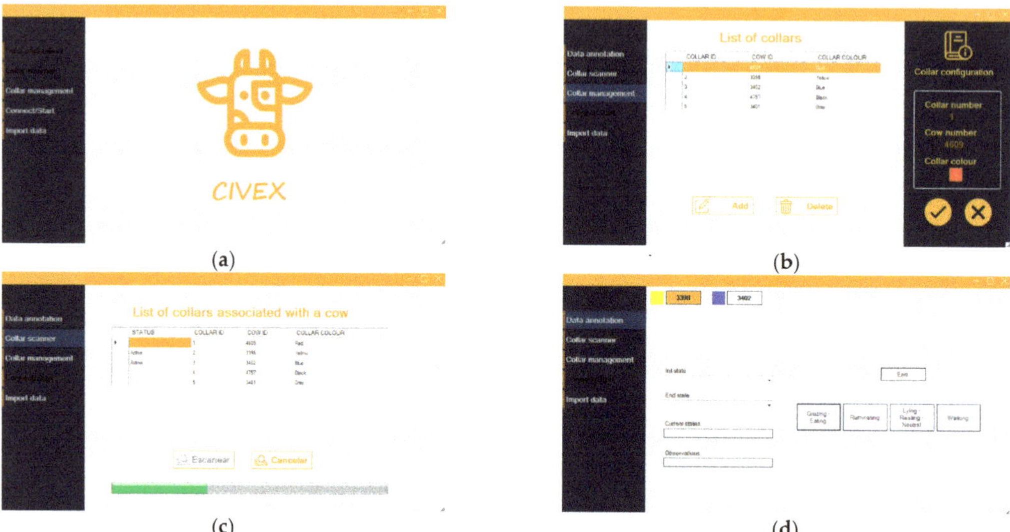

Figure 3. PC Software for collar management and data collection based on visual observation. (**a**) Startup screen; (**b**) collar management; (**c**) collar scanner; (**d**) data annotation.

During operation, it was noted that human observers sometimes had difficulty identifying cows when recording actions, especially over long distances or when animals were close to each other. To solve this, each manufactured collar was printed with a different colour (Figure 2c shows a red collar). Collar management allows registering and assigning a new collar to an animal, unregistering, or reassigning an existing one and updating the colour of the collar.

Collar scanner (Figure 3c)

This feature allows to check active collars within the BLE's range, by listening to a keepalive message containing the ID that the devices send every minute. The discovered collars are listed on an overview and their status is changed to active (Figure 3c). The application prevents the annotation of data from not active collars.

Data annotation (Figure 3d)

It allows the annotation, by a human observer, of the status and the action the cow is performing. Additional observations can also be recorded. All annotated data and observations are stored on the PC. Once the annotation activity is finished, the software automatically generates a file containing the recorded information. Two predetermined sets of states have been considered:

- General behaviour. The tags considered are: "Grazing/Eating", "Ruminating", "Neutral" and "Walking".
- Standing behaviour. The tags considered are: "Standing" and "Lying" position.

This reduction was necessary to identify transitions between lying and standing positions (lying bouts) which would not be registered in the general behaviour tag set, where the neutral state can happen both standing and lying (if the cow is doing nothing else).

Import data:

Once the operator recovers the collar and obtains the sensor data files stored in the microSD, the application allows the generation of a final file, using the timestamp in both files to combine and synchronize the data from the collar with the data from the file containing the visual observation data (stored in the PC). This synchronized data set is used to develop our own parturition prediction algorithm described in Section 2.3.2.

2.3. Calculation

2.3.1. Animals, Facility and Data Collection

For this study, ten collars were developed. However, cow monitoring was limited to a maximum of three animals simultaneously in order to improve the quality of the annotated data. The rest of the produced collars were saved as replacements, to minimize dead times and data loss in case a collar stopped working. The monitored beef cattle belong to an extensive livestock farm located in the municipality of Carrascal de Barregas, in the province of Salamanca, Spain.

During the study, collars were fitted to six cows (see Table 2). For eleven months (August 2020–July 2021) two experienced observers (one working in the morning shift and the other during the afternoon shift) annotated the actions of the animals for a total of 6 h every day. It is to be noted that although the observers were following the animals continuously, some situations introduced uncertainty in the data, for example, when cows stampede from one location to another. To mitigate the data deterioration, observers left the annotations blank when detecting these situations. To record the different states a laptop running the application previously presented was used while maintaining a clear line of sight with the animals. Every week, the data stored in the microSD card of the collar was downloaded, and the batteries were replaced. A detailed overview of the cow-wise distribution of the data is presented in Table 2. As datapoints were dumped in the SD card on an hourly basis, it is straightforward to know the number of hours we got data from. The total quantities shown in Table 2 accumulate the number of hours the collar was recording data (raw data) and the number of hours the observer labelled behaviours for every studied cow.

Table 2. Cow-wise distribution of the data.

Cow	Raw Data Collection Period(dd/mm/yyyy)	Calving Date and Hour	Hours of Raw Data	Hours of Labelled Data
01	24/08/2020–17/02/2021	01/12/2020 13 h:30′	1.634	212
02	24/08/2020–25/05/2021	24/02/2021 08 h:30′	3.417	510
03	01/03/2021–15/06/2021	05/05/2021 16 h:45′	1.720	279
04	05/10/2020–12/07/2021	25/05/2021 13 h:35′	2.957	470
05	08/02/2021–30/07/2021	11/07/2021 20 h:15′	1.130	159
06	24/08/2020–27/01/2021	-	1.887	147
Total	24/08/2020–30/07/2021	-	12.745	1.777

The schema of data acquired with the collars is shown in Table 3. In total, more than 855 million (855,319,572) raw datapoints have been recorded, of which approximately 114 million (114,167,178) are labelled.

Table 3. Schema of raw data collected by the collars.

DB18B20		GNSS			IMU						
TimestampTemp	Lon	Lat	Alt	Speed	ax	ay	az	gx	gy	gz	Temp

Previous research work related to monitoring of pregnant cows around calving, (Jensen, 2012) and (Titler et al., 2015), has been focused on the period immediately around the time of calving (one and four days, respectively). This approach however is not practical for application where monitoring is less frequent, such as extensive livestock farms where large herds are held. Therefore, our research was focused on the long-term monitoring of the pregnant animals, which extended up to two months after calving. Using this approach, we could analyse the individual behavioural change during different stages of pregnancy, which previous research has proved can differ greatly between individuals [21].

The labels used for data annotation are presented in Tables 4 and 5. These two sets of labels distinguish the two datasets introduced before, namely general behaviour and standing/lying behaviour. Considering both label sets, approximately 1.777 h of cow behaviour have been annotated by two experienced observers (working part-time in morning and afternoon shifts).

Table 4. General behaviour annotations.

ID	Action
A1	Grazing-Eating
A2	Ruminating
A3	Neutral
A4	Walking

Table 5. Standing/lying behaviour annotations.

ID	Action
B1	Standing
B2	Lying

The distribution of the annotated actions is presented in Figure 4 for general behaviour actions and in Figure 5 for standing/lying behaviour.

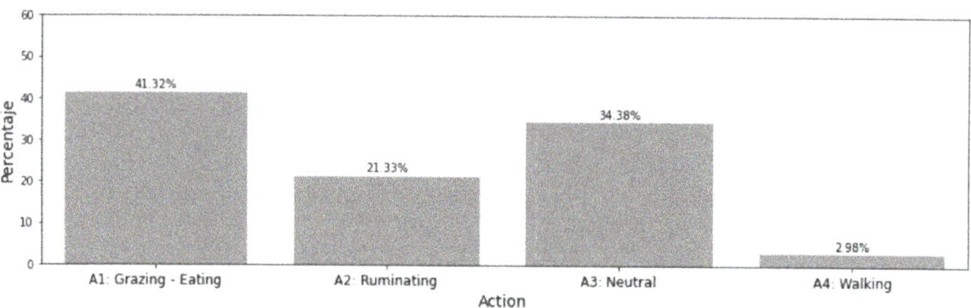

Figure 4. Distribution of general behaviour annotations.

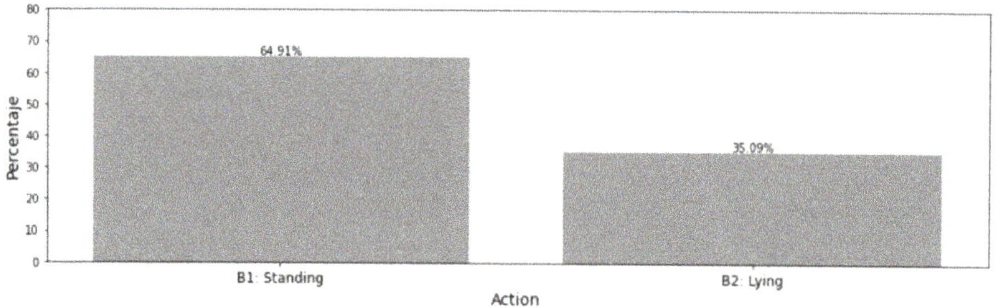

Figure 5. Distribution of standing/lying behaviour annotations.

General behaviour (Table 4) action prediction [33] has been explored during the study as additional input for calving prediction. Furthermore, data already gathered allows for further investigation in this field without the need of additional human-labelling. However, the general behaviour label set can lead to an imperfect classification of standing/lying behaviour (lying bouts) of the cow when transformed, as some actions can only be carried out in one posture (e.g., walking-standing), but other may occur both standing and lying.

For this reason, a second set of labels (Table 5) only accounting for lying and standing postures has been used to ensure a correct classification of these two postures when needed due to the importance of lying bouts detection for calving detection (as discussed in the introduction).

The annotated data from human observations show a small delay between the change of action of the cow and the annotation of this new action due to the response time of the observer and their need to operate the annotation software. For this reason, the datapoints recorded two minutes before an action change were ignored. This time period was chosen as a balance between loss of data and minimizing mislabelled data in the dataset. As cows did not change actions with high frequency in the recorded labels, two minutes resulted in enough certainty without losing a significative volume of data for each action.

As discussed previously, the sensor readings from the collars and the annotations from the observers can be joined using the timestamp of each data point to form the final dataset.

Three datasets were generated using the recorded data:

- Non-annotated data from the devices. These data have been proved to be useful for unsupervised and semi supervised learning tasks.
- General behaviour annotated data (Table 4). This dataset could be used to classify and predict the animal actions based on new reading from the devices.
- Standing/lying behaviour annotated data (Table 5). This dataset, although smaller compared to b, serves for statistical learning tasks, as well as semi-supervised learning techniques.

The experiments and algorithms have been developed using Python 3.7 [34] along several libraries, mainly: Pandas [35], Keras [36], Numpy [37], and SciPy [38]. Figures have been plotted using the Seaborn [39] library.

2.3.2. Parturition Prediction Algorithm

Calving prediction is the main objective of this pilot study. To usefully notify parturition, it is necessary to detect it with enough anticipation using a low-memory algorithm suited for the microcontroller.

The number of transitions of the animal between lying and standing has been empirically proved to be a good indicator of parturition in different studies [27,40,41]. This measure serves as an indicator of the proximity of calving due to the relative increase of its value in the 8 to 2 h before the parturition event. Furthermore, a notable decrease in the number of lying bouts in the hours after calving is also observed. However, these works study intensive dairy cattle, while our work studies extensive beef cattle, with the according significantly less restricted environment since calving barns are not used. A new low-memory algorithm based on classification of two cow postures (lying and standing), from the collar sensors readings has been developed. To classify these behaviours, accelerometer readings, commonly used to distinguish between lying and standing, as well as GPS altitude readings were initially considered. However, GPS readings were discarded due to insufficient sensor resolution.

In [21] is indicated that is more difficult to distinguish between a lying and standing position with a neck-based accelerometer since the two positions show similar accelerometer readings. Leg-based accelerometers show a distinct crossover of two axes and can easily be utilized to determine a standing or lying position. However, we have analyzed accelerometer signals read by our neck-mounted collars on the Y and Z axes (Figure 6). This figure shows that the analysis of the accelerometer signals provided by the cow's collar, allow us to clearly distinguish cow lying position (green colour), and cow standing position (blue colour). These results are similar to those presented in [29] to classify the cow posture as standing or lying.

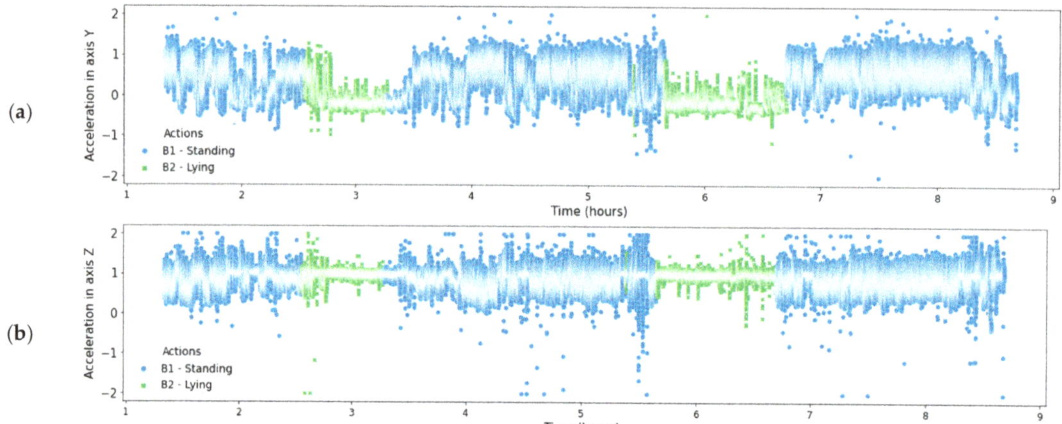

Figure 6. Y-axis and Z-axis accelerometer signal with human-labelled standing and lying annotations. (**a**) Y-axis; (**b**) Z-axis.

Our algorithm for cow posture classification (lying/standing) based on accelerometer readings has been implemented based on an heuristic threshold [27]. This approach is focused on simplicity and is based on the different distributions of acceleration recordings along the Y axis depending on the posture of the animal. Figure 7 shows the distribution of all available data labelled with the standing/lying action (11 months of data collection). Accelerometer Y-axis readings (left) indicates that standing behaviour is characterized by a larger mean (denoted by a grey triangle) than those recorded with the animal lying down. Accelerometer Z-axis readings (right), indicates that standing position have a larger interquartile range than the lying ones.

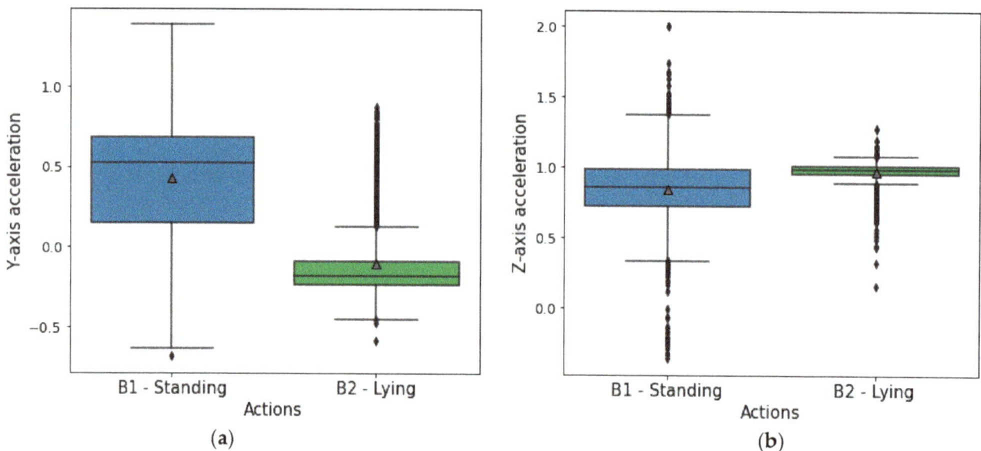

Figure 7. Y-axis and Z-axis accelerometer signal distribution while standing and lying of all data annotated with standing and lying postures. (**a**) Y-axis; (**b**) Z-axis.

To achieve "a low-memory algorithm", the accelerometer data is resampled from 17.6 Hz to 0.27 Hz as our experiments have proven that this data rate is enough for the algorithm. This way, memory requirements during the sliding window operations decreases and battery life of the device increases. Based on these appreciations and given the series of readings from the Y-axis of the accelerometer at discrete timestamps t_i, denoted by $f_{ay}(t_i)$, the following algorithm has been developed:

1. First, a rectangular sliding window of the last 15 min is used to count the number of readings in the Y-axis between a given superior and inferior threshold, thr_{sup} and thr_{inf}, respectively. The values of these thresholds are obtained from the distribution shown in Figure 7 to maximize the difference in the resulting count while standing and lying. This operation results in a function $f_{count}(t)$ (1) given by:

$$f_{count}(t) = \sum_{i=t-15\ minutes}^{t} \left[thr_{inf} < f_{ay}(i) < thr_{sup}\right], \quad (1)$$

2. Next, $f_{count}(t)$ is thresholded to obtain a binary signal $f_{standing}(t)$ (2) depending on the value of the function in each instant relative to a threshold $thr_{standing}$. This way, any value greater than $thr_{standing}$ will be denoted as 1 (standing), while values smaller than the threshold will be converted to 0 (lying).

$$f_{standing}(t) = \begin{cases} 1\ (standing) & if\ f_{count}(t) > thr_{standing} \\ 0\ (lying) & if\ f_{count}(t) \leq thr_{standing} \end{cases}, \quad (2)$$

3. This binary function $f_{standing}(t)$ is converted to a discrete transition signal $f_{lb}(t)$ (3) taking the absolute value of the difference between $f_{standing}(t)$ at any given time and its immediately previous value with each transition from standing to lying down and vice-versa represented by a 1.

$$f_{lb}(t) = \left|f_{standing}(t) - f_{standing}(t-1)\right|, \quad (3)$$

4. Finally, with this discrete transition signal $f_{lb}(t)$ computed, another rectangular sliding window is used to count the number of transitions that took place in the previous 5 h of each reading. This function $f_{parturition}(t)$, acts as a proxy to predict parturition based on the lying bouts occurrence increasement before calving.

$$f_{parturition}(t) = \sum_{i=t-5\ hours}^{t} f_{lb}(i), \quad (4)$$

3. Results and Discussion

As indicated in Table 2, cow number 03 calved on 5 May 2021 at 4:45 PM. Figure 8 shows the values of function $f_{parturition}(t)$ in the last five hours calculated with a rolling window for cow 03, for a week (from 4 May to 11 May), using the proposed algorithm for parturition prediction. A notable increase of this function is observed near the parturition instant, signalled with a vertical red dotted line.

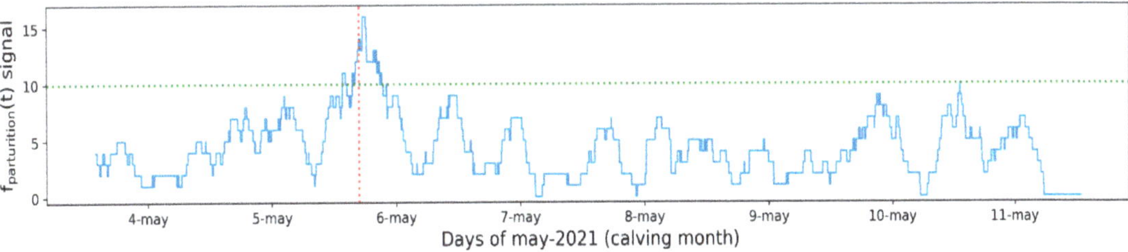

Figure 8. Number of lying bouts (function $f_{parturition}(t)$), in the last five hours calculated with a rolling window for cow 03 for a week. Dotted red line signalizes the corresponding calving instant. Dotted green line denotes a candidate value to detect the parturition event.

Figure 8 represents the function $f_{parturition}(t)$ during a week. This figure shows that it is during the hours before calving (2~3 h) when this function takes the highest values, reaching the maximum at the instant of calving. Furthermore, a horizontal dotted green line in

Figure 8 denotes a value that is only surpassed during the calving event ($f_{parturition}(t) = 10$). This value could be used as a trigger of the parturition detection for this cow 03. It is noted that this signalled value differs between individuals since there is variance between the activity and energy expenditure of each animal, and therefore this value has to be dynamically calculated (for each cow) on the collar based on previous readings.

As indicated in Table 2, during the data collected over a period of eleven months (August 2020–July 2021), five cows calved. Figure 9 shows the mean of function $f_{parturition}(t)$ in the last five hours calculated with a rolling window, and generated from the algorithm showed before (from the five calving events that took place). To calculate this mean, the values of this function have been aligned on the moment of parturition (0 h relative to calving). It is observed in Figure 9 that the mean value of the function $f_{parturition}(t)$ increases two hours before the calving of the cows. This increase allows us to determine that the cow is close to parturition. As previously mentioned, the parturition trigger value of 10 signalled in Figure 8 is only applicable to cow 03. This can be shown in Figure 9, where the $f_{parturition}(t)$ signals from the rest of the cows have brought the signal mean value slightly down.

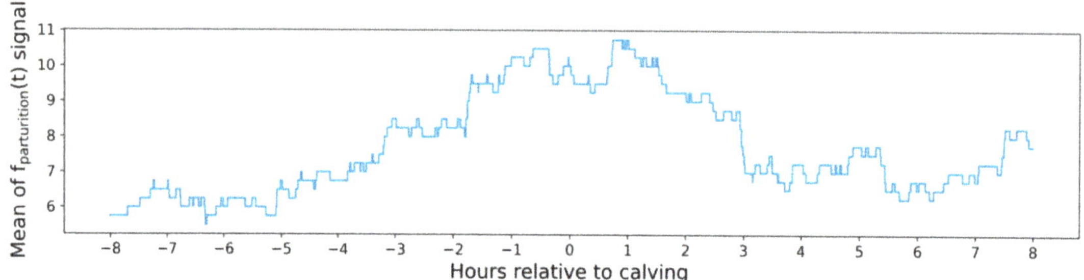

Figure 9. Mean of lying bouts (function $f_{parturition}(t)$) from the five cows which parturitions are indicated in Table 2.

The heuristic-based algorithm we have developed traces lying bouts with enough resolution to detect their increase in a time frame, which can be used to detect cow calving. This is accomplished without the need to calculate a calving indicator index that requires tracking the step count and the time spent lying down by the animal as in [27], mainly due to this variables stability when compared with the number of lying bouts near calving.

A very important aspect of the developed calving detection algorithm is its simplicity. This algorithm must be programmed in the microcontroller housed in the collar placed on the cow's neck, which represents a significant limitation in the microcontroller's computing and memory fields. Furthermore, a coarse estimate of the calving date in the receiving end of the calving prediction is also useful to use in conjunction with the algorithm, as it ensures the rejection of any naturally inviable false positive from the algorithm (i.e., calving detection one month before and after mount).

The dataset used in previous studies [27,40], usually includes data in a small temporal window around calving (1–5 days). The data collected in our study enable the back test of algorithms in a much more extensive temporal window, something essential to validate any algorithm that would run in a real-world environment, such as the proposed sensorized wearable device.

Although the proposed algorithms of this study are focused on calving prediction, the developed sensorized wearable device and the collected data enable the development of different algorithms that could be of great help to farmers both in extensive and intensive livestock. Additionally, the generalist design of the presented wearable device could be equally helpful to develop hardware solutions oriented to different animals, benefiting their caretakers with the localization data from the collar, the suite of sensors that it incorporates, and other algorithms that could be implemented within the device.

4. Conclusions and Future Work

In this paper, a low-cost neck-mounted sensorized wearable device to continuous long-term monitoring cow data has been presented. To incorporate the ability to detect cow parturitions with enough anticipation using only this device an algorithm has been developed. This algorithm can detect an increment in the number of times the cow stands up and lies down (lying bouts) using a signal calculated from the accelerometer's readings. To test the proposed algorithm, data collection using the cow neck-mounted devices and annotations from in-situ human observers has been carried out for eleven months (August 2020–July 2021). The data gathered by the neck-mounted collars correspond to six different cows monitoring in extensive livestock farming.

This preliminary study (of six cows, with five calving events), provides evidence that cow approaching parturition shows an increase in lying bouts behavioural pattern that can predict calving on average two hours before calving. To confirm these results, however, more pregnant cows need to be monitored and further research is required to refine this algorithm. The long-term character of the data acquired will allow for individualization of the thresholds for calving detection by calculating the baseline "restlessness level" of a particular animal. This could be used by the collar to generate an alert system to warn the farmer of the onset of calving.

The low-cost of this device ($\approx$100 € for small-scale production) would benefit most large livestock holdings by greatly reducing the number of hours the human experts must manually monitor individual animals. Economies of scale would allow the unitary price of the collar to be lowered even more, which would allow massive adoption even for the monitoring of large herds. However, smaller farms with less resources would also benefit from the reduced cost of the device, which lowers the entry barrier into a technology such as this, thus allowing for its adoption.

Future work will focus in two separate points. The first one, data related, requires the acquisition of calving data from a larger number of cows to validate the developed algorithm and to develop a new data driven algorithm that learns from the available data from different cows (both from the same breed and from different breeds, to deal with the variance between different species). This new algorithm could be dedicated to the classification of standing and lying behaviour or to the detection of the birth event as a time series task. Furthermore, the data already gathered from the animals long before calving and therefore related mostly to the normal activity of the animal would allow for the development of algorithms that analyse cow behaviour (grazing, ruminating, etc). A deviation of the normal behaviour of a particular animal due to sickness, heat, or an abortion could be signalised to the farmer.

The second point englobes the development of a new neck-mounted collar with IMU, GNSS and wireless low energy, and long-range communication using the LoRa protocol. This wireless communication would allow farmers to physically localize cows in extensive livestock farming, as well as receiving notifications of the detected parturition event, reducing the workload associates with parturition, and preventing dystocia in unattended calving. The energy autonomy of the device is an aspect of great importance in terms of its practical utility. In the tests carried out, it has been identified as an improvement parameter. For this, solar-powered technology is being incorporated in order to increase overall autonomy.

Author Contributions: Conceptualization, methodology and software, C.G.-S., G.S.-B. and A.C.; validation and formal analysis, J.-C.F., J.P.-T. and E.d.l.F.-L.; writing, C.G.-S., G.S.-B.; writing—review A.C.; funding acquisition J.-C.F., J.P.-T. and E.d.l.F.-L. All authors have read and agreed to the published version of the manuscript.

Funding: This research was funded by the Centre for the Development of Industrial Technology (CDTI), grant number CIVEX IDI-20180355 and CIVEX IDI-20180354 with ERDF funds, and by Spanish companies APLIFISA and COPASA.

Institutional Review Board Statement: Ethical review and approval for this study was waived since the animals were not disturbed during the study period. The image of the animal with the collar, shown in Figure 2, was taken without disturbing the behavior of the animal on the farm.

Informed Consent Statement: Not applicable.

Data Availability Statement: Not applicable.

Acknowledgments: We would like to thank Estibaliz Sánchez and Juan González-Regalado from COPASA company and Francisco Mediero, owner of the cattle farm, for their contributions to data collection, as well as for their helpful cooperation.

Conflicts of Interest: The authors declare no conflict of interest.

References

1. Gutierrez-Galan, D.; Dominguez-Morales, J.P.; Cerezuela-Escudero, E.; Rios-Navarro, A.; Tapiador-Morales, R.; Rivas-Perez, M.; Dominguez-Morales, M.; Jimenez-Fernandez, A.; Linares-Barranco, A. Embedded neural network for real-time animal behavior classification. *Neurocomputing* **2018**, *272*, 17–26. [CrossRef]
2. Liu, L.S.; Ni, J.Q.; Zhao, R.Q.; Shen, M.X.; He, C.L.; Lu, M.Z. Design and test of a low-power acceleration sensor with Bluetooth Low Energy on ear tags for sow behaviour monitoring. *Biosyst. Eng.* **2018**, *176*, 162–171. [CrossRef]
3. Borchers, M.R.; Chang, Y.M.; Tsai, I.C.; Wadsworth, B.A.; Bewley, J.M. A validation of technologies monitoring dairy cow feeding, ruminating, and lying behaviors. *J. Dairy Sci.* **2016**, *99*, 7458–7466. [CrossRef]
4. Lawrence, A.B. Applied animal behaviour science: Past, present and future prospects. *Appl. Anim. Behav. Sci.* **2008**, *115*, 1–24. [CrossRef]
5. Santman-Berends, I.M.G.A.; Schukken, Y.H.; van Schaik, G. Quantifying calf mortality on dairy farms: Challenges and solutions. *J. Dairy Sci.* **2019**, *102*, 6404–6417. [CrossRef]
6. Unold, O.; Nikodem, M.; Piasecki, M.; Szyc, K. IoT-Based Cow Health Monitoring System. *Int. Conf. Comput. Sci.* **2020**, *1*, 344–356. [CrossRef]
7. Sharma, B.; Koundal, D. Cattle health monitoring system using wireless sensor network: A survey from innovation perspective. *IET Wirel. Sens. Syst.* **2018**, *8*, 143–151. [CrossRef]
8. Quang, H.; Ngo, T.; Nguyen, T.P.; Nguyen, H. Research on a Low-Cost, Open-Source, and Remote Monitoring Data Collector to Predict Livestock′ s Habits Based on Location and Auditory Information: A Case Study from Vietnam. *Agriculture* **2020**, *10*, 180.
9. Berckmans, D. General introduction to precision livestock farming. *Anim. Front.* **2017**, *7*, 6–11. [CrossRef]
10. Arcidiacono, C.; Mancino, M.; Porto, S.M.C.; Bloch, V.; Pastell, M. IoT device-based data acquisition system with on-board computation of variables for cow behaviour recognition. *Comput. Electron. Agric.* **2021**, *191*, 106500. [CrossRef]
11. Liu, D.; He, D.; Norton, T. Automatic estimation of dairy cattle body condition score from depth image using ensemble model. *Biosyst. Eng.* **2020**, *194*, 16–27. [CrossRef]
12. Benaissa, S.; Tuyttens, F.A.M.; Plets, D.; de Pessemier, T.; Trogh, J.; Tanghe, E.; Martens, L.; Vandaele, L.; Van Nuffel, A.; Joseph, W.; et al. On the use of on-cow accelerometers for the classification of behaviours in dairy barns. *Res. Vet. Sci.* **2019**, *125*, 425–433. [CrossRef] [PubMed]
13. Vázquez Diosdado, J.A.; Barker, Z.E.; Hodges, H.R.; Amory, J.R.; Croft, D.P.; Bell, N.J.; Codling, E.A. Classification of behaviour in housed dairy cows using an accelerometer-based activity monitoring system. *Anim. Biotelemetry* **2015**, *3*, 1–14. [CrossRef]
14. Balasso, P.; Marchesini, G.; Ughelini, N.; Serva, L.; Andrighetto, I. Machine learning to detect posture and behavior in dairy cows: Information from an accelerometer on the animal's left flank. *Animals* **2021**, *11*, 2972. [CrossRef]
15. Arcidiacono, C.; Porto, S.M.C.; Mancino, M.; Cascone, G. Development of a threshold-based classifier for real-time recognition of cow feeding and standing behavioural activities from accelerometer data. *Comput. Electron. Agric.* **2017**, *134*, 124–134. [CrossRef]
16. Pratama, Y.P.; Kurnia Basuki, D.; Sukaridhoto, S.; Yusuf, A.A.; Yulianus, H.; Faruq, F.; Putra, F.B. Designing of a smart collar for dairy cow behavior monitoring with application monitoring in microservices and internet of things-based systems. In Proceedings of the 2019 International Electronics Symposium (IES), Surabaya, Indonesia, 27–28 September 2019; pp. 527–533. [CrossRef]
17. Park, M.C.; Jung, H.C.; Kim, T.K.; Ha, O.K. Design of cattle health monitoring system using wireless bio-sensor networks. In *Electronics, Communications and Networks IV, Proceedings of the 4th International Conference on Electronics, Communications and Networks (CECNET IV), Beijing, China, 12–15 December 2014*; CRC Press: Boca Raton, FL, USA, 2015; Volume 1, pp. 325–328. [CrossRef]
18. Wierig, M.; Mandtler, L.P.; Rottmann, P.; Stroh, V.; Müller, U.; Büscher, W.; Plümer, L. Recording heart rate variability of dairy cows to the cloud—Why smartphones provide smart solutions. *Sensors* **2018**, *18*, 2541. [CrossRef]
19. Saint-Dizier, M.; Chastant-Maillard, S. Methods and on-farm devices to predict calving time in cattle. *Vet. J.* **2015**, *205*, 349–356. [CrossRef] [PubMed]
20. Lammoglia, M.A.; Bellows, R.A.; Short, R.E.; Bellows, S.E.; Bighorn, E.G.; Stevenson, J.S.; Randel, R.D. Body Temperature and Endocrine Interactions before and after Calving in Beef Cows. *J. Anim. Sci.* **1997**, *75*, 2526–2534. [CrossRef] [PubMed]

21. Chang, A.Z.; Swain, D.L.; Trotter, M.G. Towards sensor-based calving detection in the rangelands: A systematic review of credible behavioral and physiological indicators. *Transl. Anim. Sci.* **2020**, *4*, 1–18. [CrossRef] [PubMed]
22. Bar, D.; Soloman, R. Rumination collars: What can they tell us. In Proceedings of the First North American Conference on Precision Dairy Management, Toronto, ON, Canada, 2–5 March 2010; p. 2.
23. Bikker, J.P.; van Laar, H.; Rump, P.; Doorenbos, J.; van Meurs, K.; Griffioen, G.M.; Dijkstra, J. Technical note: Evaluation of an ear-attached movement sensor to record cow feeding behavior and activity. *J. Dairy Sci.* **2014**, *97*, 2974–2979. [CrossRef] [PubMed]
24. Pahl, C.; Hartung, E.; Grothmann, A.; Mahlkow-Nerge, K.; Haeussermann, A. Rumination activity of dairy cows in the 24 hours before and after calving. *J. Dairy Sci.* **2014**, *97*, 6935–6941. [CrossRef] [PubMed]
25. Ouellet, V.; Vasseur, E.; Heuwieser, W.; Burfeind, O.; Maldague, X. Charbonneau Evaluation of calving indicators measured by automated monitoring devices to predict the onset of calving in Holstein dairy cows. *J. Dairy Sci.* **2016**, *99*, 1539–1548. [CrossRef]
26. Rutten, C.J.; Kamphuis, C.; Hogeveen, H.; Huijps, K.; Nielen, M.; Steeneveld, W. Sensor data on cow activity, rumination, and ear temperature improve prediction of the start of calving in dairy cows. *Comput. Electron. Agric.* **2017**, *132*, 108–118. [CrossRef]
27. Titler, M.; Maquivar, M.G.; Bas, S.; Rajala-Schultz, P.J.; Gordon, E.; McCullough, K.; Federico, P.; Schuenemann, G.M. Prediction of parturition in Holstein dairy cattle using electronic data loggers. *J. Dairy Sci.* **2015**, *98*, 5304–5312. [CrossRef] [PubMed]
28. Neave, H.W.; Lomb, J.; Weary, D.M.; LeBlanc, S.J.; Huzzey, J.M.; von Keyserlingk, M.A.G. Behavioral changes before metritis diagnosis in dairy cows. *J. Dairy Sci.* **2018**, *101*, 4388–4399. [CrossRef] [PubMed]
29. Busch, P.; Ewald, H.; Stupmann, F. Determination of standing-time of dairy cows using 3D-accelerometer data from collars. In Proceedings of the 2017 Eleventh International Conference on Sensing Technology (ICST), Sydney, Australia, 4–6 December 2017; pp. 1–4. [CrossRef]
30. Miller, G.A.; Mitchell, M.; Barker, Z.E.; Giebel, K.; Codling, E.A.; Amory, J.R.; Michie, C.; Davison, C.; Tachtatzis, C.; Andonovic, I.; et al. Using animal-mounted sensor technology and machine learning to predict time-to-calving in beef and dairy cows. *Animal* **2020**, *14*, 1304–1312. [CrossRef] [PubMed]
31. Krieger, S.; Sattlecker, G.; Kickinger, F.; Auer, W.; Drillich, M.; Iwersen, M. Prediction of calving in dairy cows using a tail-mounted tri-axial accelerometer: A pilot study. *Biosyst. Eng.* **2018**, *173*, 79–84. [CrossRef]
32. Voß, A.L.; Fischer-Tenhagen, C.; Bartel, A.; Heuwieser, W. Sensitivity and specificity of a tail-activity measuring device for calving prediction in dairy cattle. *J. Dairy Sci.* **2021**, *104*, 3353–3363. [CrossRef] [PubMed]
33. Martiskainen, P.; Järvinen, M.; Skön, J.P.; Tiirikainen, J.; Kolehmainen, M.; Mononen, J. Cow behaviour pattern recognition using a three-dimensional accelerometer and support vector machines. *Appl. Anim. Behav. Sci.* **2009**, *119*, 32–38. [CrossRef]
34. van Rossum, G. The python development team. In *The Python Language Reference—Release 3.7.11*; Network Theory Ltd.: London, UK, 2018; ISBN 0954161785.
35. McKinney, W. Data structures for statistical computing in python. In Proceedings of the 9th Python in Science Conference, Austin, TX, USA, 28 June–3 July 2010; Volume 1, pp. 56–61. [CrossRef]
36. Chollet, F. Others Keras. 2015. Available online: https://github.com/fchollet/keras (accessed on 29 November 2021).
37. Harris, C.R.; Millman, K.J.; van der Walt, S.J.; Gommers, R.; Virtanen, P.; Cournapeau, D.; Wieser, E.; Taylor, J.; Berg, S.; Smith, N.J.; et al. Array programming with NumPy. *Nature* **2020**, *585*, 357–362. [CrossRef]
38. Virtanen, P.; Gommers, R.; Oliphant, T.E.; Haberland, M.; Reddy, T.; Cournapeau, D.; Burovski, E.; Peterson, P.; Weckesser, W.; Bright, J.; et al. SciPy 1.0: Fundamental algorithms for scientific computing in Python. *Nat. Methods* **2020**, *17*, 261–272. [CrossRef] [PubMed]
39. Waskom, M. Seaborn: Statistical Data Visualization. *J. Open Source Softw.* **2021**, *6*, 3021. [CrossRef]
40. Jensen, M.B. Behaviour around the time of calving in dairy cows. *Appl. Anim. Behav. Sci.* **2012**, *139*, 195–202. [CrossRef]
41. Rice, C.A.; Eberhart, N.L.; Krawczel, P.D. Prepartum lying behavior of holstein dairy cows housed on pasture through parturition. *Animals* **2017**, *7*, 32. [CrossRef]

Article

Automated Processing and Phenotype Extraction of Ovine Medical Images Using a Combined Generative Adversarial Network and Computer Vision Pipeline

James Francis Robson *, Scott John Denholm and Mike Coffey

Scotland's Rural College (SRUC), Animal and Veterinary Sciences, Peter Wilson Building, Kings Buildings, West Mains Road, Edinburgh EH9 3JG, UK; scott.denholm@sruc.ac.uk (S.J.D.); mike.coffey@sruc.ac.uk (M.C.)
* Correspondence: james.robson@sruc.ac.uk

Abstract: The speed and accuracy of phenotype detection from medical images are some of the most important qualities needed for any informed and timely response such as early detection of cancer or detection of desirable phenotypes for animal breeding. To improve both these qualities, the world is leveraging artificial intelligence and machine learning against this challenge. Most recently, deep learning has successfully been applied to the medical field to improve detection accuracies and speed for conditions including cancer and COVID-19. In this study, we applied deep neural networks, in the form of a generative adversarial network (GAN), to perform image-to-image processing steps needed for ovine phenotype analysis from CT scans of sheep. Key phenotypes such as gigot geometry and tissue distribution were determined using a computer vision (CV) pipeline. The results of the image processing using a trained GAN are strikingly similar (a similarity index of 98%) when used on unseen test images. The combined GAN-CV pipeline was able to process and determine the phenotypes at a speed of 0.11 s per medical image compared to approximately 30 min for manual processing. We hope this pipeline represents the first step towards automated phenotype extraction for ovine genetic breeding programmes.

Keywords: generative adversarial network; machine learning; automated medical image processing; deep neural network; animal science; CT scans; computer vision

Citation: Robson, J.F.; Denholm, S.J.; Coffey, M. Automated Processing and Phenotype Extraction of Ovine Medical Images Using a Combined Generative Adversarial Network and Computer Vision Pipeline. *Sensors* **2021**, *21*, 7268. https://doi.org/10.3390/s21217268

Academic Editor: Sylvain Girard

Received: 16 August 2021
Accepted: 28 October 2021
Published: 31 October 2021

1. Introduction

Increase in global food demand has led to livestock breeders seeking to produce breeding lines more able to match economic demand which have genetic advantages to primary traits such as growth speed and reduced feed intake. With agricultural animals providing 18% of global calories and 39% of global protein intake, they are still an essential part of global nutritional requirements [1]. One of the methods in making livestock more advantageous is to selectively breed them for commercial traits such as growth rate [2], milk quality [3], weather [4] and disease resistance [5]. Recent improvements in genomic technologies such as detection of single nucleotide polymorphisms (SNPs) and whole genome sequencing [6] have allowed unparalleled insight into the driving factors which guide animal phenotypes [7] and successful genomic breeding selection has been able to identify traits which are not only desirably economically, such as improved livestock social behaviour and carcass composition [5], but also identify novel cosmetic or welfare indicators such as predicting horn phenotypes in Merino sheep [8]. As the number and biological complexity of known phenotypes are increasing, there is a call to innovate new ways to detect phenotypes faster and more accurately [9] in addition to detecting and preserving those of potential future relevance [10].

Non-invasive imaging techniques, such as computed tomography (CT), magnetic resonance imaging (MRI) and ultrasound, can provide detailed data from which phenotypes can then be extracted [11,12] and used in breeding programmes. One major benefit of using these non-invasive imaging techniques is that internal phenotypic data, such as muscle and fat distribution [13], organ size and limb morphology, can then be incorporated more swiftly into genetic breeding programmes for live breeding animals [14]. Out of the commonly used non-invasive imaging techniques, CT scanning provides the highest resolution (1–2 mm). One hurdle which can impact extraction of useful phenotypic information is the processing and analysis of these images which can be time consuming and therefore costly, especially if there is a need to re-analyse historic databases to measure newly emerging phenotypes.

Machine learning and artificial intelligence have been successfully implemented to increase phenotype detection speed and accuracy within many different medical areas including brain cancer detection, COVID status in lungs and classification of organ deformities [15–17]. Recently the same technology has been applied to areas of agricultural science such as detection of bovine tuberculosis status based upon milk spectral data [18,19]. Briefly put, these networks work by passing data such as images, or segments thereof, through a series of layers containing artificial neurones which determine the likelihood of visually similar animals such as pigs, sheep or horses on a scale of 0 (absent) to 1 (present). The type of network commonly used to perform this image to binary diagnostic is a convolutional neural network where, as the layer depth increases, many datapoints (such as pixels) are condensed into fewer datapoints (likelihood of, e.g., pig, sheep or horse presence). The subject field of artificial intelligence, machine learning and deep learning using neural networks is extremely broad, and this research article only aims to provide a broad overview in order to demonstrate its application in agriculture and to not discuss these in depth, although many excellent reviews exist for further reading [20–23].

To perform image-to-image translations a similar type of neural network is required, although rather than condensing pixel information into a few datapoints, the shape of the layers more closely resembles that of an hourglass laying on its side (Figure 1a). This hourglass shape allows the network to perform general purpose image-to-image translation and even increase resolution of blurry input images [24]. By pairing this image-to-image transforming network with a second convolutional neural network (Figure 1a), the discriminator, which compares and scores the images produced by the image transforming network and tries to discriminate between fake and ground truth results, a self-training system can be produced. These two-component image translational networks are termed generative adversarial networks (GANs) and have traditionally been used for a variety of image translational tasks including sketch-to-photo, smile-to-frown, and non-bearded-to-bearded [24–26]. More recently, GANs have been applied to medical images to remove noise from low-dose CT, generate tissue structure from blood vessel networks, correct motion artefacts, produce CT images from MR images and synthesise new image data [27–29].

By combining GANs with another machine learning technique, computer vision (CV), any images generated by the GAN can then be analysed to extract data of interest in a fully automated way (Figure 1b). Computer vison is a research field which aims to extract understanding or context from images and can use both traditional mathematical regression techniques as well as deep learning classification networks [30,31]. Application of CV can range from simple inspection of food quality and ripeness by counting the number of pixels within images of fruit and vegetables which fall within certain colour hue ranges [32,33] up to more complex tasks such as identifying road signage or pedestrians to guide automated driving systems [34].

We use both smart techniques (GANs and CV, Figure 1a,b, respectively) to aid processing and analysis of agricultural medical images of sheep. This research aims to first implement a GAN to perform ovine CT processing steps involving global information manipulation such as object and organ removal since within the image are multiple objects (scanning cradle and padding) and organs (testes) of varying size, morphology and orientation. Then, with the processed image containing only key features, attempt to extract phenotypes relevant for breeding programmes using CV techniques in an automated process.

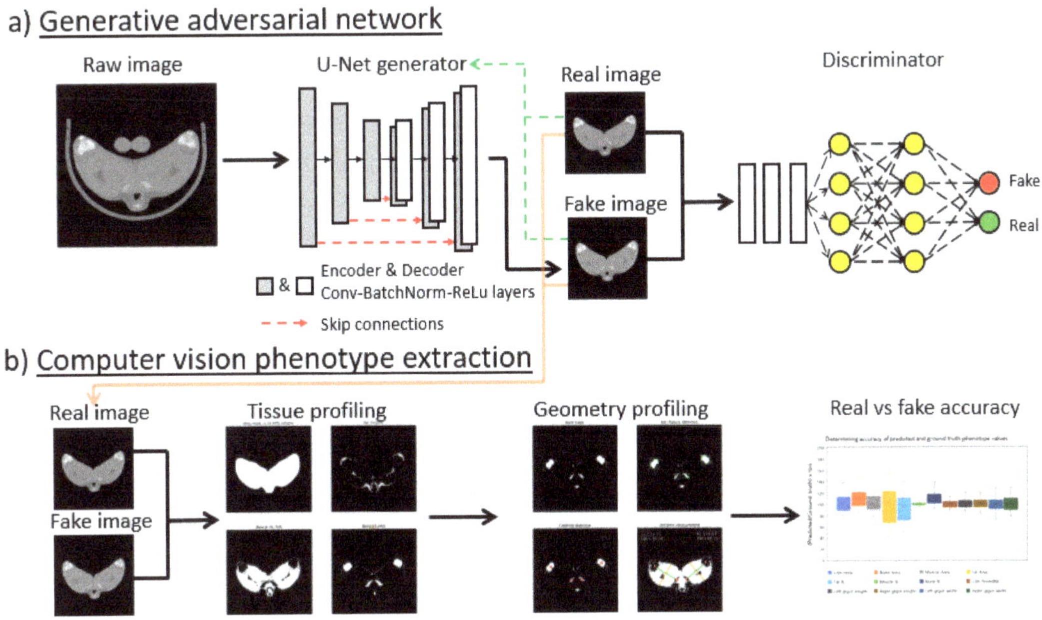

a) Generative adversarial network

b) Computer vision phenotype extraction

Figure 1. Combined GAN-CV pipeline for phenotype extraction. Neural networks can be trained to perform image-to-image translations such as in (**a**) where a raw ovine CT scan is passed through a generator network, a series of convolution, batch normalisation and ReLu activation function layers, to produce a "fake" image. Skip connections apply regions from the encoded to the encoded images and improve object border definition. By reducing differences between the real and the fake images (green dashes) the autoencoder also learns to better produce fake images independently. A second neural network, the discriminator, then determines if an image is considered real or fake. By pairing the two neural networks to work against each other, an adversarial component emerges, where the generator tries to produces images to fool the discriminator and the discriminator tries to determine if these images are real or not. Phenotype extraction is performed on both real and fake images (**b**) to determine tissue composition and shape before being compared to confirm accuracy.

2. Materials and Methods

2.1. Ovine Ischium Scan Collection

A single cross-sectional 2D image was taken through the top of the leg at the point of the ischium for each lamb using a Somatom Scope (Siemens, located at the SRUC-BioSS CT unit in Edinburgh, Scotland) with a slice thickness of 10 mm for a variety of breeds including Beltex, Blue Texel, Charollais, Hampshire Down, Meatlinc, Shropshire, Southdown, Suffolk and Texel as performed by Bunger et al. [12]. The images at this stage are referred to as "raw" images throughout the paper as they are unprocessed. All CT images produced from the scans are exported in the "Digital Imaging and Communications in Medicine" (DICOM) format, a unified filetype for medical imaging techniques. Such DICOM images contain additional data regarding the subject, such as age, sex and location, in addition to collection parameters such as equipment and scanning methodology used. Image dimensions used for this study were 512 × 512 pixels of an intensity value between 0 (black) and 2550 (white) where 0 typically represents low-density matter such as air and 2550 represents extremely dense matter such as metal.

2.2. Determination of Tissue Pixel Intensities

Pixel intensities corresponding to respective tissues of fat, muscle and bone were calculated based upon comparison with dissected tissue as explored by Bunger et al. [12]. Pixel intensity windows for fat, muscle and bone were 800–1000, 1000–1100 and 1100–1750, respectively. This group's previous research allowed us to incorporate set pixel intensity windows for each tissue type into the CV pipeline easily.

2.3. Manual Image Processing and Phenotype Analysis

All images had been previously labelled by manual phenotype extraction. Parts of the image superfluous for downstream phenotype determination including scanning cradle and testes were removed using STAR software routines [12] using the method described by Glasbey et al. [35]. Images produced from this processing are considered as "ground truth". From the ground truth images, tissue phenotype could then be extracted by calculating tissue distribution within the experimentally determined windows. Other phenotypes such as gigot length were measured manually by measuring the distance from the centre of the ischium bone cross-section to that of the femur bone cross-section in a "click and drag" fashion. Processing the images in this fashion took approximately 30 min.

2.4. GAN Model

GANs are two-component systems which have a generator component G to generate images and a discriminator component D to determine if the image is real or fake. The generator G takes an input image to translate into an output image y and can operate in either an unconditional fashion where random noise z is supplied or in a conditional fashion where an input image x or random noise z is supplied, G: $\{x \text{ or } z\} \rightarrow y$. The discriminator D determines if the image produced is "real" or "fake" and helps train the generator G to produce images which can pass as "real". GANs thus attempt to optimize the following function [36]:

$$\min_{G} \max_{D} V(G,D) = \mathrm{E}_{x,y}[\log D(x, y)] + \mathrm{E}_{x,z}[\log(1 - D(x, G(x,z))] \quad (1)$$

Further improvement of the generator G can be incorporated by including a function to minimise the absolute pixel differences between "real" and "fake" images [25].

$$\min_{G} L_{L1}(G) = \mathrm{E}_{x,y,z}[y - G(x,z)] \quad (2)$$

Which results in the following final model:

$$G* = \min_{G} \max_{D} V(G,D) + L_{L1}(G) \quad (3)$$

2.4.1. GAN Training

The GAN network trained in this study is an implementation of AUTOMAP [37] and Pix2Pix [25] which has been optimised for use with paired image datasets [38]. This particular GAN was chosen for this study as it was designed from the ground up to process paired sets of images, such as those commonly found in the medical field where an image can be altered to produce a "before" and "after" whilst maintaining the same subject ID and type, e.g., sheep–sheep, human–human, in a conditional synthesis process. This is in contrast to other popular GANs, such as CycleGAN and DCGAN, which perform unconditional synthesis by capturing key style concepts, from large batches of example images to translate images between two highly different abstract style concepts such as horse-to-zebra, photograph-to-Van Gogh or sketch-to-cat [39,40].

A dataset containing 126 raw and ground truth image pairs of mixed breed ovine CT scans taken from 2019–2020 were used for GAN training (Supplementary File S1). DICOM pairs were first split into training (n = 101) and validation (n = 25) datasets (80% and 20%, respectively). The raw and ground truth pairs of DICOM filename IDs were first given a suffix of "_0" or "_1", respectively, to act as identifiers. All file extensions were then modified to ensure compatibility with the DICOM processing libraries used in this study. The script used to train the GAN, along with the full list of GAN settings used for this study, is available within Supplementary File S2. Key settings for training the GAN were as follows: random translation = 0, epochs = 100, weight for L1 reconstruction loss = 0, weight for L2 reconstruction loss = 10.0, weight for softmax focal reconstruction loss = 1.0, weight for total variation = 10^{-3}. Following training, both the L1 (absolute pixel difference) and L2 (mean squared error) were approaching stable values (Supplementary Figure S1).

2.4.2. Image Processing Using Trained GAN on Unseen Data

Thirty-two raw CT scans (Supplementary File S3) taken from 2018–2019 and belonging to the breed Charollais were passed through the trained GAN model to produce "predicted" images that were given a suffix of "_2" to clearly differentiate between the raw and ground truth counterparts (Supplementary File S2).

2.5. CT Scan Similarity Comparison

2.5.1. CT Scan Histogram Comparison

Alternative image manipulation techniques, such as removing pixels above or below certain intensities, were not suitable for processing the CT scans in the DICOM format as the pixel intensities of image objects needing to be removed overlapped with that of the subject's tissue. Furthermore, pixels in certain areas could not be removed since subject orientation was not constant. Due to the large irregular pixel area changes needed to process the images, a deep neural network that can perform image-to-image translations was deemed to be of potential use. This can be visualised by comparing the pixel intensity histograms of both the raw and ground truth images below in Figure 2 (generated as part of the computer vision pipeline in Supplementary File S4).

2.5.2. Calculation of Image Similarity

Mean squared error (MSE) and structural similarity index (SSI) metrics were used to compare the raw and ground truth images with the resulting predicted images. Mean squared error is a full pixel-wise reference metric with values closer to zero being better; it is the sum of the accumulative mean squared difference across each pixel location between a pair of images. This technique, however, is extremely sensitive and seemingly large amounts of MSE can be accumulated by very minor shifts in the image, as perceived by the human eye, such as slight rotations or horizontal and vertical translations [41]. A newer, more holistic and subtle approach which avoids the extreme position sensitivity of MSE is calculating the SSI, which analyses local similarities in structure, luminance and contrast to more closely mimic how the human eye perceives similar images [42]. Both MSE and SSI were calculated for each pairwise comparison of image classes (raw, ground truth or predicted in this study) using the SciKit Image python image processing library as documented in Supplementary File S4 [43].

2.6. Phenotype Measurement Using Computer Vision

Automated phenotype extraction from ground truth and predicted (processed) images was performed using a pipeline which incorporated known pixel intensity value thresholding for each component of the carcass, based upon manual dissection, for each tissue type in combination. Geometric phenotypes were computed predominantly using the area, contour and perimeter functions within the CV library SciKit Image [43]. In addition, a set of bespoke functions were also written to detect probable tissue pixel intensity windows of fat, muscle and bone if no known set values were available, or if the images being analysed were from different sources. All steps of phenotype extraction using computer vision are documented in Supplementary File S4.

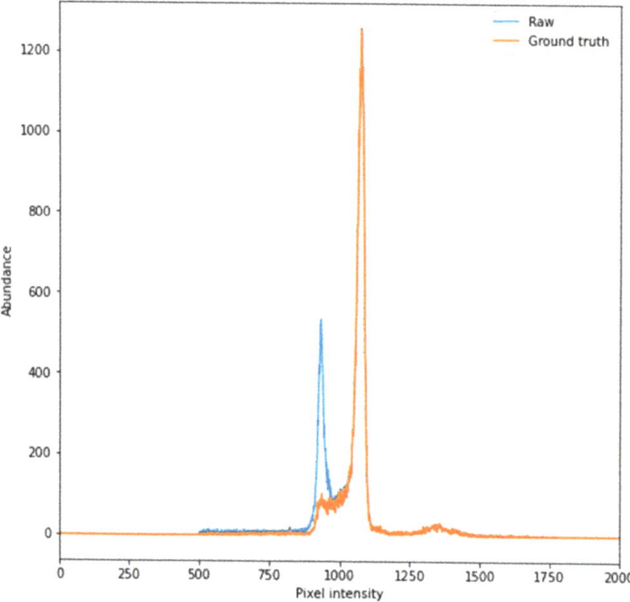

Figure 2. A representative pixel intensity histogram of raw and ground truth image shows large variance. By comparing the raw and ground truth pixel intensity histograms it can be visualised that they a) share certain areas of similarity (as seen at the peak between 1000 and 1250) but also b) contain regions which have different non-zero abundances (within the peak between 750 and 1000). As there are no regions where pixel intensity is either present or not present, images cannot be processed by simply flattening pixel intensities which lie between certain values. This type of non-linear transformation is a task in which neural networks perform well.

2.6.1. Tissue Distribution

The areas of all tissues within the ground truth and predicted images were calculated using the SciKit Image contour function for later use in determining percentage tissue composition. Tissue masks for each image were applied by first setting pixel intensity values (fat, muscle, bone) outside the respective tissue windows to zero and then setting values within the window to max (2550). Fat, muscle and bone % of each image were determined by comparing the number of pixels that fell within each of the respective tissue masks to that of the area of all tissue. By visualising each of the tissue masks independently, muscle and fat distribution could be observed in addition to locations of key physical features such as bones for further geometric phenotype analysis.

2.6.2. Skeleton Geometry

One key phenotype used for estimation of muscularity is the ratio of the length and width of the gigot muscle. These dimensions are typically measured by hand from the CT scan image but, by using CV models, we can extract this information automatically from the bone tissue mask image by implementing SciKit Image area and crofton perimeter functions [44]. Since small pieces of grit and sand may appear in the bone mask, due to high density as detected by X-rays, only bone mask objects over 200 pixels in both area and perimeter are referenced. Then, to avoid including spinal bone tissue, the four largest objects in the most +Y direction are assumed to be the features of interest and are placed into pairs according to their position along the X axis. The distance in pixels is then calculated between each pair of bones to determine gigot length. A line perpendicular to that between the bone pairs is then used to find the furthest non-zero positions within the muscle tissue mask and thus determine gigot width.

2.7. Computing Hardware and Software

The training of machine learning models can be an intensive computational task which typically requires powerful graphics processing units (GPUs). As such, all computation was performed on an NVIDIA DGX Station workgroup server [45]. The DGX workstation provided supercomputing performance with one out of a total of four TESLA V100 GPUs being used for computations underpinned by an Ubuntu operating system. All code was run within a Compute Unified Device Architecture (CUDA) 10.1 docker container which allows parallelisation of general-purpose processing to be applied to the powerful GPUs. Within this container, the open source learning framework Chainer was used to accelerate creation of the neural networks [46]. The GAN trained in this study is an implementation of AUTOMAP [37] and Pix2Pix [25] which has been optimised for use with paired image datasets [38]. Predicted images produced by the GAN were then processed using a bespoke python script run within a Jupyter notebook (Supplementary File S4). The notebook contains code within cells which can either (a) run individual steps and generate intermediary output figures (slower) or (b) calculate metrics and compare images without visualising any medical images (faster).

3. Results

The trained model was able to transform the raw images with a high degree of accuracy and perform the large image area manipulations, such as scanning cradle and testicle removal, needed to produce images similar to the manually processed ground truth images. The accuracy of these transformations was confirmed by visual inspection of predicted images and measurement of image similarity metrics including MSE and SSI. Phenotypic traits such as fat, muscle and bone tissue distribution and both gigot length and width were then automatically extracted from the predicted (transformed) images using CV techniques. All values calculated using this pipeline area are recorded in an output file (Supplementary File S5).

3.1. CT Scan Processing Using Trained GAN

Raw CT scans not previously seen by the GAN were processed using the trained model at a speed of 0.11 s per scan. Predicted and ground truth images and pixel intensity histograms were first compared visually to initially assess GAN suitability and ensure that they were visually similar (Figure 3). Quantitative metrics such as MSE and SSI were further determined to accurately assess the success of the GAN for processing the CT scans (Figure 4).

3.1.1. Images Produced from Trained Model

The trained model was able to perform the major structural alterations within the image dataset needed to transform the raw CT scans into something which, by eye, strongly resembled the ground truth images as shown below in Figure 3. Image IDs 1732, 9638 and 8353 were chosen to illustrate this transformation since, on visual inspection, they contained the largest area of features needing to be removed (large testes and a large scanning cradle).

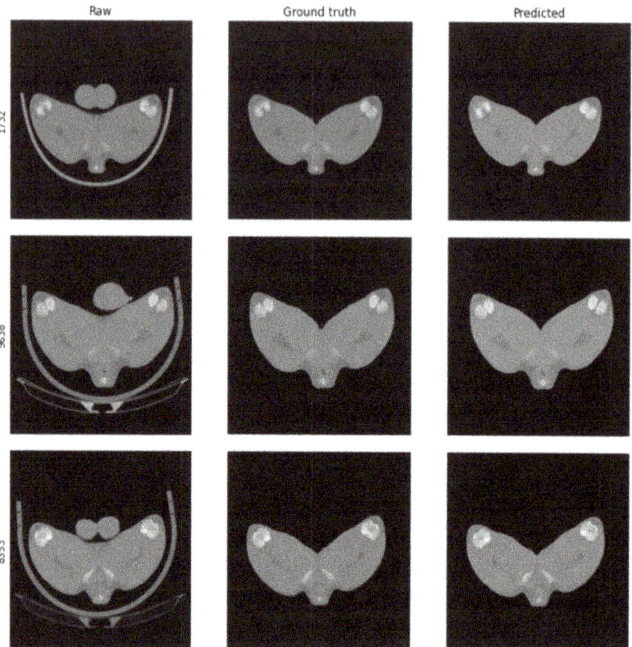

Figure 3. Representative comparison of raw, ground truth and predicted CT scan images. A trained generative adversarial network (GAN) was used to process raw CT images (**left** column) into something resembling manually processed ground truth images (**middle** column). Non-quantitative visual inspection of predicted results (**right** column) indicated that images produced by this GAN are similar to ground truth counterparts. The GAN showed good capabilities in automatically handling the large image transformations needed to remove image objects such as testes and scanning cradle.

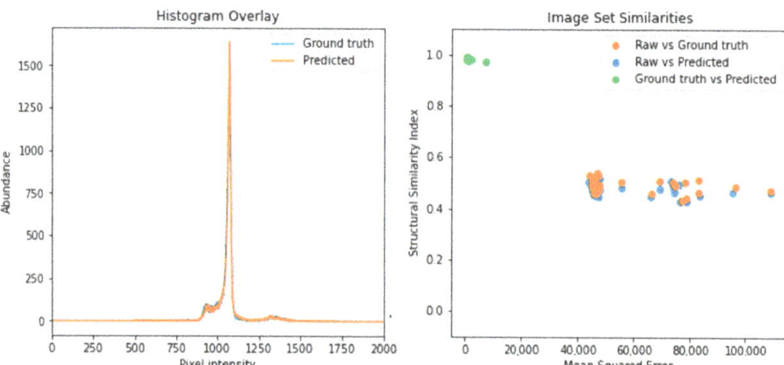

Figure 4. Quantifying a high degree of quantified similarity between ground truth and predicted images. Comparing a representative pixel intensity histogram of a ground truth and predicted image (**left**) showed a high degree of overlay and that peaks were present in similar areas at similar amplitudes, indicating a similar distribution of pixel intensities within each image. Structural components of image groups were compared (**right**) using mean squared error (MSE) and structural similarity indexes (SSIs) which revealed a) high average MSE (58,674 ± 17,766 and 58,008 ± 17,319, n = 32) with low average SSI (0.49 ± 0.025 and 0.48 ± 0.024, n = 32) between raw vs. ground truth and raw vs. predicted image groups, respectively, b) low average MSE (1028 ± 1201) and high average SSI (0.98 ± 0.0035) when comparing ground truth vs. predicted images. These high SSI and low MSE values confirm the suitability of a trained generative adversarial network to perform highly accurate ovine CT image processing.

3.1.2. Image Similarity Metrics Confirm a High Degree of Similarity

Just as raw and ground truth image histograms were compared previously, likewise the ground truth and predicted images were compared in a similar fashion which revealed two histograms, highly similar, showing a large proportion of overlap and a high degree of similarity from visual inspection. The likeness of the raw, ground truth and predicted image sets (n= 32) was compared pairwise using MSE and SSI. Both raw vs. ground truth and raw vs. predicted showed the lowest image similarity values with an average MSE of 58,674 ± 17,766 and 58,008 ± 17,319 and with average SSIs of 0.49 ± 0.025 and 0.48 ± 0.024, respectively, indicating a high degree of image dissimilarity. On the other hand, comparing images in the ground truth and predicted datasets showed a much lower average MSE (1028 ± 1201) and a far higher average SSI of 0.98 ± 0.0035, indicating a far greater similarity and indicating high accuracy of the trained model in mimicking the manual processing of CT scan images.

3.2. *Automated Phenotype Extraction*

The image processing library SciKit Image was successfully implemented to provide CV capabilities in the automated phenotype extraction pipeline. In this study, phenotypes of interest included fat, muscle and bone tissue abundance as well as leg geometry such as length and width (Supplementary File S4).

3.2.1. Leg Tissue Composition

Tissue abundance and distribution of fat, muscle and bone, within the single 2D image analysed, were calculated by counting pixels which fell within experimentally determined tissue pixel intensity windows compared to the total tissue area. Binary visualisation of these tissue value windows allowed rapid profiling of tissue distribution as seen below in Figure 5. Using this method, tissue abundances were calculated for each medical image in terms of both area and percentage composition (Figure 6). On average, the area of bone, muscle and fat across the dataset was 6488 ± 533, 44,274 ± 4051 and 5712 ± 1377 mm^2. Carcass tissue composition percentage-wise for bone, muscle and fat was 11.52 ± 0.78, 78.41 ± 1.90 and 10.07 ± 2.03%.

3.2.2. Gigot Length and Width Phenotype Extraction

By applying CV functions from the SciKit Image library such as area, perimeter and location restraints to objects in the bone tissue mask, the position and centre of key features were detected, and gigot length and width determined automatically as part of the CV script (Supplementary File S4). This process is visualised below in Figure 7. Left and right gigot lengths were 164.45 ± 8.72 mm and 166.38 ± 9.71 mm with widths being 137.55 ± 10.53 mm and 143.99 ± 12.42, respectively.

3.2.3. Phenotype Extraction Accuracy

Phenotypes from both predicted and ground truth datasets were extracted using the computer vision pipeline and compared to determine the suitability of predicted images for phenotype determination as seen below in Figure 8. Across all phenotypes, the average values were on average 101.44% that of the ground truth value with a standard deviation of 12.90% (n = 32). Muscle % was the most accurate predicted phenotype with estimated values between 93.67 and 106.65%. On the other hand, calculated fat area was the least accurate predicted phenotype with estimated values between 42.50 and 156.18% (following incomplete ovine testes removal from image ID 8346, fat-related phenotypes were not included in accuracy calculations as testes are calculated as fatty tissue. All other phenotypes for this image were recorded normally such as muscle area, bone area and gigot geometry).

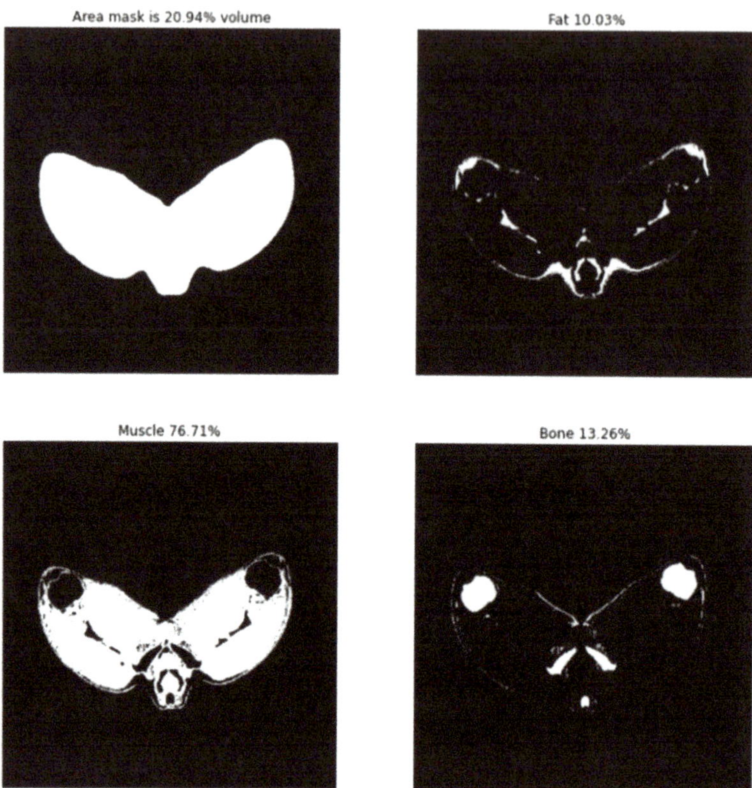

Figure 5. Representative tissue distribution of fat, muscle and bone within tissue area of predicted images. The results of the trained generative adversarial network were analysed by examining total area (**top left**) and by applying pixel intensity threshold windows to separately visualise fat (**top right**), muscle (**bottom left**) and bone (**bottom right**). The total number of pixels that fell within these pixel intensity windows determined the volume of the respective tissue types in the sample since 1 pixel = 1 mm^2.

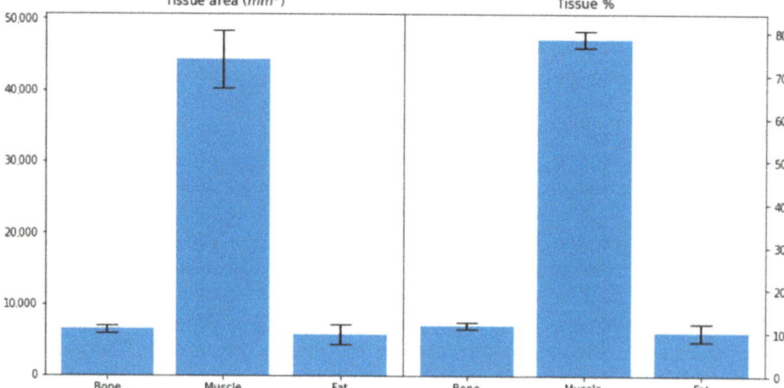

Figure 6. Area and percentage composition of tissue types within predicted ovine medical images. Using the threshold windows for each of the respective tissue types, the total area occupied was calculated for each tissue type (**left**) and what percentage this represented within each individual CT scan (**right**). On average, the area of bone, muscle and fat across the dataset was 6488 ± 533, 44,274 ± 4051, 5712 ± 1377. Carcass tissue composition percentage-wise for bone, muscle and fat was 11.52 ± 0.78, 78.41 ± 1.90 and 10.07 ± 2.03%.

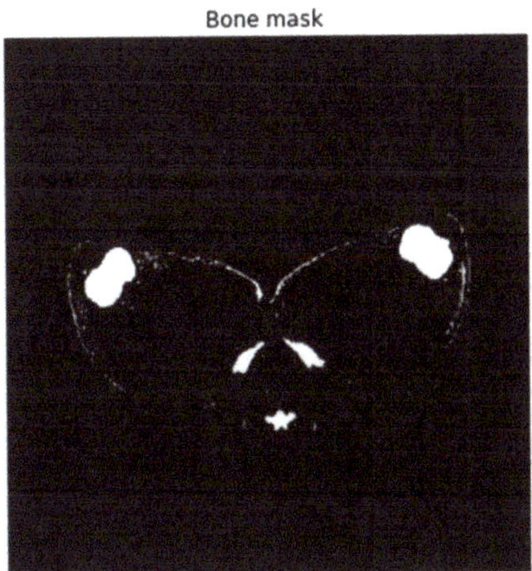

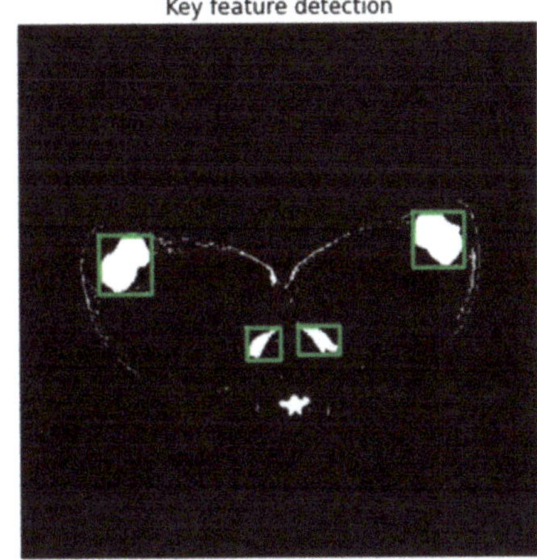

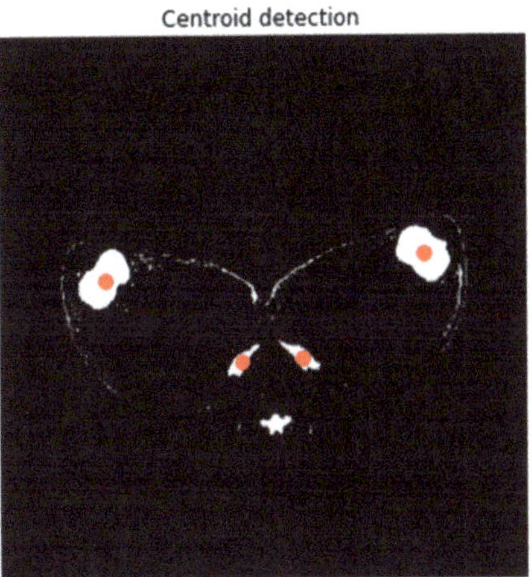

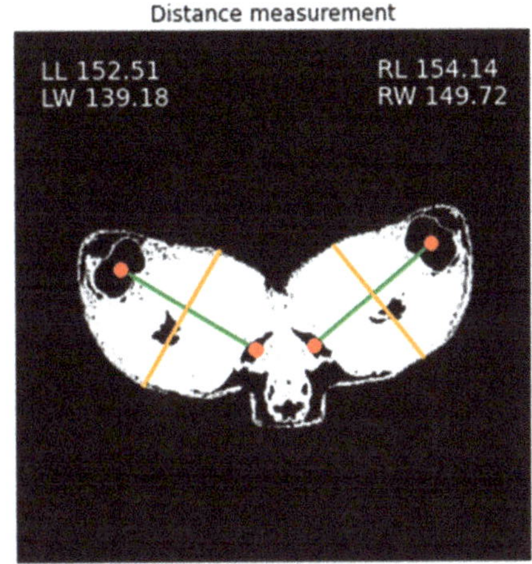

Figure 7. Automated identification of key features and determination of gigot length and width from predicted images. Using the bone tissue mask (**top left**) from the predicted image generated by the GAN, key features can be identified (**top right**). The centroid of each object within each feature area can be calculated (**bottom left**) to then measure distances and determine gigot length (LL, RL **bottom right**). By taking the perpendicular equation of the line which connects the two pairs of bones, the width of the gigot (LW, RW, **bottom right**) can be calculated by discovering the first and last non-zero values of these positions within the muscle tissue mask. Left and right gigot lengths on average were 164.45 $\pm$ 8.72 mm and 166.38 $\pm$ 9.71 mm with widths being 137.55 $\pm$ 10.53 mm and 143.99 $\pm$ 12.42, respectively.

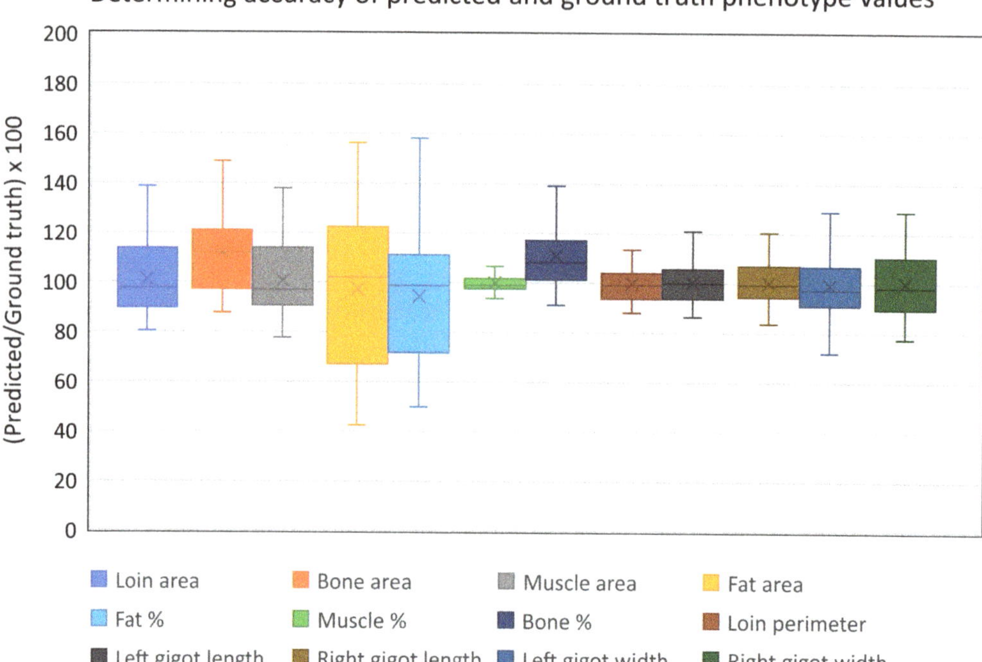

Figure 8. Comparing values of predicted and ground truth phenotypes. Prediction estimation accuracy was determined by comparing phenotype values generated from both predicted and ground truth datasets using the computer vision pipeline. Across all phenotypes, predicted values for each image were on average 101.44% that of the ground truth value with a standard deviation of 12.90% (n = 32).

4. Discussion

Continued reduction in DNA genotyping cost over time has resulted in mainstream integration of genomic selection into genetic improvement programmes for a number of domesticated animals. The increase in availability of genotypes leads to the need to identify the correlated phenotypes, as subtle or rare as they may be [10]. One technology which shows great promise in detecting these subtle phenotypes is the use of trained neural networks and CV. The processing and extraction of key data from medical images in the past have been typically performed manually by trained and experienced professionals. However, more recently, emergence of trained artificial intelligence networks has contributed to increased analysis throughput and accuracy of phenotype determination, such as the increased use and accuracy of neural networks for cancer and disease detection compared to the results of medical professionals [15,16,47,48]. By implementing similar techniques in the field of animal breeding, we hope to enhance the speed and accuracy of phenotype detection to streamline swift integration into genetic improvement programmes.

As part of this automated pipeline, a generative adversarial network was first trained to perform the necessary image-to-image translation required for automatically processing previously unseen CT scan images for subsequent phenotype extraction using CV at a speed of 0.11 s per image, this speed is far greater than the approximate 30 min required to manually process the image. The resultant images processed in this manner had an SSI of (0.98 ± 0.0035) when compared to the manually processed ground truths according to their structural similarity index and were visually indistinguishable. Automated phenotype extraction from predicted CT images was then performed by subdividing each image into the respective tissue masks to display the fat, muscle and bone volume and distribution.

By using key feature detection within the bone image mask, distances between ischium and femur bone cross-sections were calculated to determine the geometric phenotype of gigot length and width. Phenotype values determined using the computer vision pipeline were on average 101.4 % that of the ground truth value with a standard deviation of 12.90% (n = 32), indicating a high level of accuracy across the population.

One of the potential limitations of this study was the small training dataset (m = 126) as development of neural networks typically uses datasets numbering in the thousands. However, this limited dataset did not cause any major issues in accuracy as ground truth and predicted images showed an SSI of 0.98 $\pm$ 0.0035 and were indistinguishable by eye. One possible reason for such high accuracy with this limited dataset was that all subjects within the CT scans were constrained to similar postures. This hypothesis was later confirmed by re-introducing artificial random movement (such as rotation or vertical/horizontal shifts) into the images used for GAN training, resulting in a higher validation loss, poorer network performance and blurry resultant images (Supplementary Figures S1–S9).

Unfortunately, using this limited dataset resulted in one of the unseen images containing a small amount of testis tissue following processing with the GAN which was then incorrectly quantified as fat tissue. In the future, as more images are integrated into the model, we believe that the accuracy of the GAN shall improve which shall directly improve the precision of the CV phenotype determination pipeline.

5. Conclusions

In summary, we believe this research represents the first case of using an automated phenotype detection pipeline on agricultural animal medical images. This was achieved by using a combined GAN-CV pipeline to analyse agricultural medical images in a fully automated fashion. By feeding a paired image dataset into a GAN, we were able to perform the various image processing steps needed to produce a predicted image, containing only the relevant tissues, with accuracies of 98% which rivalled that of manual processing and at a fraction of the cost. Phenotypes were then extracted or calculated from these predicted images by applying CV techniques as part of an automated pipeline.

We hope to immediately expand this highly accurate GAN-CV pipeline to process and extract phenotypes from other key CT scan sections such as the 8th thoracic vertebra and 5th lumbar vertebra positions. Further on, we hope to develop a pipeline to process a complete set of layered CT images to produce an accurate 3D model from which a multitude of phenotypes can then be extracted, such as spine length and vertebra number, and detect phenotypes which are best explored in 3D space such as organ morphology [49,50]. By continuing this research we will further expand the automated extraction of phenotypes from agricultural medical imaging data and use the findings to guide genetic and genomic breeding programmes.

Supplementary Materials: The following are available online at https://www.mdpi.com/article/10.3390/s21217268/s1, Performance of generator and discriminator networks under varying degrees of random translation (Figures S1–S6), Raw, ground truth and predicted images produced from the networks with varying degrees of random translation (Figures S7–S9), Training images (File S1), Script for GAN training (File S2), Unseen images (File S3), Computer vision python notebook (File S4), Results table (File S5), Trained GAN model (File S6).

Author Contributions: Conceptualization, M.C.; methodology, J.F.R.; software, J.F.R.; validation, M.C., S.J.D.; formal analysis, J.F.R.; investigation, J.F.R.; resources, M.C., S.J.D.; data curation, M.C.; writing—original draft preparation, J.F.R.; writing—review and editing, M.C., S.J.D.; visualization, J.F.R.; supervision, M.C.; project administration, M.C.; funding acquisition, M.C. All authors have read and agreed to the published version of the manuscript.

Funding: This research was funded by Edinburgh Genetic Evaluations Services (EGENES) using data collected under a number of other funded projects (RESAS and AHDB Beef and Lamb) as part of CBS 1033316.

Institutional Review Board Statement: The study was conducted according to the guidelines of the Declaration of Helsinki, all 2019-2020 procedures involving animals were approved by the SRUC Animal Ethics Committee and were performed under UK Home Office license (PPL P90111799), following the regulations of the Animals (Scientific Procedures) Act 1986.

Data Availability Statement: All raw and ground truth training images used in this study are included as part of Supplementary File S1. All raw, ground truth and predicted images from the unseen medical images are included as part of Supplementary File S3. The trained model is provided as Supplementary File S6.

Acknowledgments: This work was supported by Nicola Lambe, Kirsty McLean and John Gordon of the SRUC CT scanning and ovine research department by collecting such excellent paired and annotated training data. NVIDIA are acknowledged for their technical support of the DGX Station and software advice.

Conflicts of Interest: The authors declare no conflict of interest.

References

1. FAO. Shaping the future of livestock; Report No. I8384EN. In Proceedings of the 10th Global Forum for Food and Agriculture, Berlin, Germany, 20 January 2018; pp. 18–20.
2. Rexroad, C.; Vallet, J.; Matukumalli, L.K.; Reecy, J.; Bickhart, D.; Blackburn, H.; Boggess, M.; Cheng, H.; Clutter, A.; Cockett, N.; et al. Genome to phenome: Improving animal health, production, and well-being—A new USDA blueprint for animal genome research 2018–2027. *Front. Genet.* **2019**, *10*, 1–29. [CrossRef]
3. Gonzalez-Recio, O.; Coffey, M.P.; Pryce, J.E. On the value of the phenotypes in the genomic era. *J. Dairy Sci.* **2014**, *97*, 7905–7915. [CrossRef]
4. Sánchez-Molano, E.; Kapsona, V.V.; Ilska, J.J.; Desire, S.; Conington, J.; Mucha, S.; Banos, G. Genetic analysis of novel phenotypes for farm animal resilience to weather variability. *BMC Genet.* **2019**, *20*, 84. [CrossRef]
5. Brito, L.F.; Oliveira, H.R.; McConn, B.R.; Schinckel, A.P.; Arrazola, A.; Marchant-Forde, J.N.; Johnson, J.S. Large-Scale Phenotyping of Livestock Welfare in Commercial Production Systems: A New Frontier in Animal Breeding. *Front. Genet.* **2020**, *11*, 793. [CrossRef] [PubMed]
6. Li, X.; Yang, J.; Shen, M.; Xie, X.L.; Liu, G.J.; Xu, Y.X.; Lv, F.H.; Yang, H.; Yang, Y.L.; Liu, C.B.; et al. Whole-genome resequencing of wild and domestic sheep identifies genes associated with morphological and agronomic traits. *Nat. Commun.* **2020**, *11*, 2815. [CrossRef] [PubMed]
7. Santos, B.F.S.; Van Der Werf, J.H.J.; Gibson, J.P.; Byrne, T.J.; Amer, P.R. Genetic and economic benefits of selection based on performance recording and genotyping in lower tiers of multi-tiered sheep breeding schemes. *Genet. Sel. Evol.* **2017**, *49*, 10. [CrossRef]
8. Duijvesteijn, N.; Bolormaa, S.; Daetwyler, H.D.; Van Der Werf, J.H.J. Genomic prediction of the polled and horned phenotypes in Merino sheep. *Genet. Sel. Evol.* **2018**, *50*, 28. [CrossRef]
9. Seidel, A.; Krattenmacher, N.; Thaller, G. Dealing with complexity of new phenotypes in modern dairy cattle breeding. *Anim. Front.* **2020**, *10*, 23–28. [CrossRef] [PubMed]
10. Leroy, G.; Besbes, B.; Boettcher, P.; Hoffmann, I.; Capitan, A.; Baumung, R. Rare phenotypes in domestic animals: Unique resources for multiple applications. *Anim. Genet.* **2016**, *47*, 141–153. [CrossRef]
11. Han, D.; Lehmann, K.; Krauss, G. SSO1450—A CAS1 protein from Sulfolobus solfataricus P2 with high affinity for RNA and DNA. *FEBS Lett.* **2009**, *583*, 1928–1932. [CrossRef] [PubMed]
12. Bunger, L.; Macfarlane, J.M.; Lambe, N.R.; Conington, J.; McLean, K.A.; Moore, K.; Glasbey, C.A.; Simm, G. Use of X-ray Computed Tomography (CT) in UK Sheep Production and Breeding. *CT Scanning-Tech. Appl.* **2011**, 329–348. [CrossRef]
13. Lee, S.; Lohumi, S.; Lim, H.S.; Gotoh, T.; Cho, B.K.; Jung, S. Determination of intramuscular fat content in beef using magnetic resonance imaging. *J. Fac. Agric. Kyushu Univ.* **2015**, *60*, 157–162. [CrossRef]
14. McLaren, A.; Kaseja, K.; McLean, K.A.; Boon, S.; Lambe, N.R. Genetic analyses of novel traits derived from CT scanning for implementation in terminal sire sheep breeding programmes. *Livest. Sci.* **2021**, *250*, 104555. [CrossRef]
15. Savage, N. How-ai-is-improving-cancer-diagnost. *Nature* **2020**, *579*, S14. [CrossRef]
16. Lin, L.; Qin, L.; Xu, Z.; Yin, Y.; Wang, X.; Kong, B.; Bai, J.; Lu, Y.; Fang, Z.; Song, Q.; et al. Using Artificial Intelligence to Detect COVID-19 and Community-acquired Pneumonia Based on Pulmonary CT: Evaluation of the Diagnostic Accuracy. *Radiology* **2020**, *296*, E65–E71. [CrossRef]
17. Lim, L.J.; Tison, G.H.; Delling, F.N. Artificial Intelligence in Cardiovascular Imaging. *Methodist Debakey Cardiovasc. J.* **2020**, *16*, 138–145. [CrossRef]
18. Denholm, S.J.; Brand, W.; Mitchell, A.P.; Wells, A.T.; Krzyzelewski, T.; Smith, S.L.; Wall, E.; Coffey, M.P. Predicting bovine tuberculosis status of dairy cows from mid-infrared spectral data of milk using deep learning. *J. Dairy Sci.* **2020**, *103*, 9355–9367. [CrossRef]

19. Brand, W.; Wells, A.T.; Smith, S.L.; Denholm, S.J.; Wall, E.; Coffey, M.P. Predicting pregnancy status from mid-infrared spectroscopy in dairy cow milk using deep learning. *J. Dairy Sci.* **2021**, *104*, 4980–4990. [CrossRef]
20. Soffer, S.; Ben-Cohen, A.; Shimon, O.; Amitai, M.M.; Greenspan, H.; Klang, E. Convolutional Neural Networks for Radiologic Images: A Radiologist's Guide. *Radiology* **2019**, *290*, 590–606. [CrossRef]
21. Hou, X.; Gong, Y.; Liu, B.; Sun, K.; Liu, J.; Xu, B.; Duan, J.; Qiu, G. Learning based image transformation using convolutional neural networks. *IEEE Access* **2018**, *6*, 49779–49792. [CrossRef]
22. Bod, M. A guide to recurrent neural networks and backpropagation. *Rnn Dan Bpnn* **2001**, *2*, 1–10.
23. Benos, L.; Tagarakis, A.C.; Dolias, G.; Berruto, R.; Kateris, D.; Bochtis, D. Machine learning in agriculture: A comprehensive updated review. *Sensors* **2021**, *21*, 3758. [CrossRef]
24. Parmar, N.; Vaswani, A.; Uszkoreit, J.; Kaiser, L.; Shazeer, N.; Ku, A.; Tran, D. Image transformer. In Proceedings of the 35th International Conference on Machine Learning ICML 2018, Stockholm, Sweden, 10–15 July 2018; pp. 6453–6462.
25. Isola, P.; Zhu, J.Y.; Zhou, T.; Efros, A.A. Image-to-image translation with conditional adversarial networks. In Proceedings of the 2017 IEEE Conference on Computer Vision and Pattern Recognition, CVPR 2017, Honolulu, HI, USA, 21–26 July 2017; pp. 5967–5976. [CrossRef]
26. Zhong, G.; Gao, W.; Liu, Y.; Yang, Y.; Wang, D.H.; Huang, K. Generative adversarial networks with decoder–encoder output noises. *Neural Networks* **2020**, *127*, 19–28. [CrossRef]
27. Singh, N.K.; Raza, K. Medical Image Generation using Generative Adversarial Networks. *arXiv* **2020**, arXiv:2005.10687.
28. Armanious, K.; Jiang, C.; Fischer, M.; Küstner, T.; Hepp, T.; Nikolaou, K.; Gatidis, S.; Yang, B. MedGAN: Medical image translation using GANs. *Comput. Med. Imaging Graph.* **2020**, *79*, 101684. [CrossRef] [PubMed]
29. Frid-Adar, M.; Diamant, I.; Klang, E.; Amitai, M.; Goldberger, J.; Greenspan, H. GAN-based synthetic medical image augmentation for increased CNN performance in liver lesion classification. *Neurocomputing* **2018**, *321*, 321–331. [CrossRef]
30. Voulodimos, A.; Doulamis, N.; Doulamis, A.; Protopapadakis, E. Deep Learning for Computer Vision: A Brief Review. *Comput. Intell. Neurosci.* **2018**, *2018*, 7068349. [CrossRef]
31. Meer, P.; Mintz, D.; Rosenfeld, A.; Kim, D.Y. Robust regression methods for computer vision: A review. *Int. J. Comput. Vis.* **1991**, *6*, 59–70. [CrossRef]
32. Wu, D.; Sun, D.W. Colour measurements by computer vision for food quality control—A review. *Trends Food Sci. Technol.* **2013**, *29*, 5–20. [CrossRef]
33. Brosnan, T.; Sun, D.W. Improving quality inspection of food products by computer vision—A review. *J. Food Eng.* **2004**, *61*, 3–16. [CrossRef]
34. Deva Koresh, J. Computer Vision Based Traffic Sign Sensing for Smart Transport. *J. Innov. Image Process.* **2019**, *1*, 11–19. [CrossRef]
35. Glasbey, C.A.; Young, M.J. Maximum a posteriori estimation of image boundaries by dynamic programming. *J. R. Stat. Soc. Ser. C Appl. Stat.* **2002**, *51*, 209–221. [CrossRef]
36. Goodfellow, I.; Pouget-Abadie, J.; Mirza, M.; Xu, B.; Warde-Farley, D.; Ozair, S.; Courville, A.; Bengio, Y. Generative adversarial networks. *Commun. ACM* **2020**, *63*, 139–144. [CrossRef]
37. Zhu, B.; Liu, J.Z.; Cauley, S.F.; Rosen, B.R.; Rosen, M.S. Image reconstruction by domain-transform manifold learning. *Nature* **2018**, *555*, 487–492. [CrossRef]
38. Kaji, S.; Kida, S. Overview of image-to-image translation by use of deep neural networks: Denoising, super-resolution, modality conversion, and reconstruction in medical imaging. *Radiol. Phys. Technol.* **2019**, *12*, 235–248. [CrossRef]
39. Radford, A.; Metz, L.; Chintala, S. Unsupervised representation learning with deep convolutional generative adversarial networks. In Proceedings of the 4th International Conference on Learning Representations, ICLR 2016, San Juan, Puerto Rico, 2–4 May 2016; pp. 1–16.
40. Zhu, J.Y.; Park, T.; Isola, P.; Efros, A.A. Unpaired Image-to-Image Translation Using Cycle-Consistent Adversarial Networks. *Proc. IEEE Int. Conf. Comput. Vis.* **2017**, *2017*, 2242–2251. [CrossRef]
41. Sara, U.; Akter, M.; Uddin, M.S. Image Quality Assessment through FSIM, SSIM, MSE and PSNR—A Comparative Study. *J. Comput. Commun.* **2019**, *7*, 8–18. [CrossRef]
42. AGandhi, S.; Kulkarni, C.V. MSE vs. SSIM. *Int. J. Sci. Eng. Res.* **2013**, *4*, 930–934.
43. Van Der Walt, S.; Schönberger, J.L.; Nunez-Iglesias, J.; Boulogne, F.; Warner, J.D.; Yager, N.; Gouillart, E.; Yu, T. Scikit-image: Image processing in python. *PeerJ* **2014**, *2014*, e453. [CrossRef]
44. Séverine, R. Analyse D'image Géométrique et Morphométrique par Diagrammes de Forme et Voisinages Adaptatifs Généraux. Ph.D. Thesis, ENSMSE, Saint-Etienne, France, 2011.
45. NVIDIA NVIDIA DGX Station: AI Workstation for Data Science Teams. Available online: https://www.nvidia.com/en-gb/data-center/dgx-station-a100/ (accessed on 21 May 2021).
46. Tokui, S.; Okuta, R.; Akiba, T.; Niitani, Y.; Ogawa, T.; Saito, S.; Suzuki, S.; Uenishi, K.; Vogel, B.; Vincent, H.Y. Chainer: A deep learning framework for accelerating the research cycle. In Proceedings of the 25th ACM SIGKDD International Conference on Knowledge Discovery & Data Mining, Anchorage, AK, USA, 4–8 August 2019; pp. 2002–2011. [CrossRef]
47. Lassau, N.; Ammari, S.; Chouzenoux, E.; Gortais, H.; Herent, P.; Devilder, M.; Soliman, S.; Meyrignac, O.; Talabard, M.P.; Lamarque, J.P.; et al. Integrating deep learning CT-scan model, biological and clinical variables to predict severity of COVID-19 patients. *Nat. Commun.* **2021**, *12*, 634. [CrossRef] [PubMed]

48. Saood, A.; Hatem, I. COVID-19 lung CT image segmentation using deep learning methods: U-Net versus SegNet. *BMC Med. Imaging* **2021**, *21*, 19. [CrossRef] [PubMed]
49. Nguyen-Phuoc, T.; Li, C.; Theis, L.; Richardt, C.; Yang, Y.L. HoloGAN: Unsupervised learning of 3D representations from natural images. In Proceedings of the 2019 International Conference on Computer Vision Workshop, ICCVW 2019, Seoul, Korea, 27–28 October 2019; pp. 2037–2040. [CrossRef]
50. Öngün, C.; Temizel, A. Paired 3D model generation with conditional generative adversarial networks. In Proceedings of the European Conference on Computer Vision, ECCV 2018 Workshops, Munich, Germany, 8–14 September 2018; Springer: Cham, Switherland, 2019; Volume 11129, pp. 473–487. [CrossRef]

 sensors

MDPI

Article

A Cascaded Model Based on EfficientDet and YOLACT++ for Instance Segmentation of Cow Collar ID Tag in an Image

Kaixuan Zhao, Ruihong Zhang and Jiangtao Ji *

College of Agricultural Equipment Engineering, Henan University of Science and Technology, Luoyang 471023, China; kx.zhao@haust.edu.cn (K.Z.); rh_zhang@stu.haust.edu.cn (R.Z.)
* Correspondence: jjtao@haust.edu.cn

Abstract: In recent years, many imaging systems have been developed to monitor the physiological and behavioral status of dairy cows. However, most of these systems do not have the ability to identify individual cows because the systems need to cooperate with radio frequency identification (RFID) to collect information about individual animals. The distance at which RFID can identify a target is limited, and matching the identified targets in a scenario of multitarget images is difficult. To solve the above problems, we constructed a cascaded method based on cascaded deep learning models, to detect and segment a cow collar ID tag in an image. First, EfficientDet-D4 was used to detect the ID tag area of the image, and then, YOLACT++ was used to segment the area of the tag to realize the accurate segmentation of the ID tag when the collar area accounts for a small proportion of the image. In total, 938 and 406 images of cows with collar ID tags, which were collected at Coldstream Research Dairy Farm, University of Kentucky, USA, in August 2016, were used to train and test the two models, respectively. The results showed that the average precision of the EfficientDet-D4 model reached 96.5% when the intersection over union (IoU) was set to 0.5, and the average precision of the YOLACT++ model reached 100% when the IoU was set to 0.75. The overall accuracy of the cascaded model was 96.5%, and the processing time of a single frame image was 1.92 s. The performance of the cascaded model proposed in this paper is better than that of the common instance segmentation models, and it is robust to changes in brightness, deformation, and interference around the tag.

Keywords: cow identification; EfficientDet; YOLACT++; cascaded model; instance segmentation

Citation: Zhao, K.; Zhang, R.; Ji, J. A Cascaded Model Based on EfficientDet and YOLACT++ for Instance Segmentation of Cow Collar ID Tag in an Image. *Sensors* **2021**, *21*, 6734. https://doi.org/10.3390/s21206734

Academic Editor: Cosimo Distante

Received: 25 July 2021
Accepted: 7 October 2021
Published: 11 October 2021

1. Introduction

Using machine vision and video surveillance equipment to automatically analyze the behavior of dairy cows, obtain their physiological and health information, and provide data support and decision-making bases for precision breeding and management has gradually become a research hotspot [1–4]. Machine vision systems have the advantages of no contact, low cost, and low stress [5]. However, the lack of stable and reliable image-based individual identification methods and technologies seriously restricts the promotion and use of these machine vision systems [6].

Radio frequency identification (RFID) is an individual identification method commonly used on large commercial dairy farms. RFID enables the recording of individual information, feeding information [7,8], and milk production [9,10] of dairy cows by reading tags attached to their bodies (usually ear tags) with wireless transmission technology. Compared with traditional methods, the reliability of information and the real-time information acquisition performance of RFID are improved. However, the workload of wearing and maintaining ear tags is substantial, which causes a stress response in the animals [11]. The RFID working performance is affected by label power, iron fences, and electromagnetic environments [12]. Additionally, when there are multiple cows in a recognition scene, individual information from multiple cows cannot be simultaneously obtained. Individual

RFID recognition devices add complexity to machine-vision-based intelligent information perception systems. Therefore, scholars have begun to study individual biometrical identification methods based on machine vision, and this recognition is realized mainly through the extraction and classification of the biological characteristics of cows [13]. The muzzle, iris, and face of the head of dairy cows contain various biological information that can be used as recognizable biological characteristics of individual cows. Different algorithms have been used to extract the descriptive features from images of muzzles [14,15], irises [16,17], and faces [18], and machine learning technology is used to classify the feature vectors to achieve individual recognition. However, the acquisition of head images requires a special shooting environment. The shooting angle, the quality of light, and the matching degree of cows all affect the details in the image and can reduce recognition accuracy.

Holstein cows are the most common cows on farms. The black and white patterns on the bodies can be used as biological features to identify individual animals. Zhao and He [19] proposed an individual cow recognition method based on a convolutional neural network. A 48×48 matrix from the trunk image of a dairy cow was extracted as a feature value, and a recognition model based on a convolutional neural network was constructed and trained. In the test, 90.55% of the images were correctly identified. Zhao et al. [6] proposed a cow recognition method based on template matching. The feature template library was generated by extracting the trunk image features of all cows, and individual cows were identified by matching their trunk image features with the features in the template library. Okura et al. [11] proposed a method for individual identification of dairy cows based on RGB-D video. The RGB images were used to obtain the texture features of dairy cows, and the depth image videos were used to obtain the features of the cows' gaits. These two complementary features were used to identify the cows. He et al. [20] proposed an individual identification method based on an improved YOLO v3 model. Images of cow backs were obtained with video frame decomposition technology, and a recognition model of optimizing anchors and improving network structure was constructed based on the Gaussian YOLO v3 algorithm. Yukun et al. [21] obtained images of the backs of dairy cows with moving cameras. While constructing an automatic system for scoring cow body condition, these authors established an individual recognition model of dairy cows based on a YOLO model and a convolution neural network. Side-view or top-view images of walking dairy cows are easy to obtain. Generally, a video or image acquisition device can be placed in the vicinity of dairy cows during feeding, drinking, or milking, and camera focus can be adapted to different recognition distances. However, this method is only applicable to Holstein cows with black and white patterns, which does not solve the problem of identifying cows with uniform colors on their bodies. In addition, the output dimension of the network corresponds to the number of cows in the herd. When the number of cows increases, the scale of the network increases exponentially. Once new cows join the herd, the entire network needs to be retrained.

An individual dairy cow identification model that functions through the detection and recognition of the ID number on their collar tag requires simple description features and a small network scale compared to biometric identification. The tagging and maintenance processes of this method impose a lower workload than RFID and do not affect the welfare of the dairy cows compared. Specifically, the ID tag worn on the cow's neck is first located, and then the ID numbers on the tag are recognized to identify the dairy cow. Zhang et al. [22] proposed a method of cow individual identification based on collar ID tags. This method first locates the tag by cascade detector combined with multi-angle detection, and then performs character segmentation and character recognition on the tag image. However, this location method cannot well-adapt to distortion deformation of ID tags. Zin et al. [23] proposed a tracking system for individual cows using visual ear tag analysis. First, the head and ear tag are detected. Then, the ear tag is recognized by finding the four-digit area, digit segmentation and digit recognition. However, the wearing process of ear tags requires punching a hole in the ear of cows, which can easily cause a stress response and affects the welfare of cows. The cascaded instance segmentation method we propose can adapt to the various deformations of ID tags, and its wearing process does not have much of an effect on the physiology and psychology of dairy cows. A comparison of different identification methods for cows is provided in Table A1.

The detection of ID tags is the first and key step to identifying individual cows, and its results directly affect the subsequent character recognition accuracy. If the ID tag is accurately segmented according to its contour, the digital recognition task becomes similar to license plate character recognition. According to existing research, it has achieved high recognition accuracy [24,25]. At present, few studies have been conducted on cow collar tag detection, but there are many studies on and applications for license plate detection. Xie et al. [26] proposed a multidirectional license plate detection framework based on CNN, which predicts the rectangular box and corresponding rotation angle to the license plate. This method can solve the problem of license plate rotation in a plane, but it cannot accommodate the tilt of the plate in three-dimensional space caused by the shooting angle. Xu et al. [27] proposed a method for locating irregular quadrilateral license plates. The proposed algorithm has two prediction branches: one is to predict the bounding box containing the license plate area and the other is to predict four groups of vertex offset values corresponding to the four bounding box corners, so as to get the vertex of irregular quadrilateral. This method is implemented based on YOLOv3 that extends the output dimensions. Kim et al. [28] proposed a two-step license plate location method that first detects the vehicle area and then locates the license plate in each vehicle area. This method can quickly filter out the complex background in an image. The license plate that is detected by these methods is a rigid object, but the four digital blocks on the ID tag we aimed to detect are attached to a flexible neck ring (to reduce the foreign body sensation experienced by the cow).

The flexible collar ID tag detection task is required to solve the following two key problems: First, in a side-looking image of a cow walking, the cow is in a continuous state of activity, so the tag is rotated and distorted in different planes, causing different degrees of deformation. Second, when the cow is far from the image acquisition equipment, the pixel area of the tag is relatively small, which causes difficulty in accurate detection. Therefore, we constructed a cascaded model for instance segmentation of the targets. First, the EfficientDet-D4 [29] model is used to detect the bounding box surrounding the ID tag, which effectively filters out most of background in the image and makes the segmentation task more targeted. Then, the image in the bounding box is sent to the YOLACT++ [30] model, and the ID tag is accurately segmented according to its contour to solve the tag deformation problem.

To accurately segment collar ID tags of cows, we conducted the following work: (1) To address the detection and recognition of a cow collar ID tag, we propose a high-precision cascaded model based on EfficientDet and YOLACT++ for instance segmentation, which overcomes the detection difficulty caused by the small area and large deformation of the tag. (2) We tested the performance of EfficientDet-D0–D5 model in the ID tag detection task, and analyzed the ability of different models to detect small targets. (3) The YOLACT++ model with different backbone networks (ResNet50/ResNet101) and different numbers of prototype masks was used to segment ID tags, and the effects of different parameters on the accuracy and speed of a single target segmentation task were analyzed. (4) The common two-stage segmentation models Mask RCNN [31], Mask Scoring RCNN [32], and one-stage instance segmentation model Solov2 [33] were used to segment ID tags, and the accuracy and speed of our proposed method and the above methods were compared. (5) The robustness of the cascaded instance segmentation model to changes in area, ID tag deformation, and brightness was analyzed.

The main contributions of this paper can be described as follows:

(I) A cascaded model is proposed based on EfficientDet and YOLACT++ to accurately detect and segment small targets in images.

(II) The structure and parameters of the model are optimized to improve the detection accuracy and efficiency.

2. Materials and Methods

2.1. Data Acquisition

Experimental images were collected at Coldstream Research Dairy Farm, University of Kentucky, USA, in August 2016, and the subjects were Holstein cows during lactation. When a cow returned to her shed after milking, she passed through a flat straight passage that had four electric fences (two before and after each) to limit the active area of the cow, and the width of the passage was 2 m. A Nikon D5200 camera was mounted on a tripod 3.5–5 m from the passage and 1.5 m from the ground. The camera used a 35 mm lens and was set to ISO 400, autoexposure, and autofocus. As cows passed through the camera's field of view, it continuously captured pictures at fixed time intervals. The resolution of the images was 6000 (horizontal) × 4000 (vertical) pixels. Images were captured from 16:00 to 18:00 on sunny days and was performed under natural light. The images were stored in the camera's local memory card. One of the original images is shown in Figure 1a. The cows were all wearing collars, as shown in Figure 1b. Each collar contained four square blue plastic blocks with white numbers; the four-digit numbers were the only identity labels of the cows.

Figure 1. (**a**) An original image. (**b**) A schematic diagram of a neck collar: 1, buckle; 2, neck band; 3, character block; 4, digital label; 5, weight.

To verify the background robustness of the cascaded model, we collected images of dairy cows wearing collar ID tags in the feeding bank at Sheng Sheng Farm, in Luoyang, Henan province, China. The images were captured from 9:30 to 11:30 under natural light on 16 September 2021. A cellphone (Xiaomi 10, Xiaomi Inc., Beijing, China) was used for hand-held shooting. The camera was set to autofocus and autoexposure mode. Images were captured from different angles when dairy cows were fixed on fences. A total of 200 images were captured, where 20 cows were involved and four different collar IDs were used. The resolution of the images was 5792 (horizontal) × 4344 (vertical) pixels.

The images were screened to exclude those with no cow or those that were overexposed, leaving 1344 images for the experiment. Due to the different moving speeds of the cows through the field of view, the number of samples of the individuals differed. A total of 670 images of 36 cows were randomly selected as the training set, which included 788 tags. A total of 268 images of 16 cows were randomly selected as the validation set, which included 321 tags. The 406 images of the remaining 26 cows were used as the test set, which included 492 tags. The ratio of images in the training set, validation set, and test set was approximately 5:2:3, and there was no cross-duplication between individuals in different data sets. The training set was used to fit the ID tag detection and segmentation model. The validation set was used to preliminarily evaluate the model to adjust its hyperparameters. The test set was used to evaluate the generalization ability of the final model.

2.2. Data Labelling

Labelme software (https://github.com/CSAILVision/LabelMeAnnotationTool (accessed on 22 October 2020) was used to annotate the data and build data sets in COCO format. Because the activities of the cows led to different degrees of deformation of their tags, the polygon mode was selected to label the target in an image. For the cascaded instance segmentation model in this paper, two steps of image annotation were required. Step (1): The tags in the original image were labelled to train and test the ID tag detection model. Step (2): The detection model trained in step (1) was used to detect ID tags in the training set and crop the detected bounding boxes. The cropped images were labelled and taken as the training set for the segmentation model. We performed the same for the validation set and test set of the detection model to obtain the validation set and test set of the segmentation model, respectively. Because the resolution of the original image was too high (4000 × 6000 pixels), the memory requirement of the model training was very high, so the images of the training set and the validation set were compressed to 1200 × 800 pixels when training the ID tag detection model.

2.3. Cascaded Model for Instance Segmentation

To solve the problem of the small area and the deformation of the cow collar ID tag in the images, a cascaded detection method was developed in this study. First, the detection model was used to detect the ID tag, and the image in the bounding box surrounding the ID tag was cropped as the input to the segmentation model. Then, the ID tag was accurately segmented according to its contour using the instance segmentation model. For the detection model in the first step, since the area of the target contained a small portion of the whole image, the feature extraction network was required to obtain both high-level semantic information and low-level spatial information. We wanted the model to allow the input image resolution to be as large as possible to retain more feature information. EfficientDet is a scalable model architecture for object detection based on EfficientNet. EfficientDet-D0–D7 were obtained by the composite scaling of each part of the detection network. This composite scaling method enabled us to balance accuracy and speed and to choose a better model. The BiFPN structure in the EfficientDet model enabled the network to obtain rich semantic and spatial information about the target through the upsampling, downsampling, and weighted fusion of different feature layers. Therefore, we chose EfficientDet as the ID tag detection model and tested the performance of different EfficientDet models to identify and select the optimal model.

For the ID tag segmentation task, the segmentation result was the final result, which directly affected the accuracy of the subsequent character recognition. Therefore, we hoped that the mask along the edge of the tag could completely contain all the ID numbers and did not contain redundant background. Because the image to be segmented was relatively small and the target area generally occupied more than 1/2 of the whole image, the difficulty of segmentation was low, and the fully convolutional network could efficiently segment the ID tags. Therefore, we chose the real-time instance segmentation model YOLACT++ based on a fully convolutional network to complete the tag segmentation task.

2.3.1. EfficientDet Detection Model

EfficientDet uses EfficientNet as its backbone to extract feature maps. EfficientNet obtains EfficientNet-B0–B7 by scaling the baseline model while adjusting the depth, width, and resolution of the input image. As the baseline, EfficientNet-B0 is composed of 1 stem and 7 blocks, as shown in Figure 2a. The stem structure functions to adjust the number of channels through convolution. The block includes several mobile inverted bottleneck convolution (MBConv) block modules. The design concept of the MBConv block modules involves inverted residuals and ResNet. First, a 1 × 1 convolution is performed to upgrade the dimension of the feature maps and a 3 × 3 or 5 × 5 depthwise separable convolution is performed, then a simple attention mechanism is added after this structure. Finally, 1 × 1 convolution is used to reduce the dimensionality of the feature maps, which are connected to the input side to form a residual structure. The channel attention mechanism effectively reduces the redundant channel feature information in the image, accelerates the network training speed, and reduces the memory required for training. Based on EfficientNet-B0, EfficientNet-B1–B7 are obtained by changing the width coefficient, depth coefficient, and input image size.

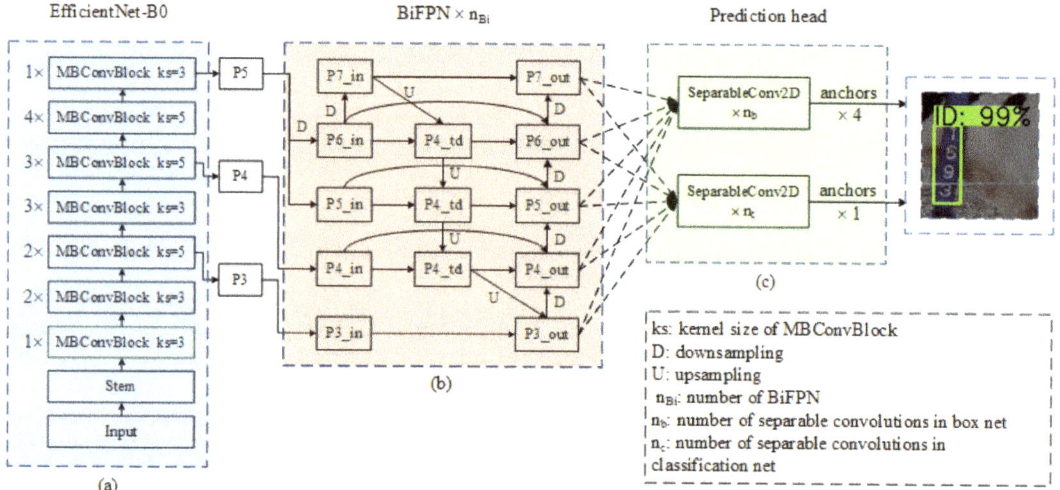

Figure 2. The structure of EfficientDet. (**a**) EfficientNet-B0, (**b**) BiFPN, (**c**) prediction head.

Simply, BiFPN is an enhanced version of FPN. The feature extraction process of BiFPN is shown in Figure 2b. BiFPN mainly includes two parts: the first is feature upsampling and feature weighted fusion; the second is feature downsampling and feature-weighted fusion. After downsampling and adjusting the number of channels, the feature maps extracted by EfficientNet are used as the input to BiFPN. First, upsampling and stacking of input features are performed, and then downsampling and stacking are performed. In the next BiFPN, the feature layers of the previous stage are used as the input, and up and downsampling and feature fusion are carried out again. This feature extraction method of upper and lower circulation sampling and weighted fusion retains the spatial information of the ID tag and obtains semantic information. From EfficientDet-D0 to EfficientDet-D7, BiFPN has an increasing number of cycles, which means that the depth of the network increases and the extracted feature information is richer. However, with the deepening of the network, the speed of training and reasoning is reduced.

The prediction head consists of two parts: the classification network and the prediction box regression network. The former assesses the category of the target, and the latter regresses the location of the target, as shown in Figure 2c. Before prediction, anchors are generated on the feature layers extracted by BiFPN. Through repeated separable convolutions, the classification branch and the prediction box regression branch generate 1 category parameter and 4 position adjustment parameters for each anchor, and finally obtain the location of the prediction box and the category of the target in the prediction box. From EfficientDet-D0 to EfficientDet-D7, the classification branch and the prediction box regression branch have different depths. When the EfficientDet head uses more separable convolutions, it may be less sensitive to small targets while acquiring deep semantic information.

Compared with other EfficientDet network structures, the number of parameters of EfficientDet-D6 and EfficientDet-D7 is significantly larger. Considering the image resolution and detection efficiency, we did not consider the use of EfficientDet-D6 or EfficientDet-D7 in the ID tag detection task.

2.3.2. YOLACT++ Segmentation Model

In the YOLACT++ instance segmentation model [30], a series of prototype masks and mask coefficients are generated by a fully convolutional network and fully connected layers, respectively, and the final mask is obtained by a linear combination of the two. As a one-stage model, YOLACT++ also adds a fast mask rescoring network to improve the segmentation accuracy of the mask so the model has excellent detection speed and high segmentation accuracy.

The YOLACT++ model uses ResNet as its backbone and FPN to construct feature maps P3, P4, P5, P6, and P7 with different sizes and advanced semantic information, as shown in Figure 3a. To adapt to the different scales and deformations of the target, a deformable convolutional network (DCN) is introduced into ResNet. The prototype generation branch (Protonet) takes the P3 layer of FPN (feature pyramid net) as its input (Figure 3c) because the P3 layer, as the deep backbone feature, has high resolution and can produce high-quality masks. Protonet is a fully convolutional network (FCN) composed of 3×3 and 1×1 convolution layers. Protonet predicts k prototype masks for the image, and all the final predicted masks are the linear combination of these k prototype masks. The prediction head takes the five feature maps (Pi) output by FPN as its input and uses the fully connected layer to generate three branches. One branch is used to predict the confidence of the target belonging to c categories, the second branch is used to predict the four position regression parameters of the bounding box, and the third branch is used to predict k mask coefficients (k corresponds to the number of prototype masks), as shown in Figure 3b. Then, non-maximum suppression (NMS) is carried out according to the predicted bounding box and the corresponding category confidence. The linear combinations of prototype masks and corresponding mask coefficients are the results of instance segmentation. These operations can be efficiently implemented using a single matrix multiplication and sigmoid:

$$\mathrm{M} = \sigma\left(PC^T\right) \tag{1}$$

where P is an $h \times w \times k$ matrix of prototype masks and C is an $n \times k$ matrix of mask coefficients for n instances that survive NMS and score thresholds. Finally, the masks are cropped with the predicted bounding box.

2.4. *Training Platform and Parameter Settings*

2.4.1. Training Platform

The software environment of our experimental platform was an Ubuntu 18.04 LTS 64 bit system. The programming language was Python 3.7. CUDA10.1 and cuDNN 7.6.5 were used as the parallel computing architecture of the deep neural network and GPU acceleration library. We selected Pytorch 1.4 as the deep learning framework. The GPU was a NVIDIA GeForce GTX 1080Ti, and the memory was 11 GB. The CPU had a 3.50 GHz Intel(R) Core(TM) i7-7800X processor, and its working memory was 32 GB.

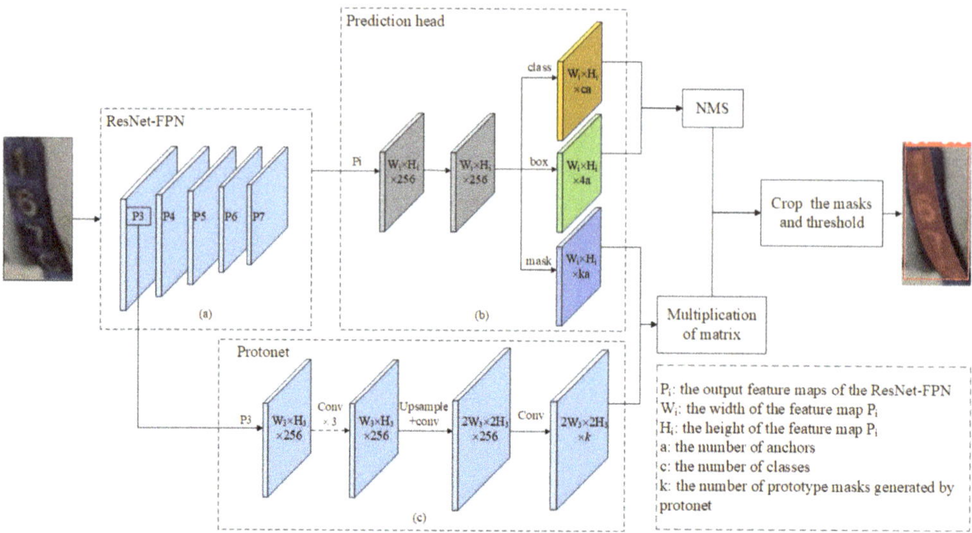

Figure 3. The structure of YOLACT++. (**a**) ResNet-FPN, (**b**) prediction head, (**c**) protonet.

2.4.2. Training Parameters of EfficientDet and YOLACT++

First, the training set and validation set constructed in step (1) of Section 2.2 were used to train the ID tag detection model EfficientDet. AdamW was selected as the optimizer for model training, and the batch size was set to 2. The initial learning rate was set to 1×10^{-3}. If the loss of the validation set was less than 0.1 in three epochs, the learning rate would have become 0.1 times that of the original. The weight decay coefficient and momentum coefficient were set to 1×10^{-4} and 0.9, respectively. The maximum number of iterations of all models was 1×10^{4}. According to the statistics of the tag size in the scaled image, the anchor sizes were determined to be 4, 8, 16, 32, and 64. The K-means clustering algorithm was used to calculate the anchor ratios suitable for our dataset, which were (0.7, 1.4), (1.0, 1.0), and (1.4, 0.7). The same training environment and training parameters were used to train Efficient-D0–D5 based on the pretraining model. After training, the performance of different EfficientDet models was evaluated with test sets.

The training set and validation set constructed in step (2) of Section 2.2 were used to train the ID tag segmentation model YOLACT++. SGD was selected as the optimizer for model training, and the batch size was set to 4. The initial learning rate was set to 1×10^{-4}. The weight decay coefficient and the momentum coefficient were set to 1×10^{-4} and 0.9, respectively. In the training process, the maximum number of iterations of the model was 1×10^{4}. ResNet50 and ResNet101 were selected as the backbone of YOLACT++ for training to compare the effects of different backbones on the accuracy and speed of the ID tag segmentation model. To study the influence of the generated prototype masks number k on the segmentation effect and speed of a single target, the YOLACT++ models were trained with k = 4, 16, and 32.

2.5. Precision Evaluation Index of Model

In this study, COCO detection evaluation indexes were used to evaluate the precision of the model. The intersection over union (IoU) is a value used to measure the degree of overlap between a prediction box and a groundtruth box, and its formula is:

$$\text{IoU} = \frac{S_p \cap S_g}{S_p \cup S_g} \quad (2)$$

where S_p represents the area of the predicted bounding box and S_g represents the area of the groundtruth bounding box. The IoU threshold is used to determine whether the content in the prediction box is a positive sample.

For the target detection model, the commonly used evaluation indices are precision (P) and recall (R), and their calculation formulas are:

$$P = \frac{TP}{TP + FP} \tag{3}$$

$$R = \frac{TP}{TP + FN} \tag{4}$$

where TP represents the number of correctly predicted targets; FP represents the number of falsely predicted targets, that is, the background is mistaken for a positive sample; FN represents the number of missed targets, that is, a positive sample is mistaken as the background. Confidence is an important indicator in target detection algorithms. For each prediction box, a confidence value was generated, indicating the credibility of the prediction box. Different combinations of P and R were obtained by setting different confidence thresholds. Taking P and R as vertical and horizontal coordinates, respectively, the PR curve could be drawn. When the IoU threshold was set to 0.5, the area under the PR curve was $\mathrm{AP}^{\mathrm{IoU}\,=\,0.50}$ (AP50). When the IoU threshold was set to 0.75, the area under the PR curve was $\mathrm{AP}^{\mathrm{IoU}\,=\,0.75}$ (AP75). AP was averaged over multiple intersection over union (IoU) values. Specifically, we used 10 IoU thresholds of 0.50:0.05:0.95. The average of multiple IoU thresholds more comprehensively reflects the performance of the model.

From the statistics of the test results, for the first step of the ID tag detection task, only when the IoU of the prediction and groundtruth bounding box was greater than 0.5 could the prediction box contain all the numbers on a tag. Therefore, the $\mathrm{AP}^{\mathrm{IoU}\,=\,0.50}$ (AP50) and AP of the detected bounding boxes were selected as the evaluation indices of the accuracy of the tag detection model. For the second step of the ID tag segmentation task, only when the IoU of the prediction and groundtruth mask was greater than 0.75 could the prediction mask contain all the numbers on a tag without background. Therefore, the $\mathrm{AP}^{\mathrm{IoU}\,=\,0.75}$ (AP75) and AP of the segmented masks were selected as the evaluation indices of the accuracy of the tag segmentation model. For the proposed cascaded instance segmentation method, we multiplied the AP50 of the detection model and the AP75 of the segmentation model to obtain the final accuracy of the ID tag detection model.

3. Results

3.1. Training and Testing of EfficientDet

During EfficientDet training, in the first 1000 iterations, the loss decreased rapidly. In 1000–6000 iterations, the loss had no obvious convergence trace but continuously oscillated. After 6000 iterations, due to the reduction in the learning rate, the loss started to converge again and finally reached a stable state. Therefore, for EfficientDet, reducing the learning rate at the late training stage effectively inhibited the loss oscillation of the model and accelerated the convergence rate. From EfficientDet-D0 to EfficientDet-D5, the training time gradually increased from the initial 3 h to 43 h, indicating that the complexity of the network structure significantly affected the training time of the model.

To test the performance of different EfficientDet models in the ID tag detection task, the original images (6000 × 4000 pixels) in the test set, which were constructed in step (1) in Section 2.2, were input to the trained Efficient-D0–D5 models for detection. According to the detection results, we aimed to find the best EfficientDet model that achieved a balance between accuracy and speed. The $\mathrm{AP}^{\mathrm{IoU}\,=\,0.50}$ (AP50) and AP of the detected bounding boxes and inference time per image were used as he evaluation indices. The test results are shown in Figure 4.

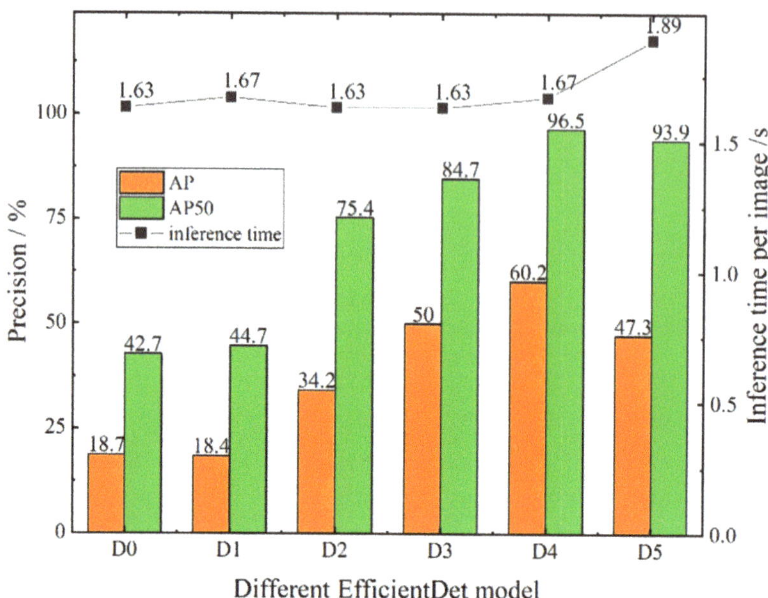

Figure 4. The precision and efficiency of EfficientDet-D0–D5. D0–D5 represent EfficientDet-D0–EfficientDet-D5, respectively.

As shown in Figure 4, from EfficientDet-D0 to EfficientDet-D4, the accuracy increased, indicating that increased network depth and multiple BiFPN cycles significantly improved the extraction and expression of image features, and the reasoning time for a single image did not significantly increase. The main factor affecting the reasoning speed of different EfficientDets was the complexity of the model. Although EfficientDet-D0–D4 had different complexities, their parameters were within 5–20 million. For our ID tag detection task, these differences had less influence on the reasoning speed than the high resolution of the image. Thus, the reasoning time of the EfficientDet-D0–D4 models for a single image had no obvious change.

Although EfficientDet-D5 has a wider and deeper network than EfficientDet-D4, its accuracy in the tag detection task was lower than that of EfficientDet-D4. This shows that for our small target detection task, the spatial information of small targets gradually reduced when the network reached a certain depth, which led to a decrease in detection accuracy. However, the number of parameters of the EfficientDet-D5 model was 30 million, which is approximately 1.5 times that of EfficientDet-D4, so its inference time was longer than that of the previous model. Therefore, we finally adopted the EfficientDet-D4 model with its high accuracy and efficiency as the ID tags detection model.

Figure 5 shows the detection results of EfficientDet-D0–D5 for some of the images in the test set. Due to the high resolution, only the image content related to the prediction box is cropped.

Figure 5 shows that for EfficientDet-D0 and EfficientDet-D1, problems of missing targets and inaccurate location often occurred, indicating that the shallow network structure could not effectively extract the features of small targets in the image. For EfficientDet-D2 and EfficientDet-D3, there were few missed targets but many false detections. This indicates that improvement in the network depth, width, and input image resolution increased the ability to extract features from small targets, but semantic information sufficient to accurately classify anchors was not extracted. For EfficientDet-D4, the model could not only accurately classify and locate small targets but also had higher confidence in correct classification than the previous model, which accurately and efficiently completed the

ID tags detection task. The confidence of the detection boxes of EfficientDet-D5 was high, but there were false samples near the target. This shows that high-level semantic information could correctly classify anchors when the network depth increased, but the low-level spatial information of small targets decreased, resulting in false detection boxes near targets. Thus, for small target detection tasks, reasonable network depth and width are the keys to simultaneously obtaining accurate semantic information and complete position information.

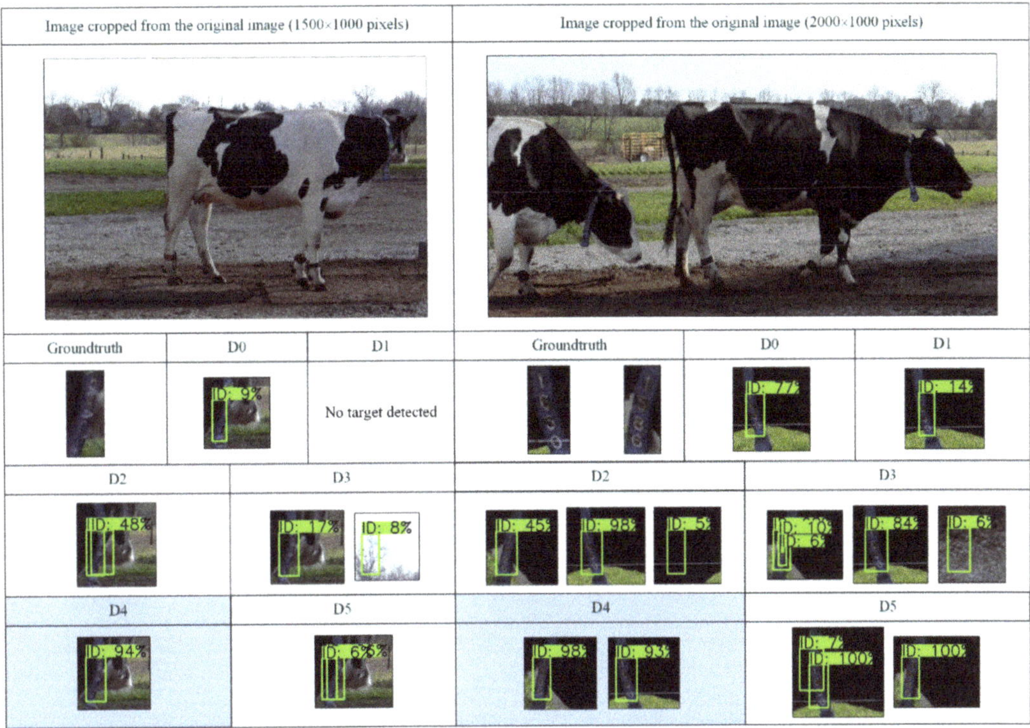

Figure 5. Some of detection results of EfficientDet-D0–D5. Groundtruth represents the true bounding box in the image to be detected; D0–D5 represent EfficientDet-D0–EfficientDet-D5, respectively; the green boxes in the detection image represent the prediction results of the model; and ID represents the class of detected targets. For our ID tag detection task, ID is the only class. The number behind ID represents the confidence of the corresponding detection box, and the unit is % (not shown in some black background images).

3.2. Training and Testing of YOLACT++

During training, the model converged rapidly in the first 500 iterations. From 500 to 6000 iterations, although it stabilized overall, there were still some large loss values. The losses stabilized after the 6000th iteration. The model with the ResNet101 backbone had a slightly longer training time than the model with the ResNet50 backbone. The greater the *k* value, the longer the training time. Compared with the *k* value, the backbone had more influence on the training time.

To study the influence of different parameters on the accuracy and detection speed of the ID tag segmentation model, after the training was completed, the images of the test set constructed in step (2) in Section 2.2 were input to the YOLACT++ models with different parameters for segmentation. The $AP^{IoU = 0.75}$ (AP75) and AP of the segmented masks and detection speed were used as test indices. The test results are shown in Figure 6.

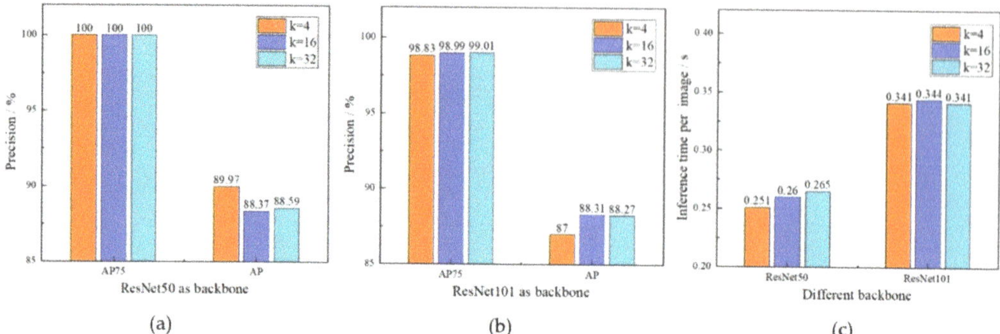

Figure 6. The precision and efficiency of YOLACT++. (**a**) The detection accuracy of the model with ResNet50 as backbone. (**b**) The detection accuracy of the model with ResNet101 as backbone. (**c**) The detection speed of the model with different parameters.

As shown in Figure 6a,b, the accuracy index AP75 of the models with different parameters reached nearly or exactly 100%, and the detection time of a single image was 0.25–0.34 s, indicating that the YOLACT++ model could quickly and accurately segment the ID tag through a linear combination of the prototype masks and the mask coefficients. The accuracy of the YOLACT++ model using ResNet50 as the feature extraction network was higher than when using ResNet101 as shown in Figure 6a,b. This result indicates that for simple segmentation tasks, the depth of ResNet50 was sufficient to extract the features of the target in the image. When ResNet50 was used as the backbone, reducing the number of prototype masks generated slightly improved the segmentation accuracy. This shows that for single-target segmentation, due to the reduced background in the image, too many protomasks will interfere with accurate segmentation of a tag.

In terms of detection speed, the speed of the YOLACT++ model with ResNet50 as its backbone was higher than that of ResNet101, as shown in Figure 6c. When the backbone of YOLACT++ was ResNet50, reducing the k value slightly improved the detection speed. However, when the backbone of YOLACT++ was ResNet101, reducing the k value had little effect on the detection speed. This result indicates that compared with the k value, the backbone had a greater impact on the detection speed. According to the test results, we finally decided to use ResNet50 as the backbone for feature extraction and chose to generate four prototype masks. As a result, the overall accuracy of the cascaded model based on EfficientDet-D4 and YOLACT++ was 96.5%, and the total detection time for a single image was 1.92 s. Figure 7 shows some test results for the YOLACT++ model. The predicted masks cleanly surrounded the number on the tag with relatively high confidence. The result had good robustness to the rotation of the tag, the change in brightness, and the interference around the label.

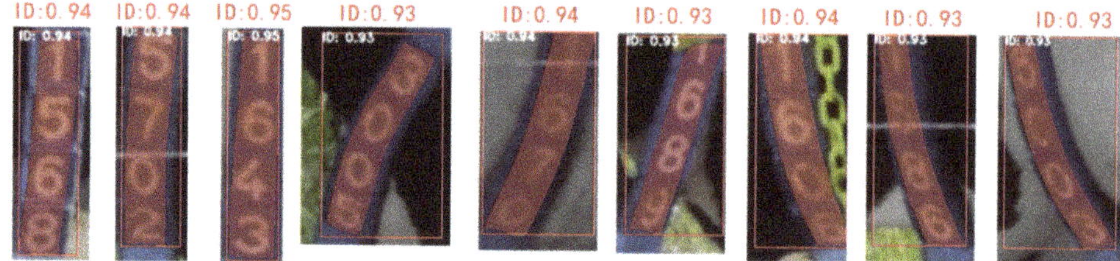

Figure 7. Some of segmentation results of YOLACT++. ID represents the identified class name, and the number after ID represents the confidence for the predicted mask.

4. Discussion

4.1. Comparison with Common Instance Segmentation Model

The proposed cascaded detection model was compared with the common two-stage models Mask RCNN, Mask Scoring RCNN, and the one-stage model SOLOv2. Mask RCNN and Mask Scoring RCNN are two-stage detection models based on a region proposal network. The detection accuracy of these algorithms is high, but their detection speed is slow. SOLOv2 is a one-stage detection model based on anchor box regression. The detection accuracy of this algorithm is slightly low, but its speed is fast. We used the same training set, validation set, and test set to train and test the accuracy and detection time of different models in the same operating environment. The results are shown in Table 1.

Table 1. The precision and efficiency of cascaded model and other models.

Model		Backbone	AP75 (%)	Inference Time (s/Iteration)
Cascaded model	EfficientDet-D4	EfficientNet-B4	96.5	1.92
	YOLACT++	ResNet50		
Mask RCNN		ResNet101	85.3	2.63
Mask Scoring RCNN		ResNet101	58.2	3.39
SOLOv2		ResNet101	18.5	1.21

The overall accuracy of our proposed cascaded instance segmentation model is 96.5%, and its detection time for a single image is 1.92 s. The two-stage instance segmentation model Mask RCNN accurately locates and segments most ID tags with high accuracy, and its segmentation index AP75 is 85.3%, but its detection time for a single image is 2.63 s, which is slightly longer. Mask Scoring RCNN with a Re-score branch performs worse than Mask RCNN in our ID tag segmentation task, and its segmentation index AP75 is 58.2%. The detection time per image is 3.39 s, which is longer than that of Mask RCNN. As a one-stage instance segmentation model, SOLOv2 has a short detection time of 1.21 s. However, its segmented mask is rough along the edge of the tag but tortuous, so its segmentation accuracy is low, at 18.5%. In most cases, the masks with tortuous contours contain some background outside the tags.

The detection results for some images are shown in Figure 8. Three situations with high segmentation difficulty are depicted in the figure. The first is when the brightness of the tag is too low and there are multiple targets in the image. The second is when interference around the tag has similar characteristics to characters (such as white chains). Third, when the brightness of the tag is too high, the character block borders are also displayed. In the above three cases, our method accurately segments the ID tag from the complex background. However, other models are prone to location offsets; missing some characters, including redundant backgrounds and even being unable to detect the tag. Therefore, compared with existing two- and one-stage segmentation models, our proposed cascaded instance segmentation method achieves high-precision ID tag segmentation in complex environments, which has strong robustness and solves the problem of detection difficulty caused by the small area and large deformation of the tag.

4.2. Deformation and Brightness Robustness

To analyze the performance in detecting targets with different areas, we quantified the results of detecting ID tags with different areas with EfficientDet-D4, as shown in Figure 9. As seen from Figure 9, the proportion of the ID tag area to the whole image was only 0.02–0.09%, which is representative of a small target that was difficult to detect, but the model still achieved a high detection rate. By observing the tags of different areas, we found that the rotation and distortion of the tag were the main reasons for the change in area, and the distance between the cow and the image acquisition equipment was the secondary reason. The larger the area of the ID tag, the larger the deformation of the tag. The detection accuracy of ID tags in intervals (5) and (6) was low. There were two main reasons: (1) the total number of samples in these two intervals was small, so even a small amount of false detection had a relatively large impact on the results; (2) the deformation of the tag led to an increase in false detection boxes that overlapped with the target but did not fully contain the numbers on the tag. There were only three ID tags in interval (7), which had fewer samples. When drawing its *P*–*R* curve, the accuracy and recall rate were both 100% with the confidence threshold set to 0.7, so its AP50 was 100%.

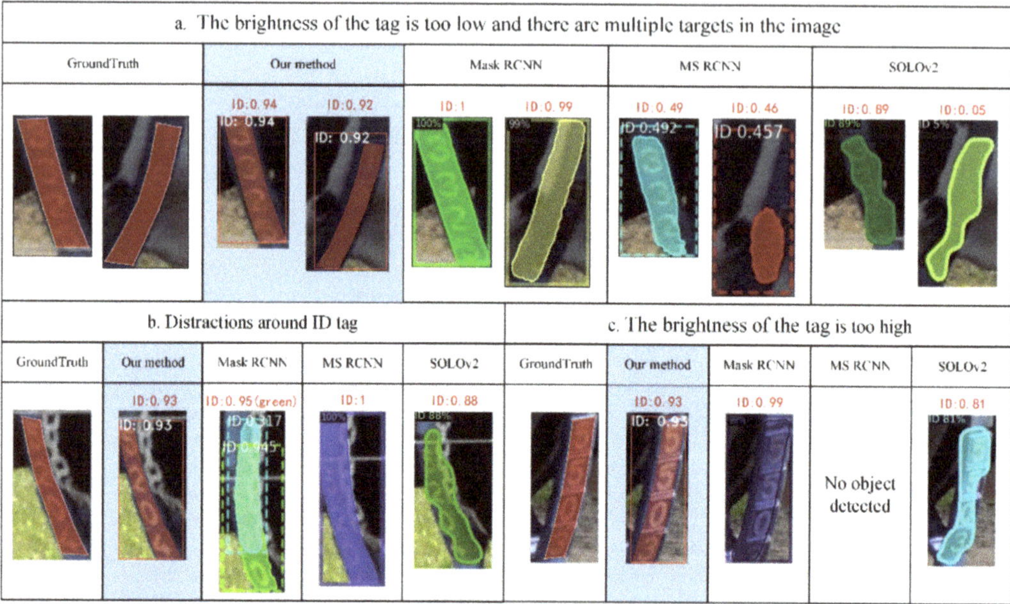

Figure 8. Some of the segmentation results of cascaded model and other models. The ID above the detection results and the number after the ID represent the class and the confidence of the prediction mask, respectively. (**a**–**c**) in the figure correspond to three difficult situations. (**a**) The brightness of the tag is too low and there are multiple targets in the image. (**b**) Distractions around ID tag. (**c**) The brightness is too high.

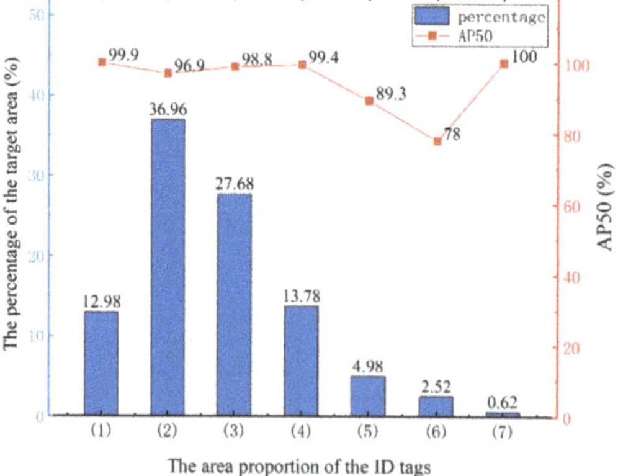

Figure 9. The precision of ID tags with different areas. The abscissa represents the proportion of the bounding box area of the ID tag to the whole image; (**1**) to (**7**) represent seven intervals from 0.02% to 0.09% in increments of 0.01%.

To analyze whether the model is robust to different types of deformation, we divided the deformation of the ID tag into three types (Figure 10): (1) the cow was walking slowly or static, and the tag was only rotated or slightly distorted; (2) the cow was walking quickly or had a lowered head, and the tag was rotated; (3) the cow's head was twisted, and the tag was both rotated and distorted. The detection results of EfficientDet-D4 for ID tags with different types of deformation were statistically analyzed (Table 2). Table 2 shows that the accuracy was the highest when the tag only slightly rotated or distorted. The accuracy was lower when the tag was rotated. When the tag was both rotated and distorted, the accuracy was the lowest. However, rotation and distortion only reduced the accuracy by 2.9%. Therefore, regardless of the state of the ID tag, the model achieves a high detection rate and has high robustness to different types of deformation.

Figure 10. The three types of deformation of ID tags: (**a**) slight rotation or distortion; (**b**) rotation; (**c**) rotation and distortion.

Table 2. ID tags detection accuracy with different types of deformation.

State of the Tag	The Number of Images	AP50 (%)
Slight rotation and distortion	94	99.4
Rotation	192	97.8
Rotation and distortion	120	96.5

The cascaded instance segmentation method proposed in this paper can also adapt to the variant brightness of the target. Figure 11 shows the detection results for some ID tags under different light conditions.

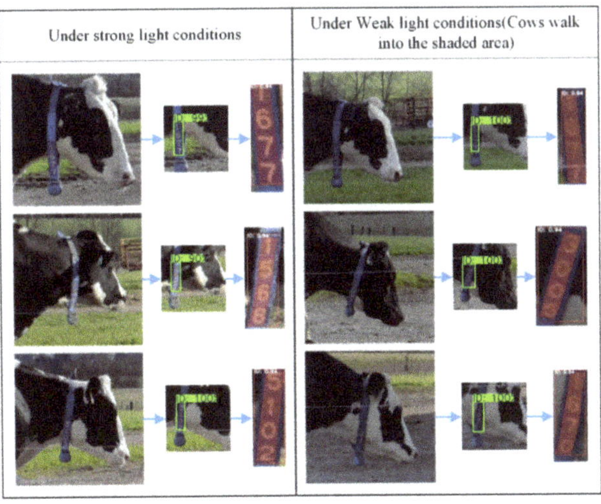

Figure 11. Some detection results under different light conditions.

4.3. Background Robustness

Without retraining the model, the dairy cows' images collected at Sheng Sheng Farm were passed through the cascaded model for detection and segmentation. The results showed that the AP50 of EfficientDet-D4 model is 94.1%, and the AP75 of YOLACT++ model is 100%. Some test results are shown in Figure 12. Even in different scenarios, the model has achieved high accuracy. Therefore, we concluded that the constructed and trained cascaded instance segmentation model has strong robustness with different backgrounds and has promising application prospects.

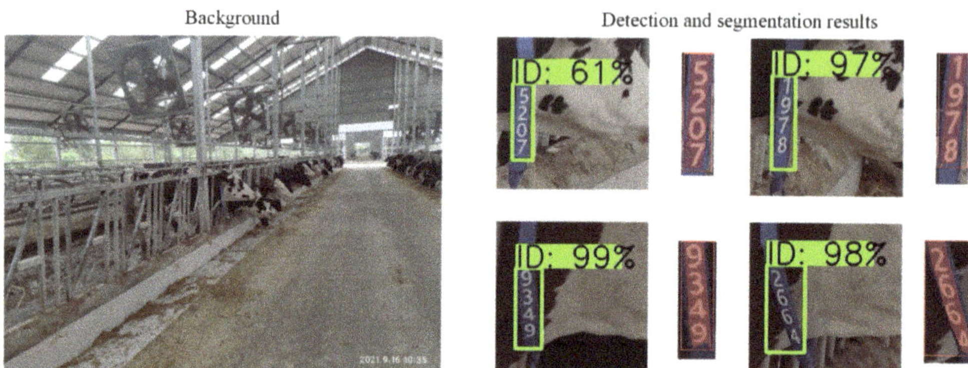

Figure 12. Detection and segmentation results with different backgrounds.

4.4. Analysis of False and Missed Detections

Since the segmentation index AP75 of the YOLACT++ model was 100%, we only analyzed the false and missed detection of ID tags by EfficientDet-D4. After statistics were compiled, there were no missing ID tags. False detection mainly included two cases: (1) a tree branch in the background was mistaken for the target, and the confidence was slightly high, as shown in Figure 13a; (2) a bounding box that overlapped with the tag but did not contain the numbers on the tag completely, as shown in Figure 13b. The reason for the first type of false detection may be that the high-level semantic features of the branches in this region were coincidentally similar to the ID tag, which led to the misjudgment of the branches as the target by the model. The reason for the second kind of misdetection may be that part of the ID tag was also included in the bounding boxes of these false detections, which led to the network failing to make correct judgements. Alternatively, these false bounding boxes were not filtered out when NMS was carried out. The confidence of false bounding boxes was generally low and could have been filtered by setting a confidence threshold.

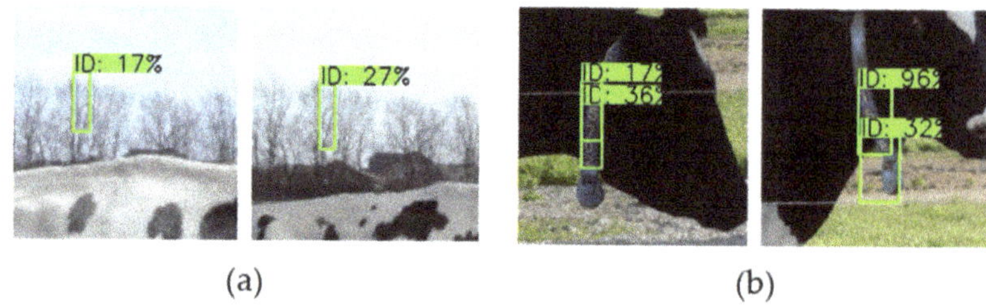

Figure 13. Some of the false detection results. (**a**) A tree branch mistaken as a target. (**b**) The bounding box partly overlapped with the target.

4.5. ID Number Recognition

ID number recognition is completed by character segmentation and character recognition. The purpose of character segmentation is to segment the color tag image into four binary images containing only a single character, which was implemented through the following steps (as shown in Figure 14). The character recognition model was constructed based on a simple convolutional neural network, with the purpose to classify the single character images. The unsegmented images detected by the EfficientDet-D4 model and the segmented images by the cascaded model proposed in this paper were passed through character segmentation model and character recognition model, respectively. The character segmentation model consists of several simple image processing methods, which are illustrated in Figure 14. The character recognition model is constructed based on LeNet-5 [34]. We changed the C5 layer of LeNet-5 from fully connected layer to convolution layer to obtain the character recognition model. The reason for this change is to reduce the redundant parameters and enrich the features.

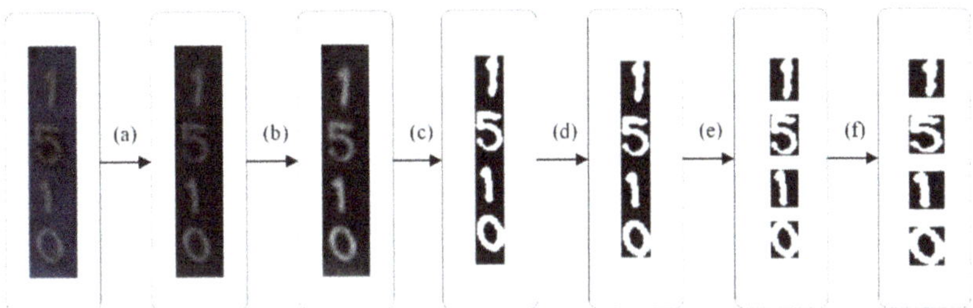

Figure 14. The process of character segmentation: (**a**) graying; (**b**) grey-level transformation; (**c**) binary segmentation; (**d**) morphological processing and removal of redundant background; (**e**) character cropping; (**f**) character normalization.

Table 2 shows that about 76% of ID tags in the dataset are rotated and twisted, so the corresponding detected bounding box will contain different background areas. In the binarization of pixels, the pale white body area of cows and grass in the background are misjudged as characters, which considerably interferes with the implementation of subsequent steps. If the brightness of the background exceeds that of the character, some characters will be lost due to the high threshold in the binary segmentation. Figure 15a depicts the character segmentation results of partially unsegmented images, and Figure 15b displays the character segmentation result of partially segmented images. In Figure 15, the images of each group from left to right are the images after detection/segmentation, the images after binarization, and the images after character normalization. It can be seen from the figure that the character segmentation results of the unsegmented images are very poor due to the influence of the background in the detection bounding box. Its character recognition result is obviously lower than that of segmented images. The accuracy of the character recognition of the segmented image is 95.4%, which is 2.05% higher than that in [22]. This proves that the segmentation of the tags image to remove the redundant background can effectively strengthen character recognition.

4.6. Future Studies

Although the cascaded model based on EfficientDet-D4 and YOLACT++ can achieve 96.5% segmentation accuracy, there is still room for improvement. For the false detection bounding boxes that overlap with the target but do not fully contain the target, their union set can be calculated as the detection results of the tag area, then the segmentation model can be used to remove the redundant background in the detection results. Alternatively, these false positives can be suppressed by better NMS methods, such as Fast NMS [35], which creates the highest confidence bounding box through mutual suppression of all detection boxes. Compared with the traditional method, it allows already-removed detections to suppress other detections, and less time is required.

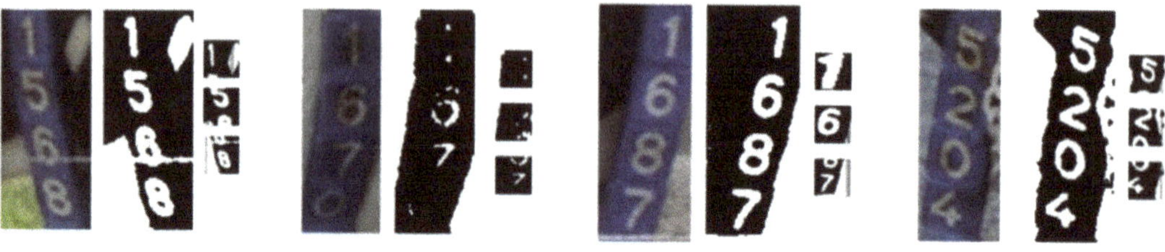

(a) unsegmented images

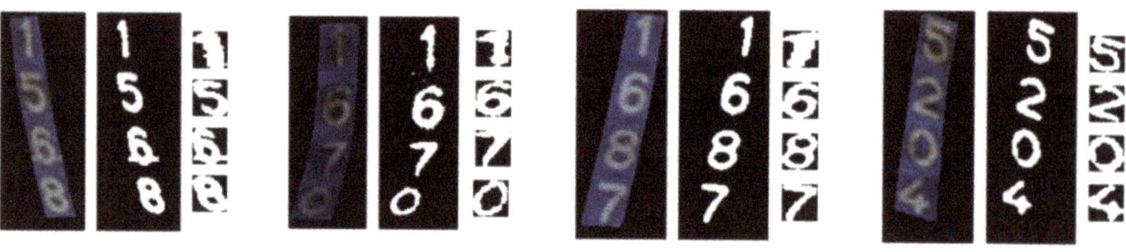

(b) segmented images

Figure 15. Character segmentation results of (**a**) unsegmented and (**b**) segmented images.

The detection speed of the cascaded model also needs to be improved. The detection time is mainly consumed in the detection process of EfficientDet-D4. Due to the high resolution of the image, it is necessary to generate many anchors at different scales and classify and regress them, which requires considerable time. In practical applications, if we know that the size of the tag is within a certain range, the number of anchors generated on each grid point can be reduced, thus effectively simplifying the detection process. Additionally, the image contains many extra background data. If the target appears only in a specified space range, adding a spatial attention mechanism to EfficientDet can cause the network to pay more attention to the areas where the ID tag may appear. This would reduce the time required to extract features from irrelevant backgrounds, thus improving the detection efficiency.

5. Conclusions

This paper proposed a cascaded method for the instance segmentation of a cow collar ID tag based on EfficientDet-D4 and YOLACT++, which accurately detects and segments the target with a small area. The detection accuracy AP50 of the EfficientDet-D4 model is 96.5%, the segmentation accuracy AP75 of the YOLACT++ model is 100%, and the overall segmentation accuracy is 96.5%. Compared with common instance segmentation models, the accuracy is improved by more than 11.2%. Changes in brightness and deformation of the tag have little effect on the detection accuracy of the proposed model. It shows high anti-interference capability and has the potential to be applied to remote and multi-target cow identification on dairy farms. In the future, we can optimize the structure of EfficientDet and propose a better NMS method to reduce the false detection. Additionally, an attention mechanism and other strategies can be considered for reducing the time used by the feature extraction process to improve the detection speed when an image has a large background area.

 sensors

MDPI

Article

An Absorbing Markov Chain Model to Predict Dairy Cow Calving Time

Swe Zar Maw [1], Thi Thi Zin [2,*], Pyke Tin [2], Ikuo Kobayashi [3] and Yoichiro Horii [4]

1 Interdisciplinary Graduate School of Agriculture and Engineering, University of Miyazaki, Miyazaki 889-2192, Japan; z3t1802@student.miyazaki-u.ac.jp
2 Graduate School of Engineering, University of Miyazaki, Miyazaki 889-2192, Japan; pyketin11@gmail.com
3 Field Science Center, Faculty of Agriculture, University of Miyazaki, Miyazaki 889-2192, Japan; ikuokob@cc.miyazaki-u.ac.jp
4 Center of Animal Disease Control, University of Miyazaki, Miyazaki 889-2192, Japan; horii@cc.miyazaki-u.ac.jp
* Correspondence: thithi@cc.miyazaki-u.ac.jp

Abstract: Abnormal behavioral changes in the regular daily mobility routine of a pregnant dairy cow can be an indicator or early sign to recognize when a calving event is imminent. Image processing technology and statistical approaches can be effectively used to achieve a more accurate result in predicting the time of calving. We hypothesize that data collected using a 360-degree camera to monitor cows before and during calving can be used to establish the daily activities of individual pregnant cows and to detect changes in their routine. In this study, we develop an augmented Markov chain model to predict calving time and better understand associated behavior. The objective of this study is to determine the feasibility of this calving time prediction system by adapting a simple Markov model for use on a typical dairy cow dataset. This augmented absorbing Markov chain model is based on a behavior embedded transient Markov chain model for characterizing cow behavior patterns during the 48 h before calving and to predict the expected time of calving. In developing the model, we started with an embedded four-state Markov chain model, and then augmented that model by adding calving as both a transient state, and an absorbing state. Then, using this model, we derive (1) the probability of calving at 2 h intervals after a reference point, and (2) the expected time of calving, using their motions between the different transient states. Finally, we present some experimental results for the performance of this model on the dairy farm compared with other machine learning techniques, showing that the proposed method is promising.

Keywords: absorbing Markov chain; cow behavior analysis; prediction of calving time

Citation: Maw, S.Z.; Zin, T.T.; Tin, P.; Kobayashi, I.; Horii, Y. An Absorbing Markov Chain Model to Predict Dairy Cow Calving Time. *Sensors* **2021**, *21*, 6490. https://doi.org/10.3390/s21196490

Academic Editors: Yongliang Qiao and Sylvain Girard

Received: 6 August 2021
Accepted: 23 September 2021
Published: 28 September 2021

1. Introduction

Even though calving is a normal physiological process, it is important to manage not only for the sake of the animals' welfare, but also for ensuring economic growth in the dairy industry [1,2]. Accurately predicting calving time helps overcome the difficulties of parturition, providing human assistance when needed, and reducing calf mortality. The problem of observing behavioral changes for predicting calving time has been widely studied [3,4]. Solutions are typically sensor-based systems that require the use of wearable or non-wearable sensors to monitor daily behavior and provide responses when calving is imminent. Wearing sensors all the time might cause great discomfort for pregnant cows, as well as risk damaging the sensors themselves while cows move around in the barn [5].

In this study, we focused on a camera-based system because it can support a smart, adjustable, time and money saving way to monitor what happens in the calving barn, and the condition of cows. Best of all, the system allows tracking everything in real time right on our PC, smartphone, or tablet. However, the system has advantages and disadvantages. Similarities between the background and the cow's body color complicates the detection of cows, and clearly identifying each individual cow is still limited in using surveillance cameras over the long-term. For this reason, we will use advanced image processing technology to solve these problems in the future. Figure 1 illustrates our research work.

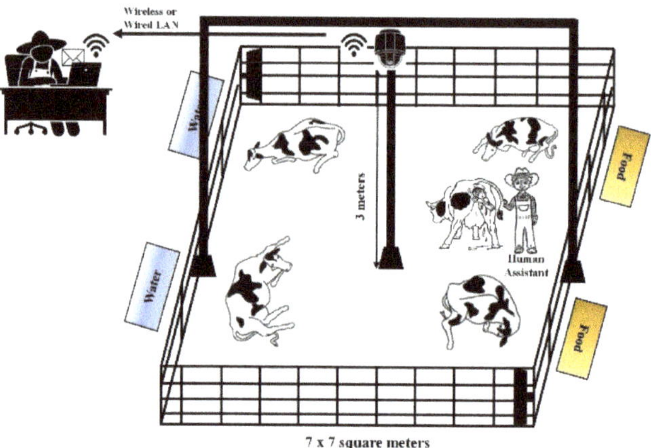

Figure 1. Demonstration of our research work.

The dairy farming industry has benefited greatly because of advances in Information and Communications Technology (ICT), as well as in Artificial Intelligence (AI) and the Internet of Things (IoT). Smart dairy farms are no longer out of our reach. The intelligent and efficient monitoring of individual cows will be a necessary part of making dairy farms smart. Among many other issues, caring for pregnant cows is crucial to dairy farm management, especially when calving occurs. Insufficient monitoring at the time of parturition can be extremely detrimental [6–8]. It can prolong the process of giving birth and increase the risk of both stillbirth and calving difficulties, causing impaired reproductive performance and increased calving-to-conception intervals.

In this research, we concentrated on monitoring daily routines for the purpose of detecting when behavioral changes increase in frequency, signaling the approach of parturition, and allowing a prediction of the exact calving time. We describe how we approached this goal by designing a system that features observations of video data collected using a 360-degree camera on an hourly basis for 72 h before the start of calving. For this reason, we propose a new method based on image processing techniques and Markov chain model to predict the time at which cow calving will occur. We also compare our proposed method with three other machine learning techniques: K-nearest Neighbors (KNN), Naïve Bayes (NB), and Support Vector Machine (SVM).

Specifically, we analyze the behavior of pregnant cows in maternity barns by embedding this behavior into a Markov chain, thus predicting the time of calving. Model performance was evaluated on video data, which were collected from 25 dairy cows at Oita Prefecture in Japan, to verify that our proposed method has potential as a method for predicting calving time. From these videos taken in the maternity barn, human observers used a reversible counter system in 5 min increments to record the number of changes in lying posture, the number of transitions from lying to standing, the number of changes in standing posture, and the number of transitions from standing to lying. All these data for statistical analysis of behavior were collected for three days before the predicted calving date.

Results of this analysis showed that the proposed methodology could be applied and achieved plausible results. Our analysis indicates that investigating behavioral activity peaks in these data will be useful in improving the prediction process. However, additional avenues should be explored in pursuing research on calving time prediction. As an alternative, the application of Markov modeling [9] to predict calving time is an appealing methodology. The most frequent applications of augmented absorbing Markov chain modeling are in predicting future stock exchange trends [10], predicting web user behavior [11], and forecasting educational attainment rates [12]. Our current concern lies in using an absorbing Markov chain to develop a prediction model from observations of the behavioral changes of the dairy cows. Specifically, we propose in this paper the use of such a model to predict when calving will occur. By comparing the results of predicting calving events using our proposed method with the results with other machine learning techniques, we see that our proposed method more accurately predicts calving events.

The organization of this paper is as follows. Section 2 follows with a description of the proposed method for calving time prediction. We show some experimental results and discussion in Sections 3 and 4. Finally, we conclude our approach in Section 5 with some suggestions and discussion of possible future research.

2. Materials and Methods

2.1. Data Collection and Preparation

The animal experimentation protocol of ethical statement and approval were granted for this study, animals were neither enforced nor uncomfortably restricted during the study period. The video data of monitoring calving process used for analysis in this study were collected by an installed camera without disturbing natural parturient behavior of animals and routine management of the farm.

The experimental design is established on a large dairy farm situated in Oita Prefecture, Japan. Three primiparous and 22 multiparous pregnant dairy cows were housed in roofed cowsheds. Four or five pregnant cows were together housed in a calving pen. which was 7×7 m^2 with sawdust flooring for when calving event was close to occurring. The cows were fed with Total Mixed Ration (TMR) twice daily for their maintenance and pregnancy, as calculated based on each cow's body weight and the expected average milk yield (35 kg/day) after giving birth. They were also provided ad libitum access to clean water and mineral supplements.

Experimented cows were continuously monitored using a 360-degree GV-FER5700 camera (Geo Vision Inc., Taiwan, China) (2560×2048 pixels, recording at 30 frames per second), which was set up 3 m above the pregnant cows located in the maternity barns. This camera can capture images within a 360-degree field of view in the horizontal plane. Using this camera, the positions and states of cows appear in different parts of the 360-degree view. All cows are clearly visible from overhead, allowing a determination of the condition of each cow. The video sequences for pregnant cows are continuously collected until calving occurs.

In this section, we describe how an augmented Markov chain model can be employed to predict a calving event. To do so, we firstly prepare the dataset from video sequences taken during the three days before calving. After collecting the video sequences, the target cow regions are manually extracted by Visual Geometry Group (VGG) annotator [13] to remove the background, and to obtain the cow contour regions by using image processing techniques.

We used an approach based on statistical analysis to predict calving time. Cow behavioral activities were recorded by human observers performing a direct visual observation of each individual cow in the calving barn. The four types of conditions include two postures and two transitions. They are defined as follows.

1. L (Posture): lying in the calving barn;
2. LS (Transition): rising from a lying state to a standing state;
3. S (Posture): standing on all four legs;
4. SL (Transition): changing from a standing state to a lying state.

Images are labeled to count the number of pairs from one state to another in making a co-occurrence matrix for the 72 h before calving. Although cows may assume many other states such as eating and drinking, all other activities are assumed to be subsets of the above-mentioned activities. Because of this, our video recording only concerns a sequence of the four activities for each individual cow, continuously monitored until calving occurs. Figure 2 illustrates a sample of the four posture conditions. In Figure 3, the system architecture of our proposed method is represented.

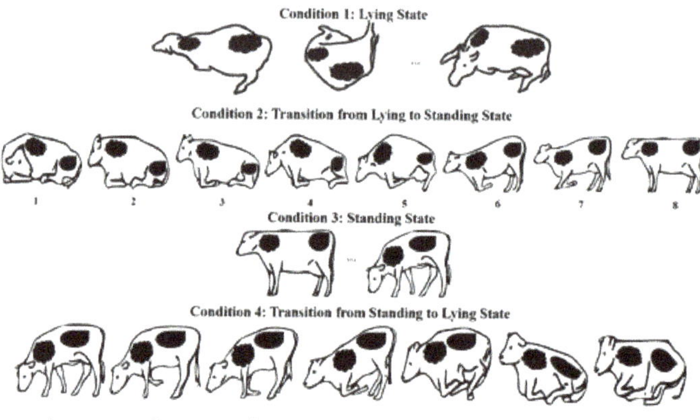

Figure 2. A sample of four posture conditions before calving.

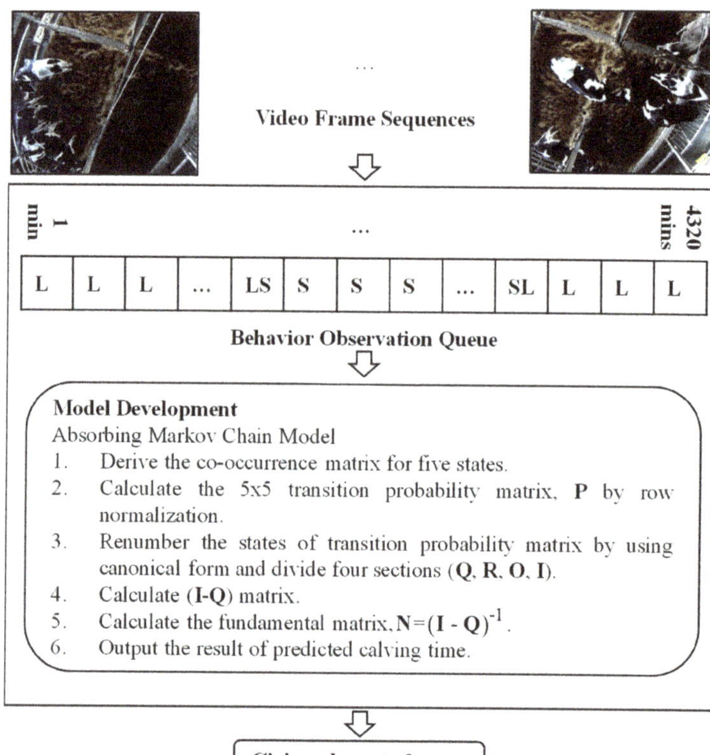

Figure 3. The system architecture of our proposed method.

Generally, absorbing Markov Chain is used to investigate behaviors of any state which eventually enter an end state or absorbing state. In our case, the absorbing state is the event at which calving occurs. In theory of Markov Chain, we can compute the time entering an absorbing state and the probability of absorbing. So, we thought that it would be tractable to apply the absorbing Markov Chain Model for the calving time of a pregnant dairy cow. Although there have been many applications in queuing and dam theory [14], we have not seen any application of absorbing Markov Chain for the prediction of dairy cow calving process. So, at this stage we have not made comparison with previous methods in this aspect.

Moreover, the nature of absorbing state from the absorbing Markov chain is similar with the nature of calving event in prediction calving time model in dairy cows. From the concept of absorbing Markov chain theory, the absorbing state is end-state, and the calving state is also end-state in the prediction of calving time model.

2.2. Creation of Co-Occurrence Matrix and Markov Chain Model

The co-occurrence and probability matrices of the Markov chain model are created using the state sequence described in Figure 2. In order to do so, we first define the number of co-occurrences of state pairs. Let $c(s_i, s_j)$ be the number of pairs of states (s_i, s_j) for $i, j = 1, 2, 3, 4, 5$, and $s_1 = \text{L}$, $s_2 = \text{LS}$, $s_3 = \text{S}$, $s_4 = \text{SL}$, $s_5 = \text{Calve}$. We can then have the corresponding co-occurrence matrix **C**, as shown below.

$$\mathbf{C} = (c(s_i, s_j)) = \begin{bmatrix} c(s_1, s_1) & c(s_1, s_2) & c(s_1, s_3) & c(s_1, s_4) & c(s_1, s_5) \\ c(s_2, s_1) & c(s_2, s_2) & c(s_2, s_3) & c(s_2, s_4) & c(s_2, s_5) \\ c(s_3, s_1) & c(s_3, s_2) & c(s_3, s_3) & c(s_3, s_4) & c(s_3, s_5) \\ c(s_4, s_1) & c(s_4, s_2) & c(s_4, s_3) & c(s_4, s_4) & c(s_4, s_5) \\ c(s_5, s_1) & c(s_5, s_2) & c(s_5, s_3) & c(s_5, s_4) & c(s_5, s_5) \end{bmatrix}.$$

The above co-occurrence matrix, **C** can be written as (1):

$$\mathbf{C} = (c_{ij}), \tag{1}$$

where, $c_{ij} = \#\{(i, j) | i, j \in S = \{1, 2, 3, 4, 5\}\}$.

We can then deduce the one step transition probabilities, p_{ij} by defining (2):

$$p_{ij} = c_{ij} / \sum_{j=1}^{5} c_{ij}, \tag{2}$$

which represents the one step transition probability of going from state i to state j in a Markov Chain. We then have the transition probability matrix, $\mathbf{P} = (p_{ij})$.

The sum of the row probabilities is equal to one, since each health state is independent of the others, and an animal must move to one of the five states. The diagonal represents the probability of staying in the same state. A state k in a Markov chain is defined as *absorbing* if $p_{kk} = 1$, in other words all $p_{kj} = 0$ for $j \neq k$. In this study, the absorbing state is the calving state. When calving occurs, further investigation stops because we have achieved the objective of predicting the calving time. Thus, five of the states are considered transient states in the Markov chain, since each one can independently transition to another state. The four transient states are lying (L), transition from lying to standing (LS), standing (S), transition from standing to lying (SL). Calving is the absorbing state. Figure 4 describes this five-state absorbing Markov chain used to predict calving time in dairy cows.

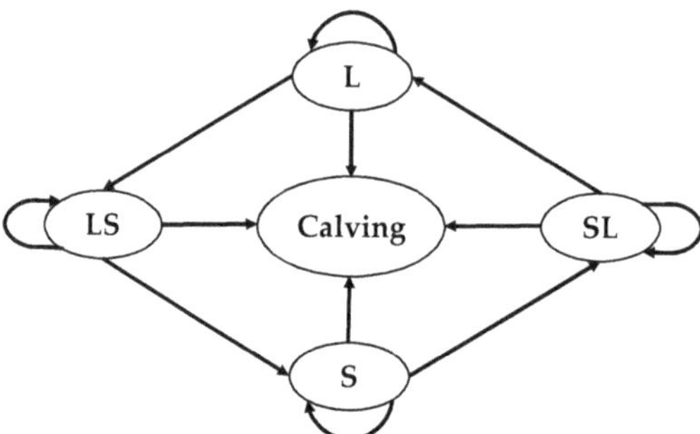

Figure 4. Five-state absorbing Markov Chain to assess the prediction of calving time in dairy cows.

2.3. Description of Calving Event

As an attractive feature of Markov models, they can describe the course of calving events over time. This is especially attractive for modeling calving since a cow's state of behavior while calving influences the prediction of calving time. The transition probability matrix **P** summarizes the probabilities of cow activities and can be used to describe the probability of calving for an individual cow with a known activity state. The elements of the probability matrix p_{ij} in the ith row and j^{th} column is denoted by $p_{ij}(t)$, which represents the probability of a transition from state i to state j during t periods or t steps, where t measures in minutes. For an m state Markov model, the probability of the system visiting state k at time t can be denoted as $p_k(t)$. Therefore, for all m states, these probabilities can be expressed as a row vector, $p(t) = [p_1(t), p_2(t), \cdots, p_m(t)]$. By using total probability, this equation can be written as (3) or (4):

$$p_k(t) = \sum_{j=1}^{m} p_{kj} p_j(t-1) \text{ for } k = 1,\ 2\ , \cdots, m, \tag{3}$$

$$p(t) = p(t-1)\mathbf{P}^1 = p(t-2)\mathbf{P}^2 = \cdots = p(0)\mathbf{P}^t, \tag{4}$$

Transient analysis may cause convergence of the probability distribution vector when t becomes very large. That is, as the number of steps t increases, the probability vector approaches a limiting value which is called the stationary distribution of the Markov chain. This stationary distribution, or as it is also called, steady state distribution, is denoted by $\pi = [\pi_1, \pi_2, \cdots, \pi_m]$ and satisfies (5).

$$\pi = \pi\mathbf{P}, \tag{5}$$

Our case features five states: L, LS, S, SL, and Calve. Therefore, we get the following probabilities. $p_i(t)$ is the probability that the cow is in one of the states after a period of t from the start. In these ways, the behavior of dairy cows can be analyzed during periods when calving is imminent. These results will be described in the next section on experimental work.

2.4. Calving Time Prediction Procedure

This section presents the method of predicting calving time using the Markov chain model discussed in the previous section. This involves adding or augmenting the calving event as an absorbing state in our four-state Markov chain model. Since the problem is to predict the time of calving, the exercise is completed when the event occurs. Because of this, the calving state is considered absorbing. This means that any of the other four states can transition directly to the calving state, but once there, no additional transitions will occur. The probability of transitioning from absorbing state to absorbing state is one; and the probability of transitioning from absorbing state to any other state is zero. The four-state

Markov chain model is transformed as described in (3) into an augmented Markov chain model of five states by adding an absorbing state (calving state).

Fundamental Matrix Solution: Absorbing Markov Chain

The matrix solution provides an exact solution for the time spent in each state, conditional on the entry state in which an individual enters the model. Such a matrix solution is only viable in time with homogeneous Markov chains with r absorbing states and m transient states. The transition probability matrix of a chain that contains an absorbing state is defined as the separation of a probability transition matrix **A** using canonical form.

$$\mathbf{A} = \begin{bmatrix} \mathbf{Q} & \mathbf{R} \\ \mathbf{O} & \mathbf{I} \end{bmatrix}, \tag{6}$$

where, **I** is an r-by-r identity matrix, **O** is an r-by-m zero matrix, **R** is a nonzero m-by-r matrix and **Q** is an m-by-m matrix.

In the proposed augmented Markov model, **Q** is the matrix that contains transition probabilities between transient states, R is column vector of the calving state probabilities, **O** is the row vector of zero matrix, and **I** in [1] is 1×1 matrix. The iterated multiplication of the augmented matrix **A** yields as follows.

$$\mathbf{A}^2 = \begin{bmatrix} \mathbf{Q} & \mathbf{R} \\ \mathbf{O} & \mathbf{I} \end{bmatrix} \times \begin{bmatrix} \mathbf{Q} & \mathbf{R} \\ \mathbf{O} & \mathbf{I} \end{bmatrix} = \begin{bmatrix} \mathbf{Q}^2 & \mathbf{QR}+\mathbf{R} \\ \mathbf{O} & \mathbf{I} \end{bmatrix}, \tag{7}$$

$$\mathbf{A}^3 = \begin{bmatrix} \mathbf{Q}^2 & \mathbf{QR}+\mathbf{R} \\ \mathbf{O} & \mathbf{I} \end{bmatrix} \times \begin{bmatrix} \mathbf{Q} & \mathbf{R} \\ \mathbf{O} & \mathbf{I} \end{bmatrix} = \begin{bmatrix} \mathbf{Q}^3 & \mathbf{Q}^2\mathbf{R}+\mathbf{QR}+\mathbf{R} \\ \mathbf{O} & \mathbf{I} \end{bmatrix}, \tag{8}$$

Hence, by induction, we obtain the following:

$$\mathbf{A}^t = \begin{bmatrix} \mathbf{Q}^t & \mathbf{Q}^{t-1}\mathbf{R}+\mathbf{Q}^{t-2}\mathbf{R}+\cdots+\mathbf{R} \\ \mathbf{O} & \mathbf{I} \end{bmatrix} \times \begin{bmatrix} \mathbf{Q} & \mathbf{R} \\ \mathbf{O} & \mathbf{I} \end{bmatrix} = \begin{bmatrix} \mathbf{Q}^t & \left(\mathbf{Q}^{t-1}+\cdots+\mathbf{I}\right)+\mathbf{R} \\ \mathbf{O} & \mathbf{I} \end{bmatrix}. \tag{9}$$

However, when t tends to infinity the transient state matrix, $\mathbf{Q}^t$ will tend to **O** (zero matrix). We then have from (6) that,

$$\mathbf{A}^\infty = \begin{bmatrix} \mathbf{O} & \mathbf{NR} \\ \mathbf{O} & \mathbf{I} \end{bmatrix}, \tag{10}$$

where, $\mathbf{N} = \mathbf{I} + \mathbf{Q}^1 + \mathbf{Q}^2 + \mathbf{Q}^3 + \cdots = (\mathbf{I} - \mathbf{Q})^{-1}$.

The matrix $\mathbf{N} = (\mathbf{I} - \mathbf{Q})^{-1}$ is called the fundamental matrix for the augmented Markov chain model. Let $\mathbf{N}(i, j)$ be the element in row i and column j. Then, we can interpret the summation of $\mathbf{N}(i, j)$ over j as the expected number of periods until *absorbing* (calving). Therefore, the expected time until the *absorbing* (calving state) occurs is shown as $\sum_j \sum_i \mathbf{N}(i, j)$. The probability of *absorbing* or calving at the expected time is $p(0) \times \mathbf{N} \times \mathbf{R}$.

3. Results

Data were collected on 25 dairy cows; 21 Holstein Black and White cows and 4 Brown Swiss cows were moved into the maternity barns from beginning 3 days before the expected calving date. We divided the 25 cows into 2 groups based on the primiparous and multiparous pregnant cows, as shown in Tables 1 and 2. They calved at ages between 21 and 26 months and the calving period was between November and December in 2017. None of the 25 test cows presented with dystocia, and assistance for newborns calves was provided, as necessary. All newborn calves were single birth. Behavior analysis data were not collected after calving. Individual cows were continuously monitored until the calving event occurred using a 360-degree camera above the pregnant cows in the maternity barns.

Table 1. Group 1: Primiparous pregnant cows data information.

Cow ID	Types of Cows	Data Collection Start Date (mm/dd/yy) and Time (h/m/s)	Calving Date (mm/dd/yy) and Time (h/m/s)
1	Black and White Holstein	11.26.2017, 17:10:00	11.29.2017, 17:10:00
3	Black and White Holstein	11.26.2017, 19:35:00	11.29.2017, 19:35:00
4	Black and White Holstein	11.30.2017, 15:10:00	12.03.2017, 15:10:00

Table 2. Group 2: Multiparous pregnant cows data information.

Cow ID	Types of Cows	Data Collection Start Date (mm/dd/yy) and Time (h/m/s)	Calving Date (mm/dd/yy) and Time (h/m/s)
2	Brown Swiss	11.30.2017, 21:15:00	12.03.2017, 21:15:00
5	Brown Swiss	12.09.2017, 17:15:00	12.12.2017, 17:15:00
6	Brown Swiss	12.01.2017, 10:25:00	12.04.2017, 10:25:18
7	Brown Swiss	12.03.2017, 02:40:00	12.06.2017, 02:41:22
8	Black and White Holstein	11.29.2017, 00:30:00	12.02.2017, 00:30:00
9	Black and White Holstein	12.04.2017, 10:05:00	12.07.2017, 10:06:35
10	Black and White Holstein	12.04.2017, 10:10:00	12.07.2017, 10:13:00
11	Black and White Holstein	12.04.2017, 16:05:00	12.07.2017, 16:09:40
12	Black and White Holstein	12.04.2017, 14:00:00	12.06.2017, 20:10:00
13	Black and White Holstein	12.11.2017, 06:00:00	12.14.2017, 05:58:50
14	Black and White Holstein	12.06.2017, 10:00:00	12.08.2017, 03:25:00
15	Black and White Holstein	12.12.2017, 04:50:00	12.15.2017, 04:53:09
16	Black and White Holstein	12.07.2017, 17:20:00	12.10.2017, 17:20:00
17	Black and White Holstein	12.13.2017, 21:00:00	12.16.2017, 21:03:29
18	Black and White Holstein	12.16.2017, 21:55:00	12.19.2017, 21:55:00
19	Black and White Holstein	12.14.2017, 17:15:00	12.17.2017, 17:19:00
20	Black and White Holstein	12.17.2017, 06:10:00	12.20.2017, 06:10:00
21	Black and White Holstein	12.14.2017, 16:15:00	12.17.2017, 16:17:12
22	Black and White Holstein	12.17.2017, 09:50:00	12.20.2017, 09:50:00
23	Black and White Holstein	12.15.2017, 00:50:00	12.18.2017, 01:25:21
24	Black and White Holstein	12.17.2017, 12:15:00	12.20.2017, 12:15:00
25	Black and White Holstein	11.29.2017, 00:30:00	12.02.2017, 00:30:00

Specifically, the four relevant activities were lying, transitions from lying to standing, standing, and transitions from standing to lying. From the collected videos, a sequence of the four activities is extracted for each cow as shown in the previous section. The co-occurrence matrix is constructed from the activity sequence for each individual cow. Sample co-occurrence matrices **C** are described below for Identity Document 2 (ID 2), Identity Document 11 (ID 11), and Identity Document (ID 27).

$$\mathbf{C}_{\mathrm{ID2}} = \begin{bmatrix} 1360 & 67 & 0 & 0 & 1 \\ 0 & 0 & 65 & 4 & 1 \\ 0 & 0 & 2686 & 68 & 1 \\ 66 & 3 & 0 & 0 & 1 \\ 0 & 0 & 0 & 0 & 1 \end{bmatrix}$$

$$\mathbf{C}_{\mathrm{ID11}} = \begin{bmatrix} 1324 & 38 & 0 & 0 & 1 \\ 0 & 0 & 38 & 4 & 1 \\ 0 & 0 & 2834 & 38 & 1 \\ 38 & 5 & 0 & 0 & 1 \\ 0 & 0 & 0 & 0 & 1 \end{bmatrix}$$

$$\mathbf{C}_{\mathrm{ID27}} = \begin{bmatrix} 1229 & 29 & 0 & 0 & 1 \\ 0 & 0 & 29 & 2 & 1 \\ 0 & 0 & 2972 & 28 & 1 \\ 28 & 2 & 0 & 0 & 1 \\ 0 & 0 & 0 & 0 & 1 \end{bmatrix}$$

By row normalization, we obtain the Markov chain probability matrices **P** for ID 2, ID 11, and ID 27 as follows.

$$\mathbf{P}_{\mathrm{ID2}} = \begin{bmatrix} 0.952 & 0.047 & 0 & 0 & 0.001 \\ 0 & 0 & 0.890 & 0.096 & 0.014 \\ 0 & 0 & 0.976 & 0.024 & 0 \\ 0.957 & 0.029 & 0 & 0 & 0.014 \\ 0 & 0 & 0 & 0 & 1 \end{bmatrix}$$

$$\mathbf{P}_{\mathrm{ID11}} = \begin{bmatrix} 0.971 & 0.028 & 0 & 0 & 0.001 \\ 0 & 0 & 0.884 & 0.093 & 0.023 \\ 0 & 0 & 0.986 & 0.013 & 0 \\ 0.864 & 0.114 & 0 & 0 & 0.023 \\ 0 & 0 & 0 & 0 & 1 \end{bmatrix}$$

$$\mathbf{P}_{\mathrm{ID27}} = \begin{bmatrix} 0.976 & 0.023 & 0 & 0 & 0.001 \\ 0 & 0 & 0.906 & 0.063 & 0.031 \\ 0 & 0 & 0.990 & 0.009 & 0 \\ 0.903 & 0.065 & 0 & 0 & 0.032 \\ 0 & 0 & 0 & 0 & 1 \end{bmatrix}$$

3.1. *Calving Event as an Absorbing State of Markov Chain Model Implementation*

The transition probability matrix defined as a Markov chain probability matrix is regular. The calving event is added as an absorbing state of the Markov chain as described above. The **Q** matrix for cow ID 2 is as follows.

$$\mathbf{Q}_{\mathrm{ID2}} = \begin{bmatrix} 0.952 & 0.047 & 0 & 0 \\ 0 & 0 & 0.890 & 0.096 \\ 0 & 0 & 0.976 & 0.024 \\ 0.957 & 0.029 & 0 & 0 \end{bmatrix}$$

$\mathbf{R}_{\mathrm{ID2}} = \begin{bmatrix} 0.001 & 0.014 & 0 & 0.014 \end{bmatrix}^{\mathrm{T}}$;
$\mathbf{O}_{\mathrm{ID2}} = \begin{bmatrix} 0 & 0 & 0 & 0 \end{bmatrix}$; and $\mathbf{I}_{\mathrm{ID2}} = [1]$;
$\mathbf{N}_{\mathrm{ID2}}$ is $(\mathbf{I}_{\mathrm{ID2}} - \mathbf{Q}_{\mathrm{ID2}})^{-1}$ which is given by:

$$\mathbf{N}_{\mathrm{ID2}} = \begin{bmatrix} 372.492 & 17.969 & 657.388 & 17.487 & 1065.335 \\ 356.900 & 18.245 & 667.502 & 17.756 & 1060.402 \\ 361.332 & 17.459 & 679.840 & 17.977 & 1076.608 \\ 366.725 & 17.720 & 648.286 & 18.245 & 1050.975 \end{bmatrix}$$

Thus, from the theory developed in our method proposed in Section 3, the sum of all entries gives the expected time at which the calving event occurs. We obtain the predicted calving time as, $\sum_j \sum_i \mathbf{N}(i,j)$ = 4253.320 min = 70.889 h from the beginning. The actual calving time is 72 h from the beginning. Therefore, our proposed method provides an accurate prediction. The probability of calving is expressed in the previous section. Thus, the probability of calving is certainly almost 1.

Similarly, we have derived the most useful statistics such as the co-occurrence matrices and their corresponding probabilities for all cows in this study. By adding the concept of an absorbing barrier state (calving), we derived the time for entering the absorbing state, determining the estimated time that calving occurs.

3.2. Patterns of Activities of Cows before Calving

We also investigated patterns in the four activities of lying, transitions from lying to standing, standing, and transitions from standing to lying. In order to do so, we raised powers to the probability matrix **P**, and look at the probabilities of diagonal elements. In other words, we researched the behavior of p_{11} , p_{22}, p_{33} and p_{44}. We found that those entries in $\mathbf{P}^t$ of cow ID 2 for $t = 1, 2, 4, 6, \ldots, 24$ are represented in Table 3. As shown, the lying state, and standing state probabilities decrease when the cow approaches the calving state. However, the transition state probabilities increase. These patterns are shown in Figures 5–7.

Table 3. State probability patterns of ID2, ID11, and ID 27.

t	ID 2				ID 11				ID 27			
	p_{11}(L)	p_{22}(LS)	p_{33}(S)	p_{44}(SL)	p_{11}(L)	p_{22}(LS)	p_{33}(S)	p_{44}(SL)	p_{11}(L)	p_{22}(LS)	p_{33}(S)	p_{44}(SL)
1	0.953	0.001	0.975	0.002	0.973	0.004	0.997	0.005	0.982	0.001	0.983	0.001
2	0.908	0.002	0.951	0.001	0.946	0.005	0.994	0.004	0.964	0	0.967	0.001
4	0.831	0.001	0.908	0.001	0.895	0.003	0.989	0.001	0.929	0.001	0.936	0.001
6	0.764	0.002	0.870	0.002	0.847	0.001	0.984	0.003	0.897	0	0.907	0
8	0.706	0.001	0.838	0.001	0.802	0.002	0.979	0.002	0.867	0.001	0.880	0.001
10	0.656	0.001	0.810	0.002	0.759	0.003	0.974	0.004	0.839	0.002	0.855	0.002
12	0.613	0.001	0.786	0	0.719	0.001	0.969	0.003	0.812	0.002	0.831	0.002
14	0.575	0.002	0.765	0.001	0.681	0.002	0.965	0.001	0.788	0.001	0.809	0.001
16	0.543	0.001	0.746	0.002	0.646	0.004	0.960	0.003	0.765	0.002	0.788	0.001
18	0.515	0.001	0.731	0	0.612	0.002	0.956	0.005	0.743	0	0.769	0.001
20	0.491	0.002	0.717	0.001	0.580	0.002	0.952	0.004	0.723	0.003	0.750	0.002
22	0.470	0.003	0.705	0.003	0.551	0.005	0.948	0.005	0.704	0.002	0.733	0.002
24	0.452	0.011	0.695	0.01	0.522	0.01	0.944	0.009	0.686	0.008	0.718	0.008

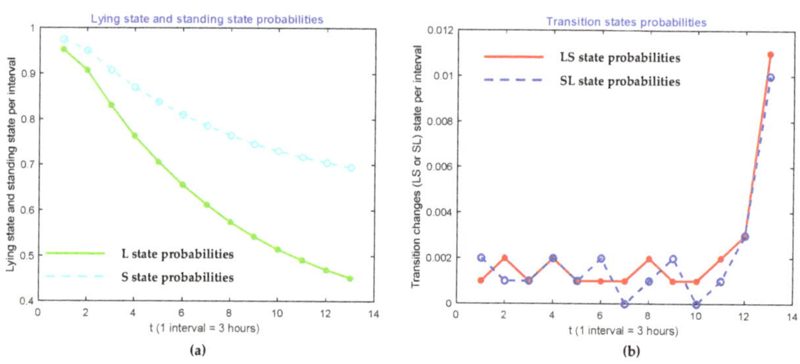

Figure 5. Comparison of (**a**) lying state and (**b**) transition state probabilities of ID 2.

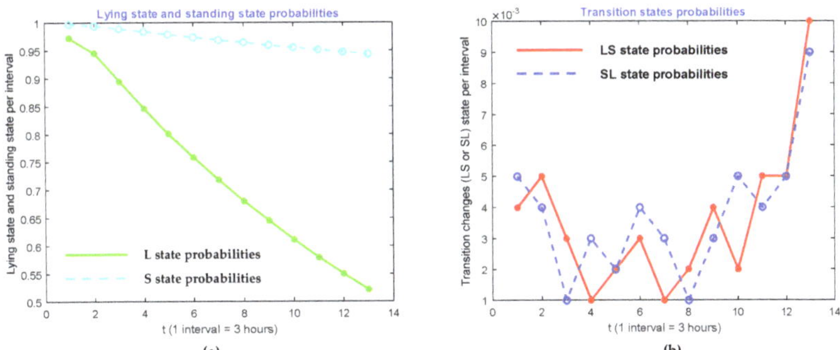

Figure 6. Comparison of (**a**) lying state and (**b**) transition state probabilities of ID 11.

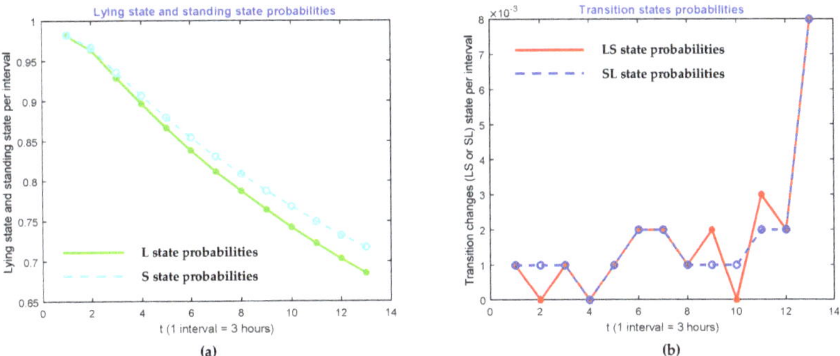

Figure 7. Comparison of (**a**) lying state and (**b**) transition state probabilities of ID 27.

4. Discussion

4.1. Discussion on Calving Time Prediction Approaches in the Literature Surveys

Though we conducted a thorough review of the literature, we have not found any method resembling our approach, and cannot make comparisons with other methods of Markov chain analysis. However, we did find some appealing approaches in the literature, such as machine learning [15], online image analysis [16], indications of posture changes [17], and investigations of farm devices [18]. However, we feel that our approach is much easier to implement and promises comparatively favorable outcomes.

Monitoring cow behavior to predict calving events is not superficial work. In fact, no one approach could cover all aspects of monitoring cow behavior. A sizable amount of research has appeared in the literature involving the development of methods and models to predict calving time, and results have been quite promising. This section concerns a brief explanation on the topic of predicting calving time based on cow behavior monitoring, focusing on the augmented absorbing Markov chain model used to build our predictive model and analyzing the performance of our proposed method by measuring some machine learning techniques' prediction results.

The effort of monitoring the calving process is a matter of assessing whether human assistance is required in the upcoming hours or overnight, or whether difficulties in giving birth are likely. Again, such difficulties may adversely affect production, and could even risk the life of mother and calf. Thus, an accurate and efficient method of predicting a calving event will continue to play a central role in precision dairy farming. It is no wonder that much research involving multiple disciplines has focused on predicting calving times and related research. However, we have yet to see satisfactory accomplishments in the literature.

Although a variety of unavoidable stressors continue to affect cows through calving and dry off (stopping milk production), our increased knowledge of events leading to calving should have a positive impact on milk production, as well as on cow health and overall wellbeing. The calving time prediction methods and devices can be divided into the following three categories based on:

(a). Hormonal changes;
(b). Clinical signs; and
(c). Behavioral changes before calving.

Since the first two categories are beyond the scope of the current focus, we shall review some research that falls in the third category. The video cameras or accelerometers recording the behavior of cows can be integrated in systems using image analysis or locomotive activity to alert the dairy farmer when calving is imminent. The four comparable predictive models had been established for calving difficulty in dairy heifers and cows using four machine learning techniques: multinomial regression, decision trees, random forests, and neural networks [19]. Among many other findings, is the use of calibration evaluation techniques that have not been frequently used in agricultural or animal health applications. Apart from these models, our discussion below will extend to some other research on calving time prediction.

Some of them utilize physical measures, such as body temperature [20,21], the blood levels of progesterone, and the relaxation of pelvic ligaments [22,23]. Recently, a combination of data from sensors detecting cumulative activity, rumination activity, feeding activity, and body temperature achieved a more accurate calving time prediction system than those based exclusively on the date of insemination [24]. However, some obstacles remain to accurately predicting the starting time of calving.

Overall, systems based on behavioral analysis seem to have the most potential, because significant changes in behavior occur on the day of calving. Analyses of behavior changes normally begin several days before delivery and last until calving time. From the literature review, the most important facts and figures are as follows: searching for isolation, moving the tail, walking aimlessly, turning the head towards the abdomen, reducing rumination time, reducing the time spent lying, sniffing the ground, and frequently changing posture [25,26]. The most distinctive trends that precede calving were also noted in Santegoeds' work [27]. In summary, these trends include the following: (1) the number of steps taken increases very slightly but significantly 10 days before calving, and more significantly over the 2 last days; (2) the time spent lying decreases slightly but significantly 10 days before calving, more significantly 3 days to 12 h before calving, and increases thereafter until after calving.

Moreover, the standing pattern is almost perfectly opposite to the lying pattern. This difference is due to an increase in time spent walking around, which interrupts periods of standing, rather than periods of lying. Walking time rises notably from two days before calving. The number of times standing up radically increases in the last6 h. Therefore, several researchers believed that close observation of cattle in the last gestation period is essential to detect the onset of calving and to reduce neonatal losses [28]. The calving time prediction is performed by using time series analysis of data on posture changes collected from video sequences recorded in the maternity barn [29–31]. He determined the number of transitions every hour before actual calving events using this time series analysis and could thereby predict the time of calving. Similarly, video cameras or accelerometers recording cow behavior should be integrated in systems using image analysis [32–34].

4.2. Discussion on Proposed Method

We have tested the proposed calving time prediction model using video data for pregnant dairy cows. The results are shown in Table 4. This table provides both predicted times and the actual calving times and also shows the results of using our method on data collected over a period of just 48 h. The majority of these predictions were accurate within a range of3 h. These results show great promise for practical applications in managing precision dairy farms. The results also reveal that prediction times and actual times were almost the same. These results indicate that only two days of data are needed for accurately predicting calving time. The average value of mean absolute error (MAE) for these calving time predictions is 1.101 using data collected over 72 h, and 1.229 using data collected over 48 h. We also compared our proposed method with some machine learning techniques such as K-nearest neighbors (KNN), Naïve Bayes (NB) and Support Vector Machine (SVM) by blindly testing on five cows, as shown in Table 5 and Figure 8. For the machine learning techniques, the four types of conditions comprise two postures (L, S) and two transitions (LS, SL), which are defined as four predictors. Calving and not-calving states are considered the two responses. For each cow, the calving state is defined as the response in the last 3 h before calving. During the other 69 h, the response is the not-calving state. According to the predicted calving time results of Table 4, our proposed method can accurately estimate every calving event of each cow between 69 and 73 h before the event.

Table 4. Experimental results of cows predicted calving time based on 72 h data and 48 h data before calving event occurs.

Cow ID	Predicted Calving Time on 72 h	Predicted Calving Time on 48 h
1	70.723	70.843
2	68.942	70.852
3	69.874	70.321
4	73.765	71.728
5	68.128	70.721
6	71.568	70.716
7	71.993	71.059
8	71.298	70.597
9	70.734	70.511
10	72.338	70.495
11	69.541	69.756
12	72.013	70.919
13	71.310	70.961
14	71.297	71.069
15	70.229	71.772
16	71.969	71.807
17	72.420	70.381
18	72.455	71.723
19	71.346	71.561
20	70.435	70.311
21	73.081	70.307
22	71.274	70.159
23	72.592	70.465
24	72.405	69.730
25	70.889	70.510

Table 5. Performance analysis of our proposed method by comparing with other methods.

Methods	Precision	F1 Score	Specificity	Sensitivity	Accuracy (%)
Proposed Method	1	1	1	1	100
K-nearest Neighbors (KNN)	0.890	0.846	0.811	0.811	96.100
Naïve Bayes (NB)	0.965	0.965	0.965	0.965	98.333
Support Vector Machine (SVM)	0.767	0.843	0.990	0.990	96.389

From Figure 8 of the confusion matrices, the total number of observations is 360 in the testing dataset for 5 cows. In this dataset, the 2 classes are calving and not-calving, with a total of 15 calving responses and 345 not-calving responses. The best accuracy obtained using our proposed method is 100%.

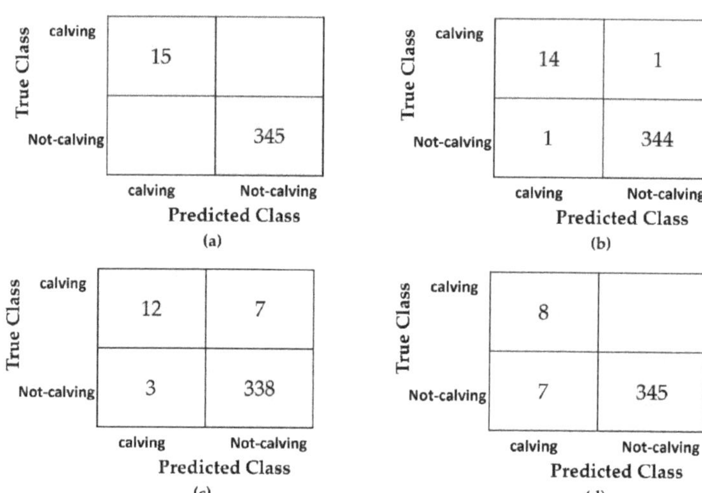

Figure 8. Confusion matrices of predicting calving events by using (**a**) Proposed Method, (**b**) NB, (**c**) KNN, and (**d**) SVM.

5. Conclusions

We have developed a five-state absorbing Markov chain model to predict calving events. Although a large number of Markov chain model applications have involved research fields such as engineering, medicine and agriculture, including livestock management and animal science, we have not seen a Markov chain application that predicts calving time. Our intention was to explore and examine how Markov Models could be applied to reproduction management for dairy cows by using them to predict calving time. In this study, we only considered four types of cow behavior. Additional activities such as head movements, rumination, and raising the tail should be considered in the future. In future research, we plan to analyze some of the above-mentioned activities using the proposed Markov model. In this paper we discuss a trial-and-error method of determining parameters for the absorbing state. However, the Monte Carlo Simulation method is also attractive as a way of determining these parameters. Much remains to be done in calving time prediction research. In the future, we will combine this stochastic model with image processing techniques to detect cows, automatically recognize their behavior, and build a better model for automatically predicting calving time.

Author Contributions: Conceptualization, S.Z.M., T.T.Z. and P.T.; methodology, S.Z.M., T.T.Z. and P.T.; software, S.Z.M.; validation, T.T.Z., P.T., I.K. and Y.H.; formal analysis, T.T.Z.; investigation, P.T.; resources, I.K. and Y.H.; data curation, S.Z.M.; writing—original draft preparation, P.T. and S.Z.M.; writing—review and editing, T.T.Z., P.T. and S.Z.M.; visualization, P.T.; supervision, T.T.Z. and P.T. All authors have read and agreed to the published version of the manuscript.

Funding: This work was supported in part by JSPS KAKENHI Grant Number 17K08066, SCOPE: Strategic and Communications R&D Promotion Program (Grant No. 172310006) and Honkawa Ranch Research Grant 2019-A-01.

Institutional Review Board Statement: Ethical review and approval were waived for this study, due to no enforced nor uncomfortable restriction to the animals during the study period. The image data of calving process used for analysis in this study were collected by an installed camera without disturbing natural parturient behaviour of animals and routine management of the farm.

Informed Consent Statement: Not applicable.

Data Availability Statement: The data presented in this study are available on request from the corresponding author. The data are not publicly available due to patent pending.

Acknowledgments: We thank K. Honkawa, D.V.M. for giving every convenience of the study in the ranch and his valuable advice.

Conflicts of Interest: The authors declare no conflict of interest.

References

1. Barrier, A.C.; Haskell, M.J. Calving difficulty in dairy cows has a longer effect on saleable milk yield than on estimated milk production. *J. Dairy Sci.* **2011**, *94*, 1804–1812. [CrossRef]
2. Mee, J.F. Why do so many calves die on modern dairy farms and what can we do about calf welfare in the future? *Animals* **2013**, *3*, 1036–1057. [CrossRef]
3. Fadul, M.; Bogdahn, C.; Alsaaod, M.; Hüsler, J.; Starke, A.; Steiner, A.; Hirsbrunner, G. Prediction of calving time in dairy cattle. *Anim. Reprod. Sci.* **2017**, *187*, 37–46. [CrossRef]
4. Bauer, J.W. The use of calving behaviours and automated activity monitors to predict and detect parturition and uterine diseases in Holstein cattle. Ph.D. Thesis, University of British Columbia, Vancouver, BC, Canada, 2020.
5. Helwatkar, A.; Riordan, D.; Walsh, J. Sensor technology for animal health monitoring. In Proceedings of the 8th International Conf. on Sensing Technology, Liverpool, UK, 2–4 September 2014; pp. 266–271.
6. Robichaud, M.V.; de Passillé, A.M.; Pearl, D.L.; LeBlanc, S.J.; Godden, S.M.; Pellerin, D.; Vasseur, E.; Rushen, J.; Haley, D.B. Calving management practices on Canadian dairy farms: Prevalence of practices. *J. Dairy Sci.* **2016**, *99*, 2391–2404. [CrossRef]
7. Zehner, N.; Niederhauser, J.J.; Schick, M.; Umstatter, C. Development and validation of a predictive model for calving time based on sensor measurements of ingestive behavior in dairy cows. *Comput. Electron. Agric.* **2019**, *161*, 62–71. [CrossRef]
8. Higaki, S.; Koyama, K.; Sasaki, Y.; Abe, K.; Honkawa, K.; Horii, Y.; Minamino, T. Calving prediction in dairy cattle based on continuous measurements of ventral tail base skin temperature using supervised machine learning. *J. Dairy Sci.* **2020**, *103*, 8535–8540. [CrossRef] [PubMed]
9. Tolver, A. *An Introduction to Markov Chains*, 2nd ed.; University of Copenhagen: Copenhagen, Denmark, 2017; p. 192.
10. Huang, W.T.; Lu, C.C. An enhanced absorbing Markov chain model for predicting TAIEX Index Futures. *Commun. Stat. Theory Methods* **2018**, *47*, 133–146. [CrossRef]
11. Park, S.; Vasudev, V. Predicting Web user's behavior: An absorbing Markov chain approach. In *Workshop on E-Business*; Springer: Cham, Switzerland, 2016; pp. 170–176.
12. Ledwith, M.C. Application of absorbing Markov chains to the assessment of education attainment rates within air force materiel command civilian personnel. In *Technological Report*; Air Force Institute of Technology: Base, OH, USA, 2019.
13. Dutta, A.; Zisserman, A. The VIA annotation software for images, audio and video. In Proceedings of the 27th ACM International Conference on Multimedia, Nice, France, 21–25 October 2019; pp. 2276–2279.
14. Phatarfod, R.M. Application of methods in sequential analysis to dam theory. *Ann. Math. Stat.* **1963**, *34*, 1588–1592. [CrossRef]
15. Borchers, M.R.; Chang, Y.M.; Proudfoot, K.L.; Wadsworth, B.A.; Stone, A.E.; Bewley, J.M. Machine-learning-based calving prediction from activity, lying, and ruminating behaviors in dairy cattle. *J. Dairy Sci.* **2017**, *100*, 5664–5674. [CrossRef] [PubMed]
16. Cangar, O.; Leroy, T.; Guarino, M.; Vranken, E.; Fallon, R.; Lenehan, J.; Mee, J. Model-Based Monitoring of Behaviour of Pregnant Cows Prior to Calving Using Online Image Analysis. Available online: https://limo.libis.be/primo-explore/fulldisplay?docid=LIRIAS1717073&context=L&vid=Lirias&search_scope=Lirias&tab=default_tab&lang=en_US&fromSitemap=1 (accessed on 20 September 2021).
17. Speroni, M.; Malacarne, M.; Righi, F.; Franceschi, P.; Summer, A. Increasing of posture changes as indicator of imminent calving in dairy cows. *Agriculture* **2018**, *8*, 182. [CrossRef]
18. Saint-Dizier, M.; Chastant-Maillard, S. Methods and on-farm devices to predict calving time in cattle. *Vet. J.* **2015**, *205*, 349–356. [CrossRef]
19. Fenlon, C.; O'Grady, L.; Mee, J.F.; Butler, S.T.; Doherty, M.L.; Dunnion, J. A comparison of 4 predictive models of calving assistance and difficulty in dairy heifers and cows. *J. Dairy Sci.* **2017**, *100*, 9746–9758. [CrossRef] [PubMed]
20. Shah, K.D.; Nakao, T.; Kubota, H. Plasma estrone sulphate (E1S) and estradiol-17β (E2β) profiles during pregnancy and their relationship with the relaxation of sacrosciatic ligament, and prediction of calving time in Holstein–Friesian cattle. *Anim. Reprod. Sci.* **2006**, *95*, 38–53. [CrossRef] [PubMed]
21. Burfeind, O.; Suthar, V.S.; Voigtsberger, R.; Bonk, S.; Heuwieser, W. Validity of prepartum changes in vaginal and rectal temperature to predict calving in dairy cows. *J. Dairy Sci.* **2011**, *94*, 5053–5061. [CrossRef] [PubMed]
22. Matsas, D.J.; Nebel, R.L.; Pelzer, K.D. Evaluation of an on-farm blood progesterone test for predicting the day of parturition in cattle. *Theriogenology* **1992**, *37*, 859–868. [CrossRef]
23. Streyl, D.; Sauter-Louis, C.; Braunert, A.; Lange, D.; Weber, F.; Zerbe, H. Establishment of a standard operating procedure for predicting the time of calving in cattle. *J. Vet. Sci.* **2011**, *12*, 177–185. [CrossRef]
24. Rutten, C.J.; Kamphuis, C.; Hogeveen, H.; Huijps, K.; Nielen, M.; Steeneveld, W. Sensor data on cow activity, rumination, and ear temperature improve prediction of the start of calving in dairy cows. *Comput. Electron. Agric.* **2017**, *132*, 108–118. [CrossRef]
25. Miedema, H.M.; Cockram, M.S.; Dwyer, C.M.; Macrae, A.I. Changes in the behaviour of dairy cows during the 24 h before normal calving compared with behaviour during late pregnancy. *Appl. Anim. Behav. Sci.* **2011**, *131*, 8–14. [CrossRef]
26. Proudfoot, K.L.; Jensen, M.B.; Weary, D.M.; Von Keyserlingk, M.A.G. Dairy cows seek isolation at calving and when ill. *J. Dairy Sci.* **2014**, *97*, 2731–2739. [CrossRef] [PubMed]

27. Santegoeds, O.J. Predicting Dairy Cow Parturition Using Real-Time Behavior Data from Accelerometers: A Study in Commercial Setting. Available online: https://www.semanticscholar.org/paper/Predicting-dairy-cow-parturition-using-real-time-A-Santegoeds/b09f439591083b4175dcd5027ca960d452af9aab, (accessed on 20 September 2021).
28. Jensen, M.B. Behaviour around the time of calving in dairy cows. *Appl. Anim. Behav. Sci.* **2012**, *139*, 195–202. [CrossRef]
29. Zin, T.T.; Sumi, K.; Tin, P. Time to dairy cow calving event prediction by using time series analysis. In Proceedings of the 12th International Conference on Computer Modeling and Simulation, Brisbane, Australia, 22–24 June 2020; pp. 143–146.
30. Zin, T.T.; Tin, P.; Hama, H. Some aspects of mathematical modeling techniques in dairy science. In *International Workshop on Frontiers of Computer Vision*; Springer Link: Ibusuki, Japan, 2020.
31. Sumi, K.; Maw, S.Z.; Zin, T.T.; Tin, P.; Kobayashi, I.; Horii, Y. Activity-Integrated Hidden Markov Model to Predict Calving Time. *Animals* **2021**, *11*, 385. [CrossRef]
32. Sumi, K.; Zin, T.T.; Kobayashi, I.; Horii, Y. A study on cow monitoring system for calving process. In Proceedings of the IEEE 6th Global Conference. on Consumer Electronics, Nagoya, Japan, 24–27 October 2017; pp. 379–380.
33. Sumi, K.; Zin, T.T.; Kobayashi, I.; Horii, Y. Framework of cow calving monitoring system using a single depth camera. In Proceedings of the International Conference on Image and Vision Computing New Zealand (IVCNZ), Auckland, New Zealand, 19–21 November 2018; pp. 1–7.
34. Maw, S.Z.; Zin, T.T.; Tin, P. Image processing and statistical analysis approach to predict calving time in dairy cows. In Proceedings of the IEEE 9th Global Conference on Consumer Electronics, Kobe, Japan, 15–16 October 2020; pp. 373–374.

Article

Cross-Modality Interaction Network for Equine Activity Recognition Using Imbalanced Multi-Modal Data †

Axiu Mao [1], Endai Huang [2], Haiming Gan [1,3], Rebecca S. V. Parkes [4,5], Weitao Xu [2] and Kai Liu [1,6,*]

1 Department of Infectious Diseases and Public Health, Jockey Club College of Veterinary Medicine and Life Sciences, City University of Hong Kong, Hong Kong, China; axmao2-c@my.cityu.edu.hk (A.M.); haimigan@cityu.edu.hk (H.G.)
2 Department of Computer Science, City University of Hong Kong, Hong Kong, China; edhuang2-c@my.cityu.edu.hk (E.H.); weitaoxu@cityu.edu.hk (W.X.)
3 College of Electronic Engineering, South China Agricultural University, Guangzhou 510642, China
4 Department of Veterinary Clinical Sciences, Jockey Club College of Veterinary Medicine and Life Sciences, City University of Hong Kong, Hong Kong, China; reparkes@cityu.edu.hk
5 Centre for Companion Animal Health, Jockey Club College of Veterinary Medicine and Life Sciences, City University of Hong Kong, Hong Kong, China
6 Animal Health Research Centre, Chengdu Research Institute, City University of Hong Kong, Chengdu 610000, China
* Correspondence: kailiu@cityu.edu.hk
† This manuscript is an extension version of the conference paper: Mao, A.X.; Huang, E.D.; Xu, W.T.; Liu, K. Cross-modality Interaction Network for Equine Activity Recognition Using Time-Series Motion Data. In Proceedings of the 2021 International Symposium on Animal Environment and Welfare (ISAEW), Chongqing, China, 20–23 October 2021 (in press).

Abstract: With the recent advances in deep learning, wearable sensors have increasingly been used in automated animal activity recognition. However, there are two major challenges in improving recognition performance—multi-modal feature fusion and imbalanced data modeling. In this study, to improve classification performance for equine activities while tackling these two challenges, we developed a cross-modality interaction network (CMI-Net) involving a dual convolution neural network architecture and a cross-modality interaction module (CMIM). The CMIM adaptively recalibrated the temporal- and axis-wise features in each modality by leveraging multi-modal information to achieve deep intermodality interaction. A class-balanced (CB) focal loss was adopted to supervise the training of CMI-Net to alleviate the class imbalance problem. Motion data was acquired from six neck-attached inertial measurement units from six horses. The CMI-Net was trained and verified with leave-one-out cross-validation. The results demonstrated that our CMI-Net outperformed the existing algorithms with high precision (79.74%), recall (79.57%), F1-score (79.02%), and accuracy (93.37%). The adoption of CB focal loss improved the performance of CMI-Net, with increases of 2.76%, 4.16%, and 3.92% in precision, recall, and F1-score, respectively. In conclusion, CMI-Net and CB focal loss effectively enhanced the equine activity classification performance using imbalanced multi-modal sensor data.

Keywords: equine behavior; wearable sensor; deep learning; intermodality interaction; class-balanced focal loss

Citation: Mao, A.; Huang, E.; Gan, H.; Parkes, R.S.V.; Xu, W.; Liu, K. Cross-Modality Interaction Network for Equine Activity Recognition Using Imbalanced Multi-Modal Data. *Sensors* **2021**, *21*, 5818. https://doi.org/10.3390/s21175818

Academic Editors: Yongliang Qiao, Lilong Chai, Dongjian He and Daobilige Su

Received: 26 July 2021
Accepted: 27 August 2021
Published: 29 August 2021

1. Introduction

The behavior of horses provides rich insight into their mental and physical status and is one of the most important indicators of their health, welfare, and subjective state [1]. However, behavioral monitoring for animals, to date, largely relies on manual observations, which are labor-intensive, time-consuming, and prone to subjective judgments of individuals [1]. The use of sensors and machine learning is well-established in monitoring gait change [2], and for lameness detection as part of the equine veterinary examination, increasing the accuracy of identification of subtle lameness, which is one of the most expensive health issues in the equine industry [3,4]. Therefore it is of significant importance to investigate and develop an automatic, objective, accurate, and quantifiable measurement system for equine behaviors. Such a system will allow caretakers to identify variations in the animal behavioral repertoire in real-time, decreasing the workloads in veterinary clinics and improving the husbandry and management of animals [5,6].

Over recent decades, automated animal activity recognition has been studied widely with the aid of various sensors (e.g., accelerometers, gyroscopes, and magnetometers) and the use of machine learning techniques. For instance, a naïve Bayes (NB) classifier was applied to recognize horse activities (e.g., eating, standing, and trotting) using triaxial acceleration and obtained 90% classification accuracy [7]. Four classifiers including a linear discriminant analysis (LDA), a quadratic discriminant analysis (QDA), a support vector machine (SVM), and a decision tree (DT) were utilized to detect dog behaviors (e.g., galloping, lying on chest, and sniffing) based on accelerometer and gyroscope data, and the results revealed that the sensor placed on the back and collar yielded 91% and 75% accuracy at best, respectively [8]. A random forest (RF) algorithm was applied to categorize cow activities using triaxial acceleration and gained high classification accuracy with 91.4%, 99.8%, 88%, and 99.8% for feeding, lying, standing, and walking events, respectively [9]. In horses, the use of receiver-operating characteristic curve analysis classified standing, grazing, and ambulatory activities with a sensitivity of 94.7–97.7% and a specificity of 94.7–96.8% [10]. However, to classify animal behaviors accurately using these machine learning methods, feature extraction and method selection are often conducted manually and separately, which requires expert domain knowledge and easily induces feature engineering issues [11]. Moreover, handcrafted features often fail to capture general and complex features, resulting in low generalization ability, i.e., these extracted features perform well in recognizing the activities of some subjects but badly for others.

Along with the recent advances in internet technology and fast graphics processing units, various deep learning approaches have been increasingly and successfully adopted in animal activity recognition with wearable sensors. Classification models based on deep learning achieve automatic feature learning through data driving and subsequent animal activity recognition. For example, feed-forward neural networks (FNNs) and long short-term memory (LSTM) models were applied to automatically recognize cattle behaviors (e.g., feeding, lying, and ruminating) using data collected from inertial measurement units (IMUs) [12,13]. Convolutional neural networks (CNNs), which accurately capture local temporal dependency and scale invariance in signals, were developed in automated equine activity classification based on triaxial accelerometer and gyroscope data [1,14,15]. FilterNet, presented based on CNN and LSTM architectures, was adopted to classify important health-related canine behaviors (e.g., drinking, eating, and scratching) using a collar-mounted accelerometer [16].

However, multi-modal data fusion has not been well handled when different sensors are used simultaneously in existing studies. Multi-modal data with different characteristics are often simply processed using common fusion strategies such as early fusion, feature fusion, and result fusion [17]. The early fusion strategy used in previous studies [12,13], i.e., extracting the same features without distinction of modalities, often caused interference between multi-modal information due to their distribution gap [18]. The result fusion scheme was suboptimal since rich modality information was gradually compressed and lost in separate processes, ignoring the intermodality correlations. As a better choice, the feature fusion strategy fuses the intermediate information of multiple modalities, which avoids the distribution gap problem and achieves intermodality interaction simultaneously [19,20]. However, feature fusion is often limited to linear fusion (e.g., simple concatenation and addition) and fails to explore deep multi-modality interactions and achieve complementary-redundant information combinations between multiple modalities [17].

In addition, the collected sensor datasets often present class imbalance problems due to the inconsistent frequency and duration of each activity resulting from specific animal physiology. Deep learning methods trained on imbalanced datasets tend to be biased toward majority classes and away from minority classes, which easily causes poor modal generalization ability and high classification error rates for rare categories [21]. Commonly used methods on imbalanced datasets mainly involve two techniques, namely, resampling and reweighting. Resampling attempts to sample the data to obtain an evenly distributed dataset, e.g., oversampling and undersampling [22]. However, oversampling and undersampling come with high potential risks of overfitting and information loss, respectively [21]. Reweighting is more flexible and convenient by directly assigning a weight for the loss function per training sample to alleviate the sensitivity of the model to data distribution [23]. This method is further divided into class-level and sample-level reweighting. The former, such as cost-sensitive (CS) loss [24] and class-balanced (CB) loss [25], depends on the prior category frequency, while the latter, such as focal loss [26] and adaptive class suppression (ACS) loss [27], relies on the network output confidences of each instance. In addition, CB focal loss, combining a CB term with a modulating factor, effectively focuses on difficult samples and considers the proportional impact of effective numbers per class simultaneously [25].

To improve the recognition performance for equine activities while tackling the above-mentioned challenges, we have developed a cross-modality interaction network (CMI-Net) which achieved a good classification performance in our previous work [28], and a CB focal loss [25] was adopted to supervise the training of CMI-Net. The CMI-Net consisted of a dual CNN trunk architecture and a joint cross-modality interaction module (CMIM). Specifically, the dual CNN trunk architecture extracted modality-specific features for accelerometer and gyroscope data, respectively, and the CMIM based on attention mechanism adaptively recalibrated the importance of the elements in the two modality-specific feature maps by leveraging multi-modal knowledge. The attention mechanism has been widely utilized in different tasks using multi-modal datasets such as RGB-D images [17,29]. It has also been adopted to focus on important elements along with channels and spatial dimensions of the same input feature [30,31]. The favorable performance presented in these studies with the attention mechanism indicated the rationality of our proposed CMIM. In our method, softmax cross-entropy (CE) loss was initially used to supervise the training of CMI-Net. However, softmax CE loss suffered from inferior classification performance, especially for monitory classes [23]. In contrast, CB focal loss, by adding a CB term to focal loss, focuses more on minor-class samples and hard-classified samples and can alleviate the class imbalance problem. Therefore, a CB focal loss [25] was also adopted. In this study, the CMI-Net was trained based on an extensively labeled dataset [32] to automatically recognize equine activities including eating, standing, trotting, galloping, walking-rider (walking while carrying a rider), and walking-natural (walking with no rider). The leave-one-out cross-validation (LOOCV) method was applied to test the generalization ability of our model, and the results were then compared to the existing algorithms. The main contributions of this paper can be summarized as follows:

- We proposed a CMI-Net involving a dual CNN trunk architecture and a joint CMIM to improve equine activity recognition performance using accelerometer and gyroscope data. The dual CNN trunk architecture comprised a residual-like convolution block (Res-LCB) which effectively promoted the representation ability and robustness of the model [33]. The CMIM based on attention mechanism enabled CMI-Net to capture complementary information and suppressed unrelated information (e.g., noise, redundant signals, and potentially confusing signals) from multi-modal data.
- We devised a novel attention module, i.e., CMIM, to achieve deep intermodality interaction. The CMIM combined spatial information from two-stream feature maps using basic CNN to produce two spatial attention maps with respect to their importance, which could adaptively recalibrate temporal- and axis-wise features in each modality. To the best of our knowledge, the attention mechanism was employed for the first time in animal activity recognition based on multi-modal data yielded by multiple wearable sensors.

- We adopted a CB focal loss to supervise the training of CMI-Net to mitigate the influence of imbalanced datasets on overall classification performance. The CB focal loss can pay more attention not only to samples of minority classes, diminishing their influence from being overwhelmed during optimization, but also to samples that are hard to distinguish. As far as we know, this is the first time the CB focal loss has been utilized in animal activity recognition based on imbalanced datasets.
- Experiments performed verified the effectiveness of our proposed CMI-Net and CB focal loss. In particular, the experimental results demonstrated that our CMI-Net outperformed the existing algorithms in equine activity recognition with the precision of 79.74%, recall of 79.57%, F1-score of 79.02%, and accuracy of 93.37%, respectively.

2. Materials and Methods

2.1. Data Description

The dataset used in this study was a public dataset created by Kamminga et al. [32]. In this dataset, more than 1.2 million 2 s data samples were collected from 18 individual equines using neck-attached IMUs. The sampling rate was set to 100 Hz for both the triaxial accelerometer and gyroscope and 12 Hz for the triaxial magnetometer. The majority of the samples were unlabeled, but data from six equines and six activities including eating, standing, trotting, galloping, walking-rider, and walking-natural were labeled extensively (87,621 2 s samples in total) and were used to classify equine activities in previous studies [7,34]. In this study, data from the triaxial accelerometer and gyroscope among the 87,621 samples were exploited separately, forming up to two tensors with a size of $1 \times 3 \times 200$ for each sample. As demonstrated in Figure 1, the activities of eating, standing, trotting, galloping, walking-rider, and walking-natural occupied 18.32%, 5.84%, 28.62%, 4.50%, 38.94%, and 3.80% of the total sample number, respectively, producing a maximum imbalance ratio of 10.25. In addition, the input sample of each axis per sensor modality was normalized by removing the mean and scaling to unit variance, which can be formulated as follows:

$$S_i = \frac{S_i - \mu_i}{\sigma_i}, \quad (1)$$

where S_i denotes all samples of a particular axis per sensor modality (i.e., X-, Y-, and Z-axis of the accelerometer, and X-, Y-, and Z-axis of the gyroscope), S_i denotes all normalized samples, and μ_i and σ_i denote mean and standard deviation values in each axis per sensor modality, respectively.

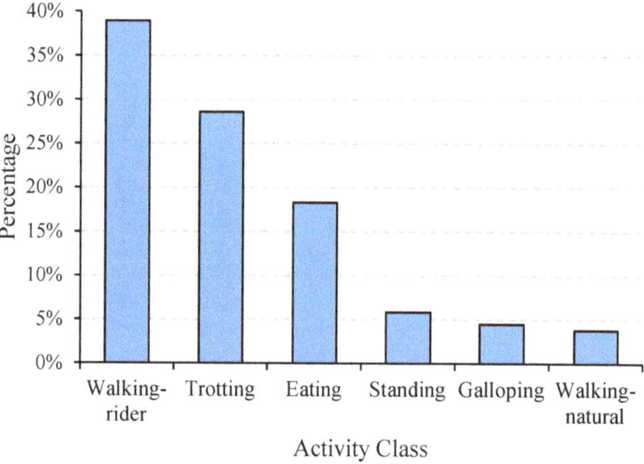

Figure 1. Histogram of class distribution.

2.2. Cross-Modality Interaction Network

Our proposed CMI-Net, where accelerometer and gyroscope data were fed into two CNN branches (represented by CNN_{acc} and CNN_{gyr}) separately, is shown in Figure 2a. The dual CNN was constructed to extract modality-specific features and concatenate these features before the final dense layer. To achieve deep interaction between the two-modality data and capture the complementary information and suppress unrelated information from them, a joint CMIM was designed and inserted in the upper layer. The details are described below.

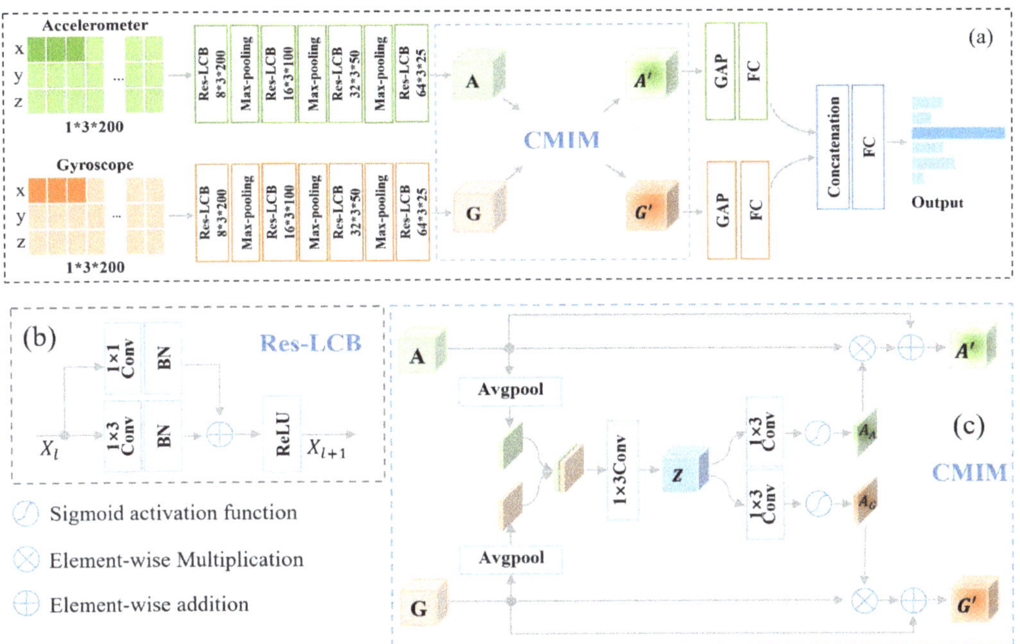

Figure 2. The architecture of our proposed cross-modality interaction network (CMI-Net). (**a**) Our proposed CMI-Net. The size of the feature maps is marked after every residual-like convolution block (Res-LCB) layer. Here, "A" and "G" denote the modality-specific features for the accelerometer and gyroscope, respectively, and "A'" and "G'" denote the refined features after modality interaction. "GAP" and "FC" are the global average-pooling layer and fully connected layer, respectively. (**b**) Res-LCB and (**c**) cross-modality interaction module (CMIM).

2.2.1. Dual CNN Trunk Architecture

The CNN_{acc} and CNN_{gyr} contained four convolution blocks, three max-pooling layers, one global average-pooling layer, and one fully connected layer, followed by concatenation and one joint fully connected layer. Inspired by the residual unit in the deep residual network that behaves like ensembles and has smaller magnitudes of responses [33], to promote the representation ability and robustness of the model, we designed a Res-LCB, as demonstrated in Figure 2b. The definition is given below.

$$X_{l+1} = RELU\left(Conv^{1\times1}(X_l) \oplus Conv^{1\times3}(X_l)\right), \tag{2}$$

where X_l and X_{l+1} denote feature maps in the l and $l+1$ layers, respectively, $Conv^{1\times1}(\bullet)$ and $Conv^{1\times3}(\bullet)$ represent 1×1 and 1×3 convolution operations, respectively, $\oplus$ denotes the elementwise addition, and $RELU\ (\bullet)$ denotes the rectified linear unit activation function [35].

2.2.2. Cross-Modality Interaction Module

Inspired by the multi-modal transfer module that recalibrates channel-wise features of each modality based on multi-modal information [36] and the convolutional block attention module that focuses on the spatial information of the feature maps [30], we devised a CMIM based on an attention mechanism to adaptively recalibrate temporal- and axis-wise features in each modality by utilizing multi-modal information. The detailed CMIM is illustrated in Figure 2c.

Let $A \in R^{C\times H\times W}$ and $G \in R^{C\times H\times W}$ represent the features at a given layer of CNN_{acc} and CNN_{gyr}, respectively. Here, C, H, and W denote the channel number and spatial dimensions of features. Specifically, H and W correspond to the axial and temporal signals, respectively. The CMIM receives A and G as input features. We first applied average-pooling operations along channels of the input features, generating two spatial maps. These two maps were then concatenated and mapped into a joint representation $Z \in R^{C'\times H\times W}$. The operation was shown as follows:

$$Z = RELU\left(Conv^{1\times 3}([Avgpool(A),\ Avgpool(G)])\right), \tag{3}$$

where C' denotes the channel number of feature Z, *Avgpool* (•) denotes the average-pooling operation, and [•] denotes the concatenation operation. Furthermore, two spatial attention maps $A_A \in R^{1\times H\times W}$ and $A_G \in R^{1\times H\times W}$ were generated through two independent convolution layers with a sigmoid function $\sigma(\bullet)$ using the joint representation Z:

$$A_A = \sigma\left(Conv^{1\times 3}(Z)\right),\ A_G = \sigma\left(Conv^{1\times 3}(Z)\right), \tag{4}$$

A_A and A_G were then used to recalibrate the input features, generating two final refined features, i.e., $A' \in R^{C\times H\times W}$ and $G' \in R^{C\times H\times W}$:

$$A' = A \otimes A_A \oplus A,\ G' = G \otimes A_G \oplus G, \tag{5}$$

where $\otimes$ denotes the elementwise multiplication. Specifically, each convolution operation under this study was followed by a batch normalization operation. The increases in channel numbers and decreases in spatial dimensions were implemented through Res-LCB and max-pooling operations, respectively.

2.3. Optimization

As the most widely utilized loss in the multiclass classification task, softmax CE loss was applied to optimize the parameters of CMI-Net. The formulation of softmax CE loss was defined as

$$L_{CE}(z) = -\sum_{i=1}^{C} y_i log(p_i) \tag{6}$$

$$\text{with } p_i = \frac{e^{z_i}}{\sum_{j=1}^{C} e^{z_j}}, \tag{7}$$

where C and $z = [z_1, \ldots, z_C]$ are the total number of classes and the predicted logits of the network, respectively. In addition, $y_i \ \ldots\ \{0,1\}, 1 \le i \le C$ is the one-hot ground-truth label. However, the models based on softmax CE loss often suffer from inferior classification performance, especially for monitory classes, due to the imbalanced data distribution [23]. Therefore, we further introduced an effective loss function to supervise the training of CMI-Net and alleviate the class imbalance problem, namely, CB focal loss.

CB focal loss, which added the CB term to the focal loss function, focused more on not only samples of minority classes, diminishing their influence from being overwhelmed during optimization, but also samples that were hard to distinguish. The CB term was related to the inverse effective number of samples per class, and focal loss added a modulating factor to the sigmoid CE loss to reduce the relative loss for well-classified samples and focused more on difficult samples. The CB focal loss was presented as

$$L_{CB_{FL}}(z) = \frac{1}{E_{n_y}} L_{FL}(z) = -\frac{1-\beta}{1-\beta^{n_y}} \sum_{i=1}^{C} \left(1-p_i^t\right)^{\gamma} log\left(p_i^t\right) \quad (8)$$

$$\text{with } p_i^t = \frac{1}{1+e^{-z_i^t}}, \quad (9)$$

$$z_i^t = \begin{cases} z_i, & if\ i = y. \\ -z_i, & \text{otherwise.} \end{cases}, \quad (10)$$

where n_y and E_{n_y} represent the actual number and the effective number of the ground-truth label y, respectively. The hyperparameter $\beta \in [0, 1)$ controlled how fast E_{n_y} grows as n_y increases, and $\gamma \geq 0$ smoothly adjusted the rate at which easy samples were down-weighted [26]. The value of β was set to 0.9999, and the search space of the hyperparameter γ was set to {0.5, 1.0, 2.0} [25] in this study. In particular, CB loss and focal loss rebalanced the loss function based on class-level and sample-level reweighting, respectively. Thus, we also utilized class-level reweighted losses, including cost-sensitive cross-entropy loss (CS_CE loss) [24], class-balanced cross-entropy loss (CB_CE loss) [25], and sample-level reweighted losses, including focal loss [26] and adaptive class suppression loss (ACS loss) [27], to validate the effectiveness of the CB focal loss.

2.4. Evaluation Metrics

The comprehensive performance of the equine activity classification model was indicated by the following four evaluation metrics, which are defined in Equations (11)–(14). Each indicator value was multiplied by 100 as the result to reflect the difference in indicator values more clearly.

$$\text{Precision} = \frac{\text{TP}}{\text{TP} + \text{FP}}, \quad (11)$$

$$\text{Recall} = \frac{\text{TP}}{\text{TP} + \text{FN}}, \quad (12)$$

$$\text{F1} - \text{Score} = \frac{2\text{TP}}{2\text{TP} + \text{FP} + \text{FN}}, \quad (13)$$

$$\text{Accuracy} = \frac{\text{TP} + \text{TN}}{\text{TP} + \text{TN} + \text{FP} + \text{FN}}, \quad (14)$$

where TP, FP, TN, and FN are the number of true positives, false positives, true negatives, and false negatives, respectively. In particular, the overall precision, recall, and F1-score were calculated by using a macro-average [37].

2.5. Implementation Details

To attain subject-dependent results, the LOOCV method was used, in which four subjects were chosen for training, one for validation, and one for testing each time and rotated in a circular manner. During training, the loss function was added by an L2 regularization term with a weight decay of 0.1 to avoid overfitting. An Adam optimizer with an initial learning rate of 1×10^{-4} was employed, and the learning rate decreased by 0.1 times every 20 epochs. The number of epochs and batch size were set to 100 and 256, respectively. The best model with the highest validation accuracy was saved and verified using test data. To evaluate the classification performance of our CMI-Net, we compared it against various existing methods, including three machine learning methods (i.e., NB, DT, and SVM) and two deep learning methods used in equine activity recognition (i.e., CNN and ConvNet7) [14,15], based on the same public dataset. Specifically, the hand-crafted features used in machine learning were the same as those used by Kamminga et al. [7]. To further explore the performance of our CMIM, we ran the network without CMIM and with it inserted after the 1st, 2nd, and 3rd max-pooling layers to obtain four different variants, i.e., Variant0, Variant1, Variant2, and Variant3, respectively. The softmax CE loss was used as the loss function for all variants. All experiments were executed using the PyTorch framework on an NVIDIA Tesla V100 GPU. The developed source code will be available at https://github.com/Max-1234-hub/CMI-Net from 1 September 2021.

3. Results and Discussion

Overall, experiments conducted on the public dataset demonstrated that our proposed CMI-Net outperformed the existing algorithms. Ablation studies were then carried out to verify the effectiveness of CMIM and that applying the CMIM in the upper layer of CMI-Net could obtain better performance. Different loss functions were adopted to validate that CB focal loss performed better than any class-level or sample-level reweighted loss used alone, and it effectively improved the overall precision, recall, and F1-score, although the overall accuracy decreased due to the imbalanced dataset used. Furthermore, recognition performance analysis was presented to help us probe the predicted performance on each activity using our CMI-Net with CB focal loss. The details are described as follows.

3.1. Comparison with Existing Methods

The comparison results of our CMI-Net with three machine learning methods (i.e., NB, DT, and SVM) and two deep learning methods (i.e., CNN and ConvNet7) [14,15] are illustrated in Table 1. The results revealed that the CMI-Net with softmax CE loss outperformed the machine learning algorithms with higher precision, recall, F1-score, and accuracy of 79.74%, 79.57%, 79.02%, and 93.37%, respectively. The reason for this superior performance was the convolution and pooling operations in CNN, which could achieve automated feature learning and aggregate more complex and general patterns without any domain knowledge [38]. The other CNN-based method [15] obtained inferior precision of 72.07% and accuracy of 82.94% compared to DT and SVM. This result is consistent with the "No Free Lunch" theorem [39] because this CNN-based method [15] was developed using leg-mounted sensor data. In addition, our CMI-Net with softmax CE loss performed better than ConvNet7 [14], which obtained lower precision, recall, F1-score, and accuracy of 79.03%, 77.79%, 77.90%, and 91.27%, respectively. This was attributed to the ability of our architecture to effectively capture the complementary information and inhibit unrelated information of multi-modal data through deep multi-modality interaction. In addition, CMI-Net with CB focal loss (γ = 0.5) enabled the values of precision, recall, and F1-score to increase by 2.76%, 4.16%, and 3.92%, respectively, compared with CMI-Net with softmax CE loss. This revealed that the adoption of CB focal loss effectively improved the overall classification performance.

Table 1. Classification performance comparison with existing methods. The best two results for each metric are highlighted in bold.

Methods	Precision (%)	Recall (%)	F1-Score (%)	Accuracy (%)
Machine learning				
Naïve Bayes	70.90	72.41	69.42	76.60
Decision tree	75.67	73.90	74.35	88.83
Support vector machine	73.92	71.30	72.19	89.65
Deep learning				
CNN [15]	72.07	76.91	73.42	82.94
ConvNet7 [14]	79.03	77.79	77.90	**91.27**
Our methods #				
CMI-Net + softmax CE loss	**79.74**	**79.57**	**79.02**	**93.37**
CMI-Net + CB focal loss (γ = 0.5) *	**82.50**	**83.73**	**82.94**	90.68

CMI-Net: cross-modality interaction network; CE: cross-entropy; CB: class-balanced; * the γ of value is 0.5, which could refer to Table 3.

3.2. Ablation Study

3.2.1. Evaluation of CMIM

To explore the effectiveness of CMIM and the impact of its position in the network on classification performance, the results corresponding to four different variants are shown in Table 2. Our proposed CMI-Net with softmax CE loss showed superior performance to Variant0 (i.e., the network without CMIM), indicating the effective performance of our interaction module. Variant1, Variant2, and Variant3 (i.e., networks with CMIM inserted after 1st, 2nd, and 3rd max-pooling layer, respectively) did not perform better in terms of precision and recall compared with Variant0, which obtained precision and recall values of 79.02% and 77.09%, respectively. This might be explained by the fact that modality-specific features learned in the shallow layer were simple and contained noise, which interfered with the process by which CMIM learned complex intermodality correlations, leading to poor predictions [40]. In addition, our architecture obtained the best performance since it applied the CMIM after a deeper layer, which enabled the network to discover more discriminative patterns and suppress irrelevant variations more effectively [41].

Table 2. Performance comparison of our CMI-Net with its variants. The best results for each metric are highlighted in bold.

Methods [&]	Precision (%)	Recall (%)	F1-Score (%)	Accuracy (%)
Variant0 [#]	79.02	77.09	76.88	91.76
Variant1 *	78.18	77.07	77.40	92.17
Variant2 *	77.50	78.44	77.91	92.92
Variant3 *	78.36	76.94	77.02	92.62
CMI-Net + softmax CE loss	**79.74**	**79.57**	**79.02**	**93.37**

[&] denotes all networks presented in this table were trained using softmax CE loss; [#] denotes the network without a cross-modality interaction module (CMIM); * denotes the network where the CMIM was inserted after the 1st, 2nd, and 3rd max-pooling layers, respectively.

The results above have proven that the inclusion of the CMIM in the network provided quantifiable improvements in identification performance. This was also reflected in the qualitative visualization of the embeddings and the corresponding clusters in Figure 3, with the help of t-distributed stochastic neighbor-embedding (t-SNE), a technique for visualizing high-dimensional data by giving each data point a location in a two- or three-dimensional map [42]. Figure 3 shows the two-dimensional embedded features from the part test dataset after the fully connected layers of both CNN branches under the network without and with CMIM by using the t-SNE technique with an init of 'pca' and perplexity of 30. Comparing the left and right columns in Figure 3, it can be observed that more compact clusters were generated under the network with CMIM by reducing the intraclass distance and enlarging the interclass distance. The core technical point was that the joint interaction module enabled adaptive amplification of salient features and suppression of unrelated features based on information from two-modality data. To further provide insights into its contribution, we presented two spatial attention maps for features extracted from the triaxial accelerometer and triaxial gyroscope data (Figure 4). As illustrated in Figure 4, the value per pixel represented the contribution degree corresponding to each temporal period and each axis, and it was adaptively recalibrated through intermodality interaction. Therefore, both quantitative and qualitative findings reinforced the suitability of our proposed CMI-Net to tasks using two-modality sensor data.

3.2.2. Evaluation of CB Focal Loss

To study the effect of CB focal loss on the optimization of CMI-Net, we show the quantitative performance in Table 3 and explore the sensitivity of its hyperparameter γ. CMI-Net with CB focal loss (γ = 0.5) achieved the best precision of 82.50%, recall of 83.73%, and F1-score of 82.94%. This indicated that CB focal loss was beneficial to the improvement of classification performance when the modulation strength was controlled appropriately, whereas negative effects occurred if the value of γ was too large or too small.

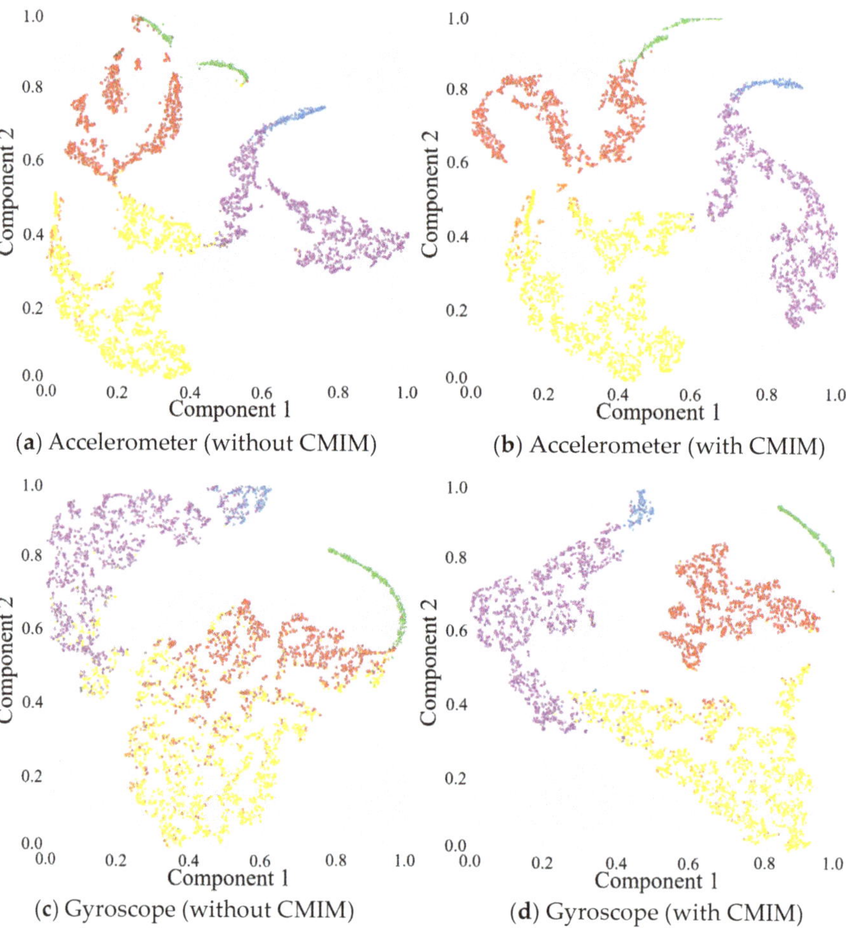

Figure 3. Embedding visualization of the features extracted from triaxial accelerometer and gyroscope data under network without and with CMIM, respectively.

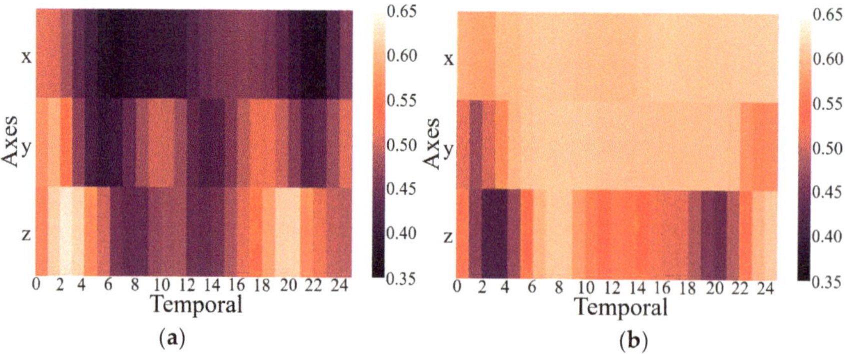

Figure 4. Attention maps for features extracted from the triaxial accelerometer (**a**) and gyroscope (**b**) data.

Table 3. Performance comparison between softmax CE loss and CB focal loss with different γ. The best results for each metric are highlighted in bold.

Loss Functions	Precision (%)	Recall (%)	F1-Score (%)	Accuracy (%)
Softmax CE Loss (baseline)	79.74	79.57	79.02	**93.37**
CB focal loss (γ = 0.1)	81.31	83.60	81.97	89.57
CB focal loss (γ = 0.5)	**82.50**	**83.73**	**82.94**	90.68
CB focal loss (γ = 1)	80.42	82.03	81.05	89.89
CB focal loss (γ = 2)	78.92	78.48	77.97	91.05

To provide further insight into the influence of CB focal loss (γ = 0.5) on the classification performance, we present the classification results of each activity under CMI-Net with CB focal loss and softmax CE loss, respectively, in Figure 5. It shows that precision, recall, and F1-score of the walking-natural were significantly improved, while other activities varied slightly when using CB focal loss. This explained that the overall classification performance increased mainly due to the increase in walking-natural, as it focused more on difficult samples and samples of minority classes. However, the overall accuracy of CMI-Net with CB focal loss decreased by 2.69% (Table 3), which was related to the different variations of recall values in different activities and the current imbalanced dataset. In particular, the overall accuracy could also be presented as the weighted average of the recall value for each activity according to the sampling frequency of each activity. As shown in Figure 5, the recall increases were 35.92% for walking-natural, 1.17% for standing, and 0.91% for galloping, and the recall decreases were 8.41% for walking-rider, 4.26% for eating, and 0.36% for trotting when using CB focal loss. It can be observed that all activities with increased recall belonged to the minority class, while the remaining activities with decreased recall belonged to the majority class, resulting in a decrease in overall accuracy. Thus, it is necessary to collect a more balanced dataset in the future.

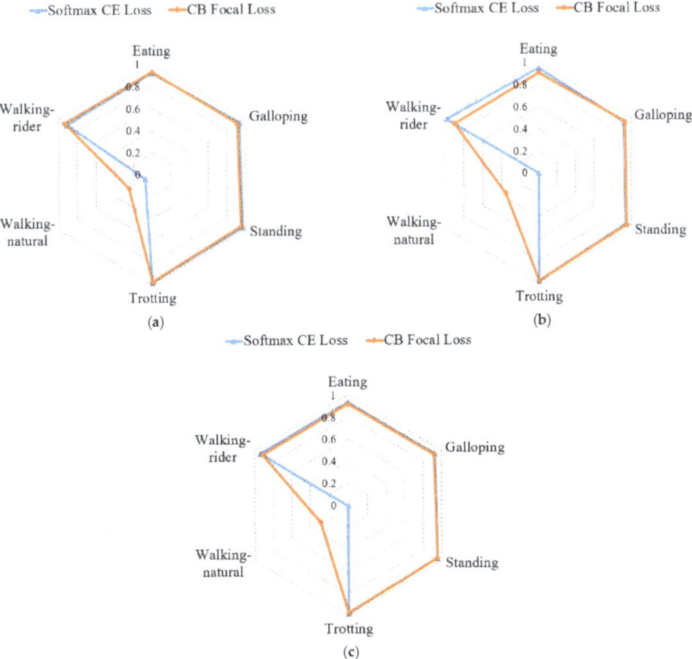

Figure 5. Precision (**a**), recall (**b**), and F1-score (**c**) comparison of each activity under softmax cross-entropy (CE) loss and class-balanced (CB) focal loss.

In addition, experiments under different loss functions were conducted to verify the effectiveness of the CB focal loss, as illustrated in Table 4. The contrasting losses mainly included CS_CE loss, CB_CE loss, focal loss, and ACS loss, as mentioned in the "Optimization" section. We found that CB focal loss combining CB loss and focal loss performed better than any of them used alone, which indicated that adding the CB term to the focal loss function improved the overall classification performance on the imbalanced dataset. In addition, the precision, recall, and F1-score of CS_CE loss and CB focal loss increased by different degrees, while both accuracies decreased compared with softmax CE loss. Specifically, the accuracy was only 83.79%, although the recall reached the highest value of 85.11%. This was because the recall of walking-rider was only 72.49%, although that of walking-natural was 69.16% (Figure 6). This result further verified that decreased accuracy occurred when using balancing techniques on the imbalanced dataset. In addition, we found that the recall of majority classes decreased while that of minority classes increased when using CS_CE loss and CB focal loss (Figure 6). This result revealed that both losses effectively focused on the samples of minority classes during training, but it is inevitable that more samples in majority classes were misclassified as minority classes so that overall accuracy would decrease.

Table 4. Classification performance comparison with different loss functions. The best two results for each metric are highlighted in bold.

Loss Functions #	Precision (%)	Recall (%)	F1-Score (%)	Accuracy (%)
Softmax CE loss	79.74	79.57	79.02	**93.37**
Class-level				
CS_CE loss [24]	**80.47**	**85.11**	**79.91**	83.79
CB_CE loss [25]	75.35	75.70	75.47	90.61
Sample-level				
Focal loss [26]	78.84	77.99	78.25	**93.30**
ACS loss [27]	77.03	76.54	76.60	92.05
CB focal loss (γ = 0.5)	**82.50**	**83.73**	**82.94**	90.68

CS_CE: cost-sensitive cross-entropy; CB_CE: class-balanced cross-entropy; ACS: adaptive class suppression.

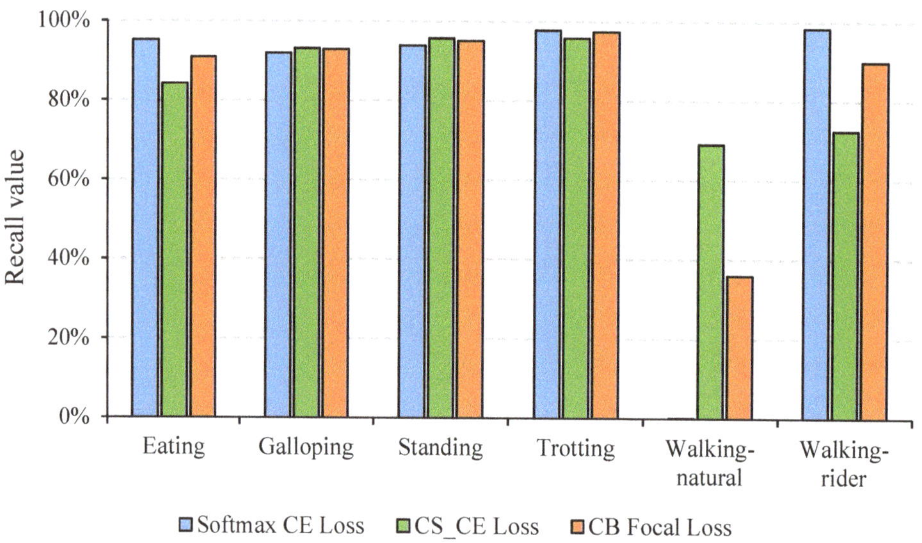

Figure 6. Recall of different activities under different loss functions including softmax CE loss, cost-sensitive cross-entropy (CS_CE) loss, and CB focal loss.

3.3. Classification Performance Analysis

In Figure 7, we show the precision and recall confusion matrix aggregating the classification results under 6-fold cross-validation when using CMI-Net with CB focal loss ($\gamma = 0.5$). Both precision and recall values of all activities had more than 90% accuracy (i.e., the precision and recall for eating were 92.86% and 90.89%, for galloping were 91.41% and 92.89%, for standing were 95.18% and 95.11%, for trotting were 97.34% and 97.46%, and for walking-rider were 93.49% and 90.01%, respectively), except for the walking-natural activity, which only obtained low precision and recall (Figure 7). This low classification precision and recall occurred for two main reasons. The first reason was class imbalance. Walking-natural as the minority class in the dataset only occupied 3.8%, which was much less than the 38.94% occupation of majority class walking-rider, which easily caused the model to be biased toward the majority classes and resulted in poor minority class recognition performance. The second reason was severe confusion with other activities, especially eating and walking-rider activities. As shown in Figure 7, 18.64% and 56.14% of the samples predicted to be class walking-natural had ground truth classes eating and walking-rider, respectively. In addition, 20.38% and 43.13% of the samples with ground truth class walking-natural were misclassified as class eating and walking-rider, respectively. This was because, during eating, the horse was slowly walking so that some samples of eating might contain walking activity [32]. The movement patterns of walking-natural and walking-rider were very similar, which interfered with the learning ability of the network for these two behavioral characteristics (Figure 8). It also revealed that there was no major variability in equine walking patterns in the presence or absence of a rider. This was consistent with a previous study that found no major changes in equine limb kinematics, although the extension of the thoracolumbar region increased during walking with a rider compared with non-ridden walking [43]. In addition, there was confusion between galloping and trotting activities with misclassification of 6.93% of galloping as trotting. This might be related to the misinterpretation by the annotator during labeling, as it was not always clear when the activity transitions occurred [32]. Additionally, a sample rate of 100Hz may limit the distinction in the transition between trotting and cantering or galloping.

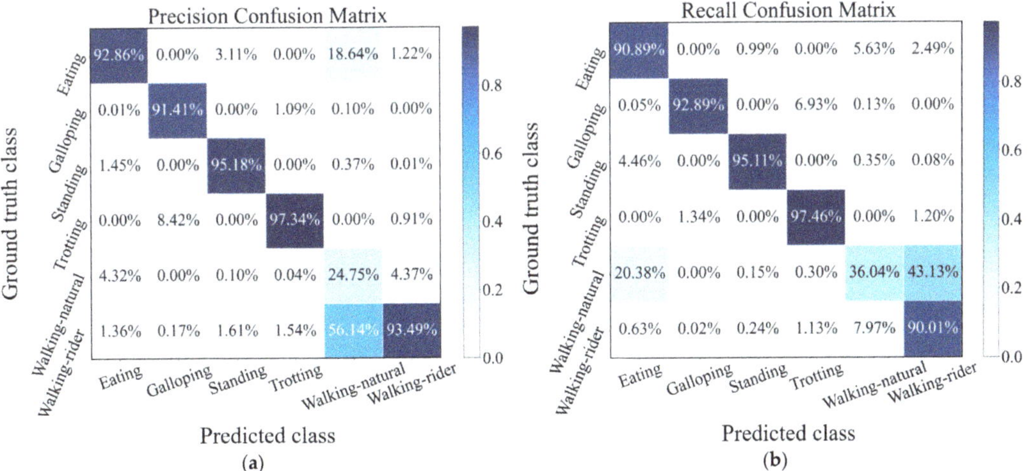

(a) (b)

Figure 7. Precision (**a**) and recall (**b**) confusion matrix of CMI-Net with CB focal loss ($\gamma = 0.5$).

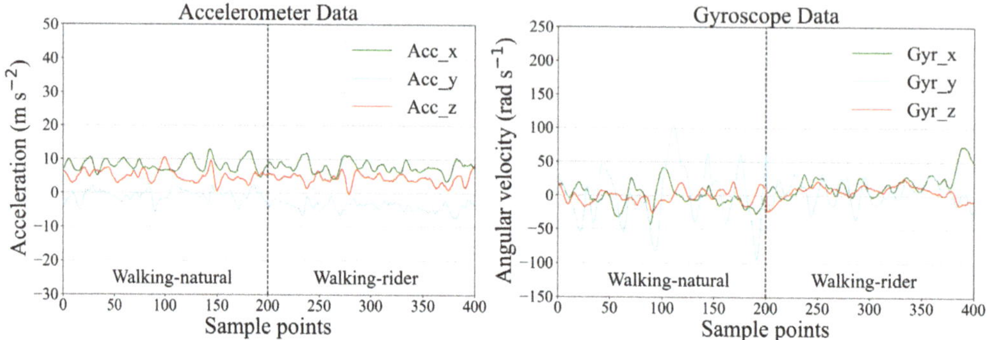

Figure 8. Example of accelerometer and gyroscope data for walking-natural and walking-rider.

3.4. Limitations and Future Works

The first limitation of our proposed method is that our model was trained on a public dataset that contained only six labeled activities, i.e., eating, standing, trotting, galloping, walking-rider, and walking-natural. Indeed, there are some other activities such as head shaking, scratch biting, rubbing, and rolling, all of which, although infrequent, are physiologically critical to equine health and welfare, and should have been labeled and included in the dataset. Due to the missing of these infrequent activities in the dataset, inevitably, as a typical open-set recognition problem [44], these unlabeled activities that occur in real behavior monitoring scenarios will be easily misclassified as the six defined activities, resulting in loss of some key information. Thus, as a next step to further improve classification performance for equine activities, we will investigate some feasible techniques such as classification-reconstruction learning and weightless neural networks [44–46] to enable our activity classifiers to not only accurately classify the defined classes appearing in training but also effectively deal with unlabeled ones generated in practice.

The second limitation is that the algorithms we developed and adopted in this study were based on supervised learning, which relied on a large number of annotated samples. Data annotation is a labor-intensive and time-consuming task, and well-annotated data is often limited as reflected by the fact that we can only find one public dataset for equine activities. With regard to the found dataset [32], in fact, there are still vast amounts of unlabeled samples that can be used to alleviate the overfitting problem and improve the generalization ability of models. Thus, how we can best use the unlabeled samples becomes a key. To this point, our work can be further expanded toward the direction of semi-supervised learning to sufficiently exploit these unlabeled data. For instance, we may first train models on the existing and well-labeled data and then apply the trained models to conduct predictions for unlabeled data. The one-hot predictions can serve as pseudo labels for those high-confidence samples, which, along with the original labels, can then be further used to train the model iteratively until the unlabeled data no longer changes.

4. Conclusions

In this study, we developed a CMI-Net involving a dual CNN trunk architecture and a joint CMIM to improve equine activity classification performance. The CMI-Net effectively captured complementary information and suppressed unrelated information from multiple modalities. Specifically, the dual CNN architecture extracted modality-specific features, and the CMIM recalibrated temporal- and axis-wise features in each modality by utilizing multi-modal knowledge and achieved deep intermodality interaction. To alleviate the class imbalance problem, a CB focal loss was leveraged for the first time to supervise the training of CMI-Net, which focused more on the difficult samples and samples of minority classes during optimization. The results revealed that our CMI-Net with softmax CE loss outperformed the existing methods, and the adoption of CB focal loss effectively improved the precision, recall, and F1-score while slightly decreasing the accuracy. In addition, ablation studies demonstrated that applying the CMIM in the upper layer of CMI-Net could obtain better performance since high-level features contained more general patterns. CB focal loss also performed better than any class-level or sample-level reweighted losses used alone. In short, the favorable classification performance indicated the effectiveness of our proposed CMI-Net and CB focal loss.

Author Contributions: Conceptualization, A.M. and K.L.; methodology, A.M.; software, A.M.; validation, A.M., E.H. and H.G.; formal analysis, A.M.; writing—original draft preparation, A.M.; writing—review and editing, K.L., W.X. and R.S.V.P.; supervision, K.L.; project administration, K.L.; funding acquisition, K.L. All authors have read and agreed to the published version of the manuscript.

Funding: This research was funded by the new research initiatives at the City University of Hong Kong.

Institutional Review Board Statement: Not applicable.

Informed Consent Statement: Not applicable.

Acknowledgments: We would like to thank Jacob W. Kamminga et al., of the Pervasive Systems Group, the University of Twente for providing the public dataset. Funding for conducting this study was provided by the new research initiatives at the City University of Hong Kong.

Conflicts of Interest: The authors declare no conflict of interest.

Abbreviations

ACS	Adaptive class suppression
CB	Class-balanced
CB_CE	Class-balanced cross-entropy
CE	Cross-entropy
CMIM	Cross-modality interaction module
CMI-Net	Cross-modality interaction network
CNN	Convolutional neural network
CS_CE	Cost-sensitive cross-entropy
DT	Decision tree
FN	False negative
FNNs	Feed-forward neural networks
FP	False positive
IMUs	Inertial measurement units
LDA	Linear discriminant analysis
LOOCV	Leave-one-out cross-validation
LSTM	Long short-term memory
NB	Naïve Bayes
QDA	Quadratic discriminant analysis
Res-LCB	Residual-like convolution block
RF	Random forest
SVM	Support vector machine
TN	True negative
TP	True positive
t-SNE	t-distributed stochastic neighbor embedding

References

1. Eerdekens, A.; Deruyck, M.; Fontaine, J.; Martens, L.; De Poorter, E.; Plets, D.; Joseph, W. A framework for energy-efficient equine activity recognition with leg accelerometers. *Comput. Electron. Agric.* **2021**, *183*, 106020. [CrossRef]
2. Parkes, R.S.V.; Weller, R.; Pfau, T.; Witte, T.H. The effect of training on stride duration in a cohort of two-year-old and three-year-old thoroughbred racehorses. *Animals* **2019**, *9*, 466. [CrossRef]
3. Van Weeren, P.R.; Pfau, T.; Rhodin, M.; Roepstorff, L.; Serra Bragança, F.; Weishaupt, M.A. Do we have to redefine lameness in the era of quantitative gait analysis? *Equine Vet. J.* **2017**, *49*, 567–569. [CrossRef]

4. Bosch, S.; Serra Bragança, F.; Marin-Perianu, M.; Marin-Perianu, R.; van der Zwaag, B.J.; Voskamp, J.; Back, W.; Van Weeren, R.; Havinga, P. Equimoves: A wireless networked inertial measurement system for objective examination of horse gait. *Sensors* **2018**, *18*, 850. [CrossRef]
5. Astill, J.; Dara, R.A.; Fraser, E.D.G.; Roberts, B.; Sharif, S. Smart poultry management: Smart sensors, big data, and the internet of things. *Comput. Electron. Agric.* **2020**, *170*, 105291. [CrossRef]
6. Rueß, D.; Rueß, J.; Hümmer, C.; Deckers, N.; Migal, V.; Kienapfel, K.; Wieckert, A.; Barnewitz, D.; Reulke, R. Equine Welfare Assessment: Horse Motion Evaluation and Comparison to Manual Pain Measurements. In Proceedings of the Pacific-Rim Symposium on Image and Video Technology, PSIVT 2019, Sydney, Australia, 18–22 November 2019; pp. 156–169. [CrossRef]
7. Kamminga, J.W.; Meratnia, N.; Havinga, P.J.M. Dataset: Horse Movement Data and Analysis of its Potential for Activity Recognition. In Proceedings of the 2nd Workshop on Data Acquisition to Analysis, DATA 2019, Prague, Czech Republic, 26–28 July 2019; pp. 22–25. [CrossRef]
8. Kumpulainen, P.; Cardó, A.V.; Somppi, S.; Törnqvist, H.; Väätäjä, H.; Majaranta, P.; Gizatdinova, Y.; Hoog Antink, C.; Surakka, V.; Kujala, M.V.; et al. Dog behaviour classification with movement sensors placed on the harness and the collar. *Appl. Anim. Behav. Sci.* **2021**, *241*, 105393. [CrossRef]
9. Tran, D.N.; Nguyen, T.N.; Khanh, P.C.P.; Trana, D.T. An IoT-based Design Using Accelerometers in Animal Behavior Recognition Systems. *IEEE Sens. J.* **2021**. [CrossRef]
10. Maisonpierre, I.N.; Sutton, M.A.; Harris, P.; Menzies-Gow, N.; Weller, R.; Pfau, T. Accelerometer activity tracking in horses and the effect of pasture management on time budget. *Equine Vet. J.* **2019**, *51*, 840–845. [CrossRef] [PubMed]
11. Nweke, H.F.; Teh, Y.W.; Al-garadi, M.A.; Alo, U.R. Deep learning algorithms for human activity recognition using mobile and wearable sensor networks: State of the art and research challenges. *Expert Syst. Appl.* **2018**, *105*, 233–261. [CrossRef]
12. Noorbin, S.F.H.; Layeghy, S.; Kusy, B.; Jurdak, R.; Bishop-hurley, G.; Portmann, M. Deep Learning-based Cattle Activity Classification Using Joint Time-frequency Data Representation. *Comput. Electron. Agric.* **2020**, *187*, 106241. [CrossRef]
13. Peng, Y.; Kondo, N.; Fujiura, T.; Suzuki, T.; Ouma, S.; Wulandari Yoshioka, H.; Itoyama, E. Dam behavior patterns in Japanese black beef cattle prior to calving: Automated detection using LSTM-RNN. *Comput. Electron. Agric.* **2020**, *169*, 105178. [CrossRef]
14. Bocaj, E.; Uzunidis, D.; Kasnesis, P.; Patrikakis, C.Z. On the Benefits of Deep Convolutional Neural Networks on Animal Activity Recognition. In Proceedings of the 2020 International Conference on Smart Systems and Technologies (SST), Osijek, Croatia, 14–16 October 2020; pp. 83–88. [CrossRef]
15. Eerdekens, A.; Deruyck, M.; Fontaine, J.; Martens, L.; de Poorter, E.; Plets, D.; Joseph, W. Resampling and Data Augmentation for Equines' Behaviour Classification Based on Wearable Sensor Accelerometer Data Using a Convolutional Neural Network. In Proceedings of the 2020 International Conference on Omni-layer Intelligent Systems (COINS), Barcelona, Spain, 31 August–2 September 2020; pp. 1–6. [CrossRef]
16. Chambers, R.D.; Yoder, N.C.; Carson, A.B.; Junge, C.; Allen, D.E.; Prescott, L.M.; Bradley, S.; Wymore, G.; Lloyd, K.; Lyle, S. Deep learning classification of canine behavior using a single collar-mounted accelerometer: Real-world validation. *Animals* **2021**, *11*, 1549. [CrossRef]
17. Liu, N.; Zhang, N.; Han, J. Learning Selective Self-Mutual Attention for RGB-D Saliency Detection. In Proceedings of the IEEE/CVF Conference on Computer Vision and Pattern Recognition, CVPR 2020, 14–19 June 2020; pp. 13753–13762. Available online: http://cvpr2020.thecvf.com/ (accessed on 27 August 2021). [CrossRef]
18. Ha, S.; Choi, S. Convolutional neural networks for human activity recognition using multiple accelerometer and gyroscope sensors. In Proceedings of the 2016 International Joint Conference on Neural Networks (IJCNN), Vancouver, BC, Canada, 24–29 July 2016; pp. 381–388. [CrossRef]
19. Mustaqeem Kwon, S. MLT-DNet: Speech emotion recognition using 1D dilated CNN based on multi-learning trick approach. *Expert Syst. Appl.* **2021**, *167*, 114177. [CrossRef]
20. Mustaqeem; Kwon, S. Optimal feature selection based speech emotion recognition using two-stream deep convolutional neural network. *Int. J. Intell. Syst.* **2021**, *36*, 5116–5135. [CrossRef]
21. Xu, X.; Li, W.; Duan, Q. Transfer learning and SE-ResNet152 networks-based for small-scale unbalanced fish species identification. *Comput. Electron. Agric.* **2021**, *180*, 105878. [CrossRef]
22. Zhang, S.; Li, Z.; Yan, S.; He, X.; Sun, J. Distribution Alignment: A Unified Framework for Long-tail Visual Recognition. In Proceedings of the IEEE/CVF Conference on Computer Vision and Pattern Recognition, CVPR 2021, 19–25 June 2021; pp. 2361–2370. Available online: http://cvpr2021.thecvf.com/ (accessed on 27 August 2021).
23. Tan, J.; Wang, C.; Li, B.; Li, Q.; Ouyang, W.; Yin, C.; Yan, J. Equalization loss for long-tailed object recognition. In Proceedings of the IEEE/CVF Conference on Computer Vision and Pattern Recognition, CVPR 2020, 14–19 June 2020; pp. 11659–11668. Available online: http://cvpr2020.thecvf.com/ (accessed on 27 August 2021). [CrossRef]
24. Khan, S.H.; Hayat, M.; Bennamoun, M.; Sohel, F.A.; Togneri, R. Cost-sensitive learning of deep feature representations from imbalanced data. *IEEE Trans. Neural Netw. Learn. Syst.* **2017**, *29*, 3573–3587. [CrossRef] [PubMed]
25. Cui, Y.; Jia, M.; Lin, T.Y.; Song, Y.; Belongie, S. Class-balanced loss based on effective number of samples. In Proceedings of the IEEE/CVF Conference on Computer Vision and Pattern Recognition, CVPR 2019, Long Beach, CA, USA, 16–20 June 2019; pp. 9260–9269. [CrossRef]
26. Lin, T.Y.; Goyal, P.; Girshick, R.; He, K.; Dollar, P. Focal Loss for Dense Object Detection. *IEEE Trans. Pattern Anal. Mach. Intell.* **2018**, *42*, 318–327. [CrossRef]

27. Wang, T.; Zhu, Y.; Zhao, C.; Zeng, W.; Wang, J.; Tang, M. Adaptive Class Suppression Loss for Long-Tail Object Detection. In Proceedings of the IEEE/CVF Conference on Computer Vision and Pattern Recognition, CVPR 2021, 19–25 June 2021; pp. 3103–3112. Available online: http://cvpr2020.thecvf.com/ (accessed on 27 August 2021).
28. Mao, A.X.; Huang, E.D.; Xu, W.T.; Liu, K. Cross-modality Interaction Network for Equine Activity Recognition Using Time-Series Motion Data. In Proceedings of the 2021 International Symposium on Animal Environment and Welfare (ISAEW), Chongqing, China, 20–23 October 2021. in press.
29. Zhang, Z.; Lin, Z.; Xu, J.; Jin, W.D.; Lu, S.P.; Fan, D.P. Bilateral Attention Network for RGB-D Salient Object Detection. *IEEE Trans. Image Process.* **2021**, *30*, 1949–1961. [CrossRef]
30. Woo, S.; Park, J.; Lee, J.Y.; Kweon, I.S. CBAM: Convolutional block attention module. In Proceedings of the European Conference on Computer Vision, ECCV 2018, Munich, Germany, 8–14 September 2018; pp. 3–19. [CrossRef]
31. Mustaqeem Kwon, S. Att-Net: Enhanced emotion recognition system using lightweight self-attention module. *Appl. Soft Comput.* **2021**, *102*, 107101. [CrossRef]
32. Kamminga, J.W.; Janßen, L.M.; Meratnia, N.; Havinga, P.J.M. Horsing around—A dataset comprising horse movement. *Data* **2019**, *4*, 131. [CrossRef]
33. He, K.; Zhang, X.; Ren, S.; Sun, J. Deep residual learning for image recognition. In Proceedings of the IEEE/CVF Conference on Computer Vision and Pattern Recognition, CVPR 2016, Las Vegas, NV, USA, 27–30 June 2016; pp. 770–778. [CrossRef]
34. Kamminga, J.W.; Le, D.V.; Havinga, P.J.M. Towards deep unsupervised representation learning from accelerometer time series for animal activity recognition. In Proceedings of the 6th Workshop on Mining and Learning from Time Series, MiLeTS 2020, San Diego, CA, USA, 24 August 2020.
35. Nair, V.; Hinton, G.E. Rectified Linear Units Improve Restricted Boltzmann Machines Vinod. In Proceedings of the 27th International Conference on Machine Learning, ICML 2010, Haifa, Israel, 21–24 June 2010. [CrossRef]
36. Joze, H.R.V.; Shaban, A.; Iuzzolino, M.L.; Koishida, K. MMTM: Multimodal transfer module for CNN fusion. In Proceedings of the IEEE/CVF Conference on Computer Vision and Pattern Recognition, CVPR 2020, 14–19 June 2020; pp. 13286–13296. Available online: http://cvpr2020.thecvf.com/ (accessed on 27 August 2021). [CrossRef]
37. Casella, E.; Khamesi, A.R.; Silvestri, S. A framework for the recognition of horse gaits through wearable devices. *Pervasive Mob. Comput.* **2020**, *67*, 101213. [CrossRef]
38. Zeng, M.; Nguyen, L.T.; Yu, B.; Mengshoel, O.J.; Zhu, J.; Wu, P.; Zhang, J. Convolutional Neural Networks for human activity recognition using mobile sensors. In Proceedings of the 6th international conference on mobile computing, applications and services, MobiCASE 2014, Austin, TX, USA, 6–7 November 2014; pp. 197–205. [CrossRef]
39. Wolpert, D.H.; Macready, W.G. No free lunch theorems for optimization. *IEEE Trans. Evol. Comput.* **1997**, *1*, 67–82. [CrossRef]
40. Wei, J.; Wang, Q.; Li, Z.; Wang, S.; Zhou, S.K.; Cui, S. Shallow Feature Matters for Weakly Supervised Object Localization. In Proceedings of the IEEE/CVF Conference on Computer Vision and Pattern Recognition, CVPR 2021, 19–25 June 2021; pp. 5993–6001. Available online: http://cvpr2021.thecvf.com/ (accessed on 27 August 2021).
41. Lecun, Y.; Bengio, Y.; Hinton, G. Deep learning. *Nature* **2015**, *521*, 436–444. [CrossRef] [PubMed]
42. Van der Maaten, L.; Hinton, G. Visualizing Data using t-SNE. *J. Mach. Learn. Res.* **2008**, *9*, 2579–2605. [CrossRef]
43. De Cocq, P.; Van Weeren, P.R.; Back, W. Effects of girth, saddle and weight on movements of the horse. *Equine Vet. J.* **2004**, *36*, 758–763. [CrossRef] [PubMed]
44. Geng, C.; Huang, S.-J.; Chen, S. Recent Advances in Open Set Recognition: A Survey. *IEEE Trans. Pattern Anal. Mach. Intell.* **2020**, *14*, 1. [CrossRef] [PubMed]
45. Yoshihashi, R.; You, S.; Shao, W.; Iida, M.; Kawakami, R.; Naemura, T. Classification-Reconstruction Learning for Open-Set Recognition. In Proceedings of the IEEE/CVF Conference on Computer Vision and Pattern Recognition, CVPR 2019, Long Beach, CA, USA, 16–20 June 2019; pp. 4016–4025.
46. Cardoso, D.O.; Gama, J.; França, F.M.G. Weightless neural networks for open set recognition. *Mach. Learn.* **2017**, *106*, 1547–1567. [CrossRef]

MDPI

Article

Classifying Ingestive Behavior of Dairy Cows via Automatic Sound Recognition

Guoming Li [1,*], Yijie Xiong [2,3], Qian Du [4], Zhengxiang Shi [5] and Richard S. Gates [6]

1 Department of Agricultural and Biosystems Engineering, Iowa State University, Ames, IA 50011, USA
2 Department of Animal Science, University of Nebraska-Lincoln, Lincoln, NE 68588, USA; yijie.xiong@unl.edu
3 Department of Biological Systems Engineering, University of Nebraska-Lincoln, Lincoln, NE 68588, USA
4 Department of Electrical and Computer Engineering, Mississippi State University, Starkville, MS 39762, USA; du@ece.msstate.edu
5 Department of Agricultural Structure and Bioenvironmental Engineering, College of Water Resources and Civil Engineering, China Agricultural University, Beijing 100083, China; shizhx@cau.edu.cn
6 Egg Industry Center, Departments of Agricultural and Biosystems Engineering, and Animal Science, Iowa State University, Ames, IA 50011, USA; rsgates@iastate.edu
* Correspondence: gmli@iastate.edu

Abstract: Determining ingestive behaviors of dairy cows is critical to evaluate their productivity and health status. The objectives of this research were to (1) develop the relationship between forage species/heights and sound characteristics of three different ingestive behaviors (bites, chews, and chew-bites); (2) comparatively evaluate three deep learning models and optimization strategies for classifying the three behaviors; and (3) examine the ability of deep learning modeling for classifying the three ingestive behaviors under various forage characteristics. The results show that the amplitude and duration of the bite, chew, and chew-bite sounds were mostly larger for tall forages (tall fescue and alfalfa) compared to their counterparts. The long short-term memory network using a filtered dataset with balanced duration and imbalanced audio files offered better performance than its counterparts. The best classification performance was over 0.93, and the best and poorest performance difference was 0.4–0.5 under different forage species and heights. In conclusion, the deep learning technique could classify the dairy cow ingestive behaviors but was unable to differentiate between them under some forage characteristics using acoustic signals. Thus, while the developed tool is useful to support precision dairy cow management, it requires further improvement.

Keywords: audio; dairy cow; deep learning; mastication; jaw movement; forage management; precision livestock management

Citation: Li, G.; Xiong, Y.; Du, Q.; Shi, Z.; Gates, R.S. Classifying Ingestive Behavior of Dairy Cows via Automatic Sound Recognition. *Sensors* **2021**, *21*, 5231. https://doi.org/10.3390/s21155231

Academic Editors: Yongliang Qiao, Lilong Chai, Dongjian He and Daobilige Su

Received: 29 June 2021
Accepted: 31 July 2021
Published: 2 August 2021

1. Introduction

Modern dairy farms continue to grow in herd size and technology adoption for maintaining or improving the production and labor efficiencies needed to feed the growing human population [1]. The U.S. is the largest dairy producer in the world with 9.39 million milking cows on farms, producing 101.25 million metric tons of milk in 2020 [2]. Despite only accounting for 6.3% of the total number of dairy farms, dairy farms containing over 500 cows have a 65.9% market share of the U.S. total milking cow inventory [3]. In these intensive production systems, forage-fed dairy production comprises over 80% [4], in which forage is a source of food and nutrients for dairy cows. Thus, sufficient provision of forage is critical to meet cow daily nutrient requirements and sustain desired milk production goals. On a short-time scale, forage intake of cows has been associated with a sequence of three jaw movements or ingestive behaviors [5], namely bites, chews, and chew-bites. A biting behavior is defined as the apprehension and severance of forage, while a chewing behavior includes the crushing, grinding, and processing of ingested grass inside the mouth [6]. A chewing-biting behavior results from the overlapping of chewing and biting events in the same jaw movement, in which the forage in the mouth is chewed, and simultaneously, a new mouthful of forage is severed [7]. The frequency and characteristics of the ingestive behaviors can change in correspondence with individual animals and surrounding environments. Therefore, efficiently and precisely monitoring and assessing the ingestive behaviors of dairy cows may provide useful insights into resource management, nutrition supply, animal health, welfare, and production.

The efficiencies and accuracies of ingestive behavior monitoring can be influenced by measurement methods. Early strategies for monitoring ingestive behavior relied on observation by technicians [8], to obtain precise individual information on a few animals but is costly and impractical for monitoring large herds [9]. Imaging methods combined with image processing algorithms or deep learning techniques may provide contactless and non-invasive measures and potentially automate the detection process [1]. However, high-quality images/videos are prerequisites for this, and appropriately recording the whole jaw movement process without occlusion within herds could be problematic. Wearable sensors including pressure sensors, accelerometers/pendulums, jaw switches, and electromyography have been examined to detect jaw movements [7]. Most of the wearable sensors focus on recognizing long-term activities (grazing or ruminating) or overall jaw movements yet have difficulties differentiating bites, chews, and chew-bites. Andriamandroso et al. [7] compared and summarized three types of sensors to classify jaw movements. They demonstrated that the accuracy of detecting jaw movement was 0.91–0.95 for noseband pressure sensors, 0.94–0.95 for microphones, and 0.65–0.90 for accelerometers, but only microphone data supported differentiating bites, chews, and chew-bites and with a relatively low accuracy of 0.61–0.95. The great potential of acoustic signals for ingestive behavior monitoring lies in the fact that ingestive sounds can be clearly transmitted from bones, skull cavities and soft tissues to recording devices typically attached to animal foreheads. Additionally, the wearable sound collecting devices do not influence cattle's natural behaviors once the animals have acclimated to the devices [6].

Several automatic recognition systems based on acoustic signals have been developed to detect and classify the three ingestive behaviors of dairy cows. Acoustic signals of dairy cows suggest that the ingestive behavior characteristics (e.g., intake rate, bite mass, bite rate, etc.) are associated with forage species (e.g., alfalfa and tall fescue) and grass heights (e.g., tall and short) [5,10], which provide critical suggestions on precision forage management for dairy cows. However, relationships between forage species/heights and key features of acoustic signals of the three ingestive behaviors remain unclear and should be explored to supplement knowledge for precision dairy management, such as grass utilization estimation, grass preference evaluation, and other forage management. In previous work, conventional machine learning models or knowledge-based algorithms have been examined [6,11–14], including random forest, support vector machine, multi-layer perceptron, decision logic algorithm, and hidden Markov model. Despite great performance, these methods require thorough designs of feature extractors to obtain appropriate features (e.g., duration, amplitude, spectrum, and power). Alternatively, deep learning techniques (e.g., convolutional neural network, CNN; and recurrent neural network, RNN) developed from conventional machine learning are representation-learning methods and can automatically discover features from raw data without extensive engineering knowledge on feature extraction [15]. Thus, these techniques may have the potential to automatically classify the ingestive behaviors of dairy cows based on acoustic signals, but the performance of various architectures requires further investigation.

The objectives of this research were to (1) examine the effects of forage species and forage height on key acoustic characteristics of bites, chews, and chew-bites for dairy cows; (2) evaluate deep learning models and optimization strategies for automatic sound recognition of these three ingestive behaviors; and (3) examine the efficacy of a deep learning model for classifying the three behaviors for various forage characteristics.

2. Materials and Methods

2.1. Dataset Description

A publicly available dataset of dairy cows was used in this research [16]. The dataset contains data collected using three microphones (Nady 151 VR, Nady Systems, Oakland, CA, USA) attached to the foreheads of three 4- to 6-year-old lactating Holstein cows weighing 608 ± 24.9 kg; and ingesting sounds (i.e., chews, bites, and chew-bites) from four types of microswards were continuously recorded for five days. The microswards consisted of sets of 4-L plastic pots with either alfalfa (*Meicago sativa*) or tall fescue (*Lolium arundinaceum*, Schreb.) with two heights, tall (24.5 ± 3.8 cm) or short (11.6 ± 1.9 cm). An illustration of the two forages used in the dataset is provided in Figure 1. The acoustic signals were saved as WAV files and labeled for observed ingestion behavior by the technicians from the same research team with a labeling agreement of over 99%. A total of 52 labeled WAV files totaling 54 min 24 s were in the dataset, and the files were processed and segmented based on ingestive behaviors, forage species, and forage height, resulting in 3038 segments with a duration of 1647.37 s being used for model evaluation and experiments (Table 1).

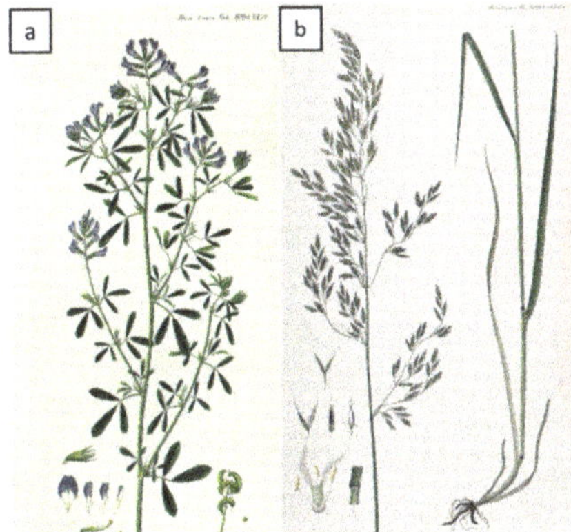

Figure 1. Generic illustration of the two forages described in this manuscript—(**a**) Alfalfa (*Meicago sativa)* and (**b**) *tall fescue (Lolium arundinaceum*, Schreb.) Sources: plantillustration.org.

Table 1. Number and duration of audio files used for model evaluation and experiments.

Forage Species	Forage Height	Number of Audio Files			Duration of Audio Files Used (s)		
		Bites	Chews	Chew-Bites	Bites	Chews	Chew-Bites
Alfalfa	Short	179	260	123	72.78	74.24	71.99
	Tall	148	416	322	175.20	184.56	182.90
Tall fescue	Short	94	454	217	143.87	144.79	141.59
	Tall	100	487	238	155.19	149.78	150.48
Total		521	1617	900	547.04	553.37	546.96

Sample illustrations of acoustic signals in frequency and time domains are presented in Figure 2 for the three ingestive behaviors. The three behaviors showed apparent differences in temporal and spatial signal patterns, and the acoustic signals of the behaviors were mostly in low frequencies (<1 kHz) [14]. Additionally, based on the observation reports from the technicians, each ingestive behavior event commonly lasted for less than 1 s [16]. The recorded acoustic signals are in 16 bit sample depth format taken at a 22.05 kHz sampling rate. There were $2^{16} = 65,536$ possible values for each recorded signal ranging from −32,768 to 32,767. For Figure 2, amplitude values in the time domain were normalized using Equation (1), and the magnitudes in the frequency domain were normalized using Equation (3).

$$Normalized\ amplitude = Recorded\ amplitude/2^{16} \quad (1)$$

$$FFT\ value = FFT.RFFT(Normalized\ amplitude) \quad (2)$$

$$ANM = ABS[(FFT\ value)_i] / \sum_{i}^{m} ABS(FFT\ value)_i \quad (3)$$

where *FFT* is Fast Fourier Transform; *RFFT* is Real Fast Fourier Transform; the function *FFT.RFFT* is to transform the discrete time domain signals into discrete frequency domain components; *ANM* is absolute normalized magnitude; *ABS* is absolutization operation; $(FFT\ value)_i$ is the ith *FFT* value normalized; and *m* is total number of FFT values converted and its length is dynamically determined by an audio segment input. Normalized amplitude and ANM range from 0 to 1. All operations in the above equations were vectorized to improve calculation efficiency.

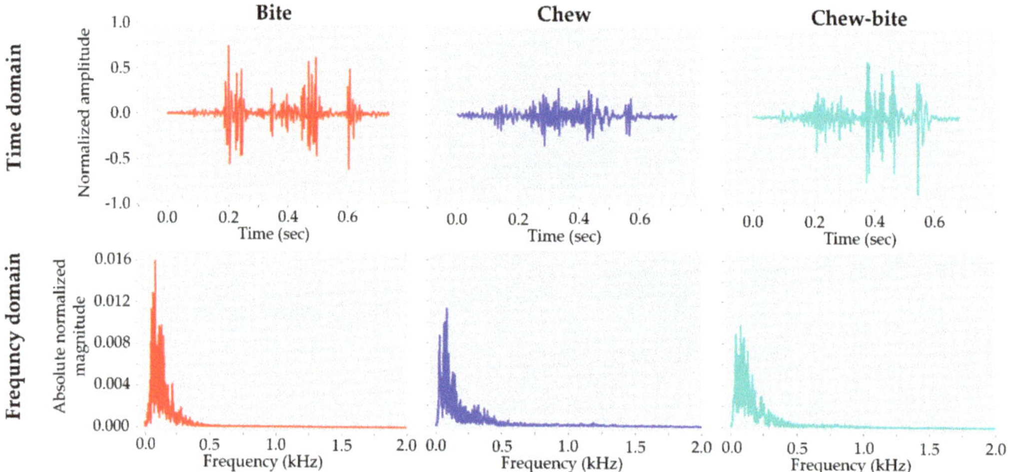

Figure 2. Sample illustrations of acoustic signals in time (**top**) and frequency (**bottom**) domains for the ingestive sounds. Higher absolute normalized values indicate higher power of the acoustic signal.

2.2. Statstical Analysis for Evaluating Effects of Forage on Acoustic Features

Statistical analyses were conducted to investigate the relationships between forage species and heights and acoustic features of the ingestive behaviors. The data used for statistical analysis were all in the time domain. The normalized amplitude and duration of the acoustic signals were extracted for each of the 3038 segments. The acoustic feature labels provided with the dataset were processed by the research team [16]. A larger normalized amplitude indicates that louder sounds were produced around cow mouths; and duration is the length of a segment, with a longer duration indicating dairy cows spent more time ingesting forage [16]. The effects of forage species, forage heights, and their interaction on the amplitude and duration of the segments corresponding to the three ingestive behaviors were analyzed with ANOVA using PROC MIXED in Statistical Analysis Software (version 9.3, SAS Institute Inc., Cary, NC, USA). Mean values were compared using Fisher's least significant difference with PDMIX800 [17], and a significant difference was considered at $p \leq 0.05$. The statistical model is the same for the three behaviors of bites, chews, and chew-bites, which can be expressed as

$$Y_{ijk} = \mu + \alpha_i + \beta_j + (\alpha\beta)_{ij} + \varepsilon_{ijk} \quad (4)$$

where Y_{ijk} is the parameter examined (i.e., amplitude and duration); μ is the least square mean of the parameter; α_i is the forage species, $i = alfalfa,\ tall\ fescue$; β_j is the forage height, $j = tall,\ short$; $(\alpha\beta)_{ij}$ is the interaction effect of forage species and height; and ε_{ijk} is the random error.

2.3. Overall Deep Learning Algorithm Workflow

As shown in Figure 3, the overall workflow consisted of four steps. The first step was to filter background noises (i.e., beeping sounds) from the input acoustic data in order to reduce interference for classification. The second step was to remove the low power signals, which can be considered as uninformative data. This may help to constrain model attention to learn important features and improve inference efficiency and accuracy. The first two steps were to clean data based on physical characteristics in the dataset. However, unwanted signals and low-power signals still existed after the initial two steps of data cleaning. The third step was to convert cleaned data into Mel-frequency cepstral coefficient (MFCC) features which are used to highlight high-power data and transform the original acoustic spectrogram into the human perception level [18]. The first three steps all involved filtering, but only the first step was named "filtering" to differentiate the data cleaning procedures. The final step was to classify the ingestive behaviors using the processed data and deep learning models. Details of the four steps are elaborated in Sections 2.4–2.6. The processing was conducted in a local machine with the processor of Intel(R) Core (TM) i9-10900KF CPU @ 3.7GHz, installed memory (RAM) of 128 GB, graphics processing unit (GPU) of NVIDIA GeForce RTX 3080, and Python-based computing environments.

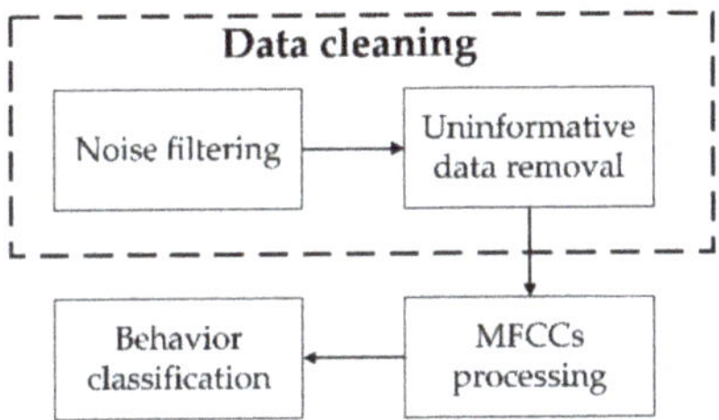

Figure 3. Overall workflow of the algorithms. MFCC is Mel-frequency cepstral coefficient.

2.4. Data Cleaning

2.4.1. Noise Filtering

A routine beeping sound associated with the recording device was produced and recorded along with the cow sounds and was randomly dispersed in the dataset, and was removed to reduce interference with detection results (Figure 4). This beeping center-frequency ranged from 3.6 to 4.5 kHz and could be effectively distinguished from cow ingesting sounds. A bandstop filter with a stopband frequency range of 3.6–4.5 kHz was used to exclude beeping sounds and maintain cow sounds.

2.4.2. Uninformative Data Removal

The noise-filtered dataset was then further processed to remove uninformative data. To maximize the removal efficiency, data in 16 bits without normalization of time domain were used (Figure 5). The input data were vectorized and averaged for every 1100 acoustic samples. Several means were obtained at various steps of averaging but only the maxima within groups were retained. If the maximal mean was smaller than a threshold, the input data with the corresponding index were set FALSE and discarded once converted to frequency domain. Based on the preliminary verification, a threshold value of 100 was used as it can cover most uninformative data while reducing signal loss. Nothing needed to be supplemented for the discarded data in time domain. Because in later step, the MFCC for behavior classification is in frequency domain, and a lack of a specific part in time domain did not influence the workflow for the behavior classification.

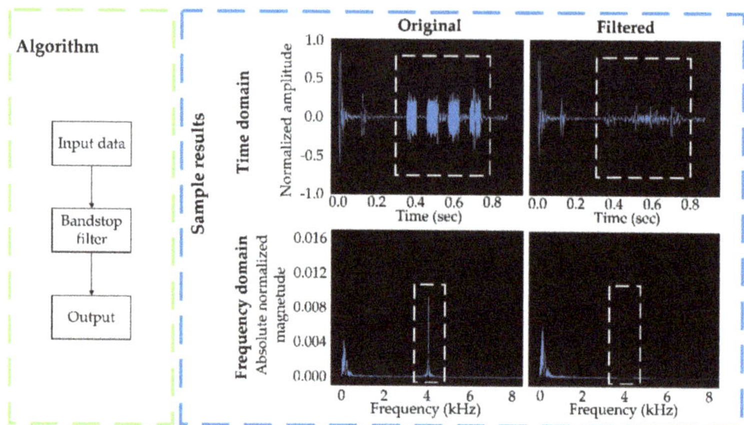

Figure 4. Illustration of the algorithm and sample results for noise removal. White dotted rectangles indicate period and frequency for device-related beeping before and after filtering.

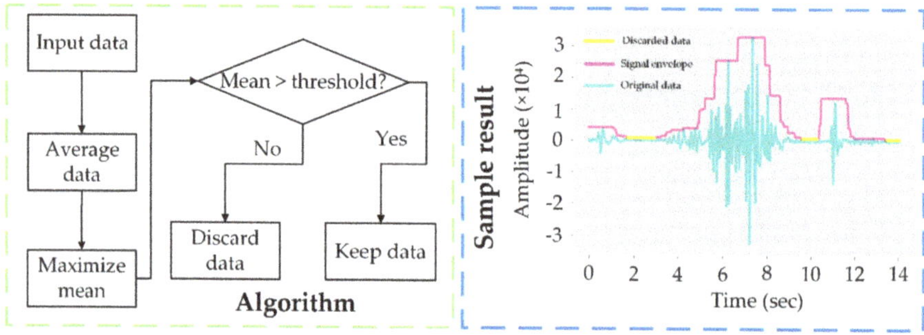

Figure 5. Illustration of the algorithm for uninformative data removal (**left**) and a sample result with a signal envelope after uninformative data removal (**right**). The amplitude values are in 16 bits and unitless.

2.5. Mel-Frequency Cepstral Coefficients Processing

The MFCC processing workflow involved several steps as summarized in Figure 6. The short-time Fourier transform was first conducted to generate a short-time amplitude spectrogram. Acoustic signals were assumed to be constant within short-time scales, and a rolling window with a length of 23 ms (512 samples) and step of 10 ms (220 samples) ran through each time-domain signal to the end of an audio file. Each windowed frame was transformed to a frequency-domain signal using a Fast Fourier transform. To disconnect adjacent overlapping frames, the Discrete Fourier Transform was operated for each frame, and a short-time amplitude spectrogram was produced accordingly. The amplitude spectrogram was converted to dB-based mel-spectrogram for human-interpretable ranges, and values in the y-axis were log-transformed for enhanced visualization, resulting in the short-time power spectrogram. A total of 26 Mel filterbanks were generated to retain more signals of lower frequency (which fits observed cow acoustic characteristics). The Mel filterbanks were mapped to the power spectrogram for building Mel-scale spectrograms, which were fed into deep learning models for behavior classification.

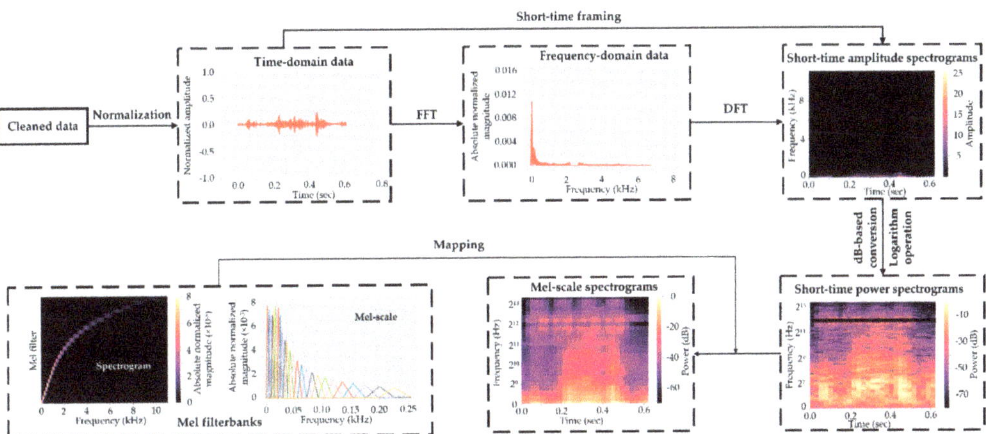

Figure 6. Overall workflow for the Mel-frequency cepstral coefficient processing. FFT is Fast Fourier transform, and DFT is Discrete Fourier Transform. Multiple mel-scale spectrograms were produced based on the short-time framing on time-domain signals.

2.6. Architectures of Deep Learning Models

Three deep learning models, one-dimensional CNN (Conv1D), two-dimensional CNN (Conv2D), and long short-term memory network (LSTM), were evaluated to classify acoustic data (Figure 7) [19].

The core component of Conv1D was the time-distributed layer. When a Mel-scale spectrogram was input into the model, the signal frequency at each time slot of the spectrogram was convolved in the time-distributed layer to extract high-level features. The model included a sequence of a time distributed layer and then a max-pooling layer to reduce acoustic signal dimensionality. Finally, dense and softmax layers were used to flatten two-dimensional features and connect target classes (the three ingestive behaviors). The model size and number of parameters were 706 kB and 348,770, respectively.

The Conv2D took the whole Mel-scale spectrogram as input and extracted major features through two-dimensional convolution. The structure was similar to that of Conv1D, in which a convolution layer followed by a max-pooling layer was repeatedly used to reduce dimensionality. The flatten, dense, and softmax layers were also used at the end of the network. The model size and number of parameters were 2056 kB and 431,290, respectively.

The third network was mainly constructed with two serial LSTM units. In each LSTM unit, a time distributed layer was to extract spectrogram features across time and skipped the next-layer connection; then a bidirectional RNN layer was to obtain features both in forward states (i.e., next frame of spectrogram) and backward states (i.e., previous frame of spectrogram); and finally, features from the time distributed layer and bidirectional RNN layer were concatenated to reinforce key components. After the two adjacent LSTM units, several dense layers and one flatten/softmax layer were built. The model size and number of parameters were 1862 kB and 392,050, respectively.

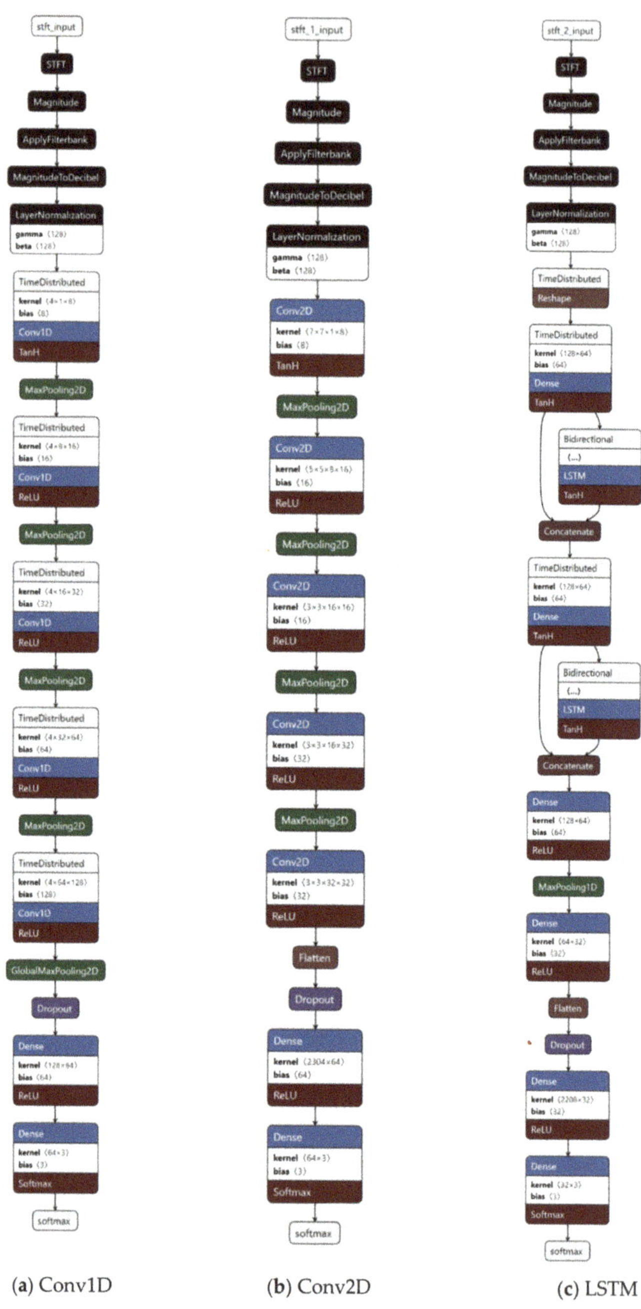

(**a**) Conv1D (**b**) Conv2D (**c**) LSTM

Figure 7. Architectures of the three proposed deep learning models for ingestive behavior classification. Conv1D represents one-dimensional convolutional neural network; Conv2D is two-dimensional convolutional neural network; LSTM represents long short-term memory network; and STFT is short-time Fourier transform.

2.7. Optimization for Classifying the Ingestive Behaviors

The classification performance of the ingestive behaviors was optimized by comparing the three models (Conv1D, Conv2D, and LSTM), two filtering strategies (original vs. filtered), and two data organization methods (imbalanced vs. balanced). The three models were trained with a dropout rate of 0.1, activation functions of relu/tanh, and training epochs of 30. The dataset was randomized into training, validation, and testing sets with a ratio of 0.7:0.1:0.2. Training and validation accuracy curves were calculated across epochs to judge whether models were underfitted/overfitted in real time, and the hold-out dataset was used for the final testing. The filtered dataset resulted from the noise filtering methods mentioned in Section 2.4.1. Although duration of audio files in Table 1 was similar for the three ingestive behaviors, the number of audio files was different (i.e., imbalanced), which may lead to biased inference for the class (i.e., chew) with a large proportion of data. The dataset was reshuffled and randomized, and the number of audio files was equalized to 521 for the three ingestive behaviors, resulting in a balanced dataset. After data reshuffling, the duration of the audio files was 547.040 s for bites, 184.460 s for chews, and 328.180 s for chew-bites.

2.8. Evaluation of Classification Performance under Various Forage Characteristics

After optimization, the optimal model, filtering strategy, and data organization method were further used to evaluate the classification performance for ingestive behaviors under various forage characteristics. Two forage species (alfalfa vs. tall fescue) and two forage heights (short vs. tall) were compared for the three ingestive behaviors. The model was trained based on forage species and heights, using the similar training hyperparameter configurations described in Section 2.7.

2.9. Evaluation Metrics

Three evaluation metrics were calculated using Equations (5)–(7), and higher values of the metrics indicate better performance.

$$Precision = \frac{True\ positive}{True\ positive + False\ positive} \tag{5}$$

$$Recall = \frac{True\ positive}{True\ positive + False\ negative} \tag{6}$$

$$F1\ score = 2 \times \frac{Precision \times Recall}{Precision + Recall} \tag{7}$$

where *true positive* is the number of cases in which models match manual labeling; *false positive* is the number of cases in which models wrongly predict behavior presence; *false negative* is the number of cases in which models wrongly predict behavior absence.

Based on the *true positive*, *false positive*, and *false negative*, confusion matrixes were calculated to indicate class-level performance. Diagonal values in a matrix indicate correct classification rates, and higher values suggest better class-level performance, whereas off-diagonal entries are related to misclassification.

Processing time was reported by Python after all audio files were processed, and processing speed was normalized by dividing the processing time by the total duration of audio files tested.

3. Results

3.1. Ingestive Sound Characteristics under Various Forage Characteristics

Table 2 shows the mean characteristics (i.e., amplitude and duration) and results of the statistical analysis for the bite, chew, and chew-bite behaviors with two forage species and heights. Overall, the amplitude and duration of the acoustic data were 0.323–0.488 and 0.152–0.212 s for bites, 0.084–0.117 and 0.073–0.148 s for chews, and 0.343–0.549 and 0.230–0.301 s for chew-bites, respectively. Except for the amplitude of the chewing sound, the amplitude and duration of the three ingestive sounds were larger for tall fescue than for alfalfa ($p < 0.01$). Between the two forage heights compared, tall forage resulted in larger sound amplitude and duration (excluding amplitudes of bites and chew-bites) ($p < 0.01$). Interaction effects of the forage species and heights on ingestive sounds were observed for all parameters examined. Alfalfa, a tender forage, had a lower value for all behaviors and for both amplitude and duration. The values were also greater for tall alfalfa than for short alfalfa, but both tended to be less than for the tall fescue regardless of its height. These results demonstrate that the three ingestive behaviors for the two forage species and two forage heights can be differentiated by acoustic sound characteristics, namely the amplitude and duration for specific ingestive behavior. The characteristics of the bite, chew, and chew-bite sounds of dairy cows were distinct under various forage characteristics, which could be indications of precision forage management, therefore, model ability to classify the sounds of ingestive behaviors under different forage characteristics should be further evaluated.

Table 2. Amplitude and duration of the bite, chew, and chew-bite sound under various forage conditions.

Factors	Bite		Chew		Chew-Bite	
	Amplitude	Duration (s)	Amplitude	Duration (s)	Amplitude	Duration (s)
Forage species						
Alfalfa	0.355b	0.176b	0.105	0.110b	0.389b	0.262b
Tall fescue	0.454a	0.208a	0.105	0.132a	0.520a	0.301a
SEM	0.012	0.004	0.002	0.003	0.008	0.004
Forage height						
Tall	0.403	0.206a	0.117a	0.138a	0.464	0.297a
Short	0.406	0.178b	0.093b	0.105b	0.446	0.266b
SEM	0.012	0.005	0.002	0.003	0.009	0.004
Interaction						
Alfalfa-Tall	0.387b	0.200a	0.127a	0.148a	0.435c	0.294a
Alfalfa-Short	0.323c	0.152b	0.084c	0.073c	0.343d	0.230b
Tall fescue-Tall	0.420b	0.212a	0.107b	0.128b	0.492b	0.301a
Tall fescue-Short	0.488a	0.205a	0.102b	0.137ab	0.549a	0.301a
SEM	0.017	0.006	0.003	0.005	0.012	0.005
p-Value						
Forage species	<0.01	<0.01	0.79	<0.01	<0.01	<0.01
Forage height	0.89	<0.01	<0.01	<0.01	0.16	<0.01
Forage species × Forage height	<0.01	<0.01	<0.01	<0.01	<0.01	<0.01

a,b,c,d Values within the same treatment groups with different letters aside indicate significant difference exists among the treatment means ($p \leq 0.05$) according to Fischer's LSD test. SEM is pooled standard error of the least square means. Amplitude is the normalized amplitude (unitless).

3.2. Performance for Classifying the Ingestive Behaviors

The performance for classifying the ingestive behaviors using three deep learning models is summarized in Table 3 and Figure 8. The *precision*, *recall*, and *F1 score* ranged from 0.615–0.941, 0.533–0.932, and 0.599–0.932 for classifying the ingestive behaviors. The *precision*, *recall*, and *F1 score* of the LSTM were averagely 0.08 higher than those of Conv1D, and 0.02 higher than those of Conv2D. Overall, the classification performance for the original dataset was similar to that of the filtered dataset with a <0.01 difference on average, while the average performance for the imbalanced dataset was 0.07 higher than that for the balanced dataset. Based on the confusion matrixes (Figure 8), chewing behavior was more accurately classified than biting and chewing-biting behaviors. The Conv1D, Conv2D, and LSTM spent 70.048–74.419, 59.408–75.120, 85.366–88.035 ms for processing 1-s acoustic data. The LSTM with the filtered and imbalanced dataset was selected for further development because of better performance and comparably faster processing speed for classifying the three ingestive behaviors.

Table 3. *Precision*, *recall*, and *F1 score* of the three deep learning models for classifying the ingestive behaviors in various datasets.

Model	Behavior	Original-Imbalanced			Original-Balanced			Filtered-Imbalanced			Filtered-Balanced		
		Precision	*Recall*	*F1 Score*	*Precision*	*Recall*	*F1 Score*	*Precision*	*Recall*	*F1 Score*	*Precision*	*Recall*	*F1 Score*
Conv1D	Bite	0.782	0.819	0.800	0.827	0.819	0.823	0.783	0.790	0.786	0.837	0.781	0.808
	Chew	0.893	0.901	0.897	0.753	0.638	0.691	0.865	0.932	0.897	0.667	0.857	0.750
	Chew-bite	0.779	0.744	0.761	0.615	0.714	0.661	0.818	0.700	0.754	0.683	0.533	0.599
	Overall	0.840	0.841	0.840	0.731	0.724	0.725	0.837	0.839	0.838	0.729	0.724	0.726
Conv2D	Bite	0.810	0.810	0.810	0.851	0.819	0.835	0.728	0.867	0.791	0.844	0.876	0.860
	Chew	0.900	0.920	0.910	0.823	0.752	0.786	0.941	0.886	0.913	0.820	0.867	0.843
	Chew-bite	0.821	0.789	0.805	0.712	0.800	0.753	0.821	0.817	0.819	0.821	0.743	0.780
	Overall	0.861	0.862	0.861	0.795	0.790	0.792	0.869	0.862	0.865	0.828	0.829	0.828
LSTM	Bite	0.829	0.829	0.829	0.855	0.895	0.874	0.820	0.867	0.843	0.881	0.848	0.864
	Chew	0.935	0.929	0.932	0.767	0.848	0.805	0.935	0.895	0.915	0.864	0.905	0.884
	Chew-bite	0.841	0.850	0.845	0.831	0.705	0.763	0.824	0.861	0.842	0.837	0.829	0.833
	Overall	0.889	0.888	0.888	0.818	0.816	0.817	0.883	0.880	0.881	0.860	0.860	0.860

Notes: Conv1D is one-dimensional convolutional neural network; Conv2D is two-dimensional convolutional neural network; and LSTM represents long short-term memory network. "Original" indicates the original dataset without any filtering; "Filtered" indicates the original dataset was filtered with the bandstop filter to remove background beeping sounds; "imbalanced" indicates the dataset with unequal audio file sizes for the three ingestive behaviors; and "balanced" indicates the dataset with equal audio file sizes for the three ingestive behaviors.

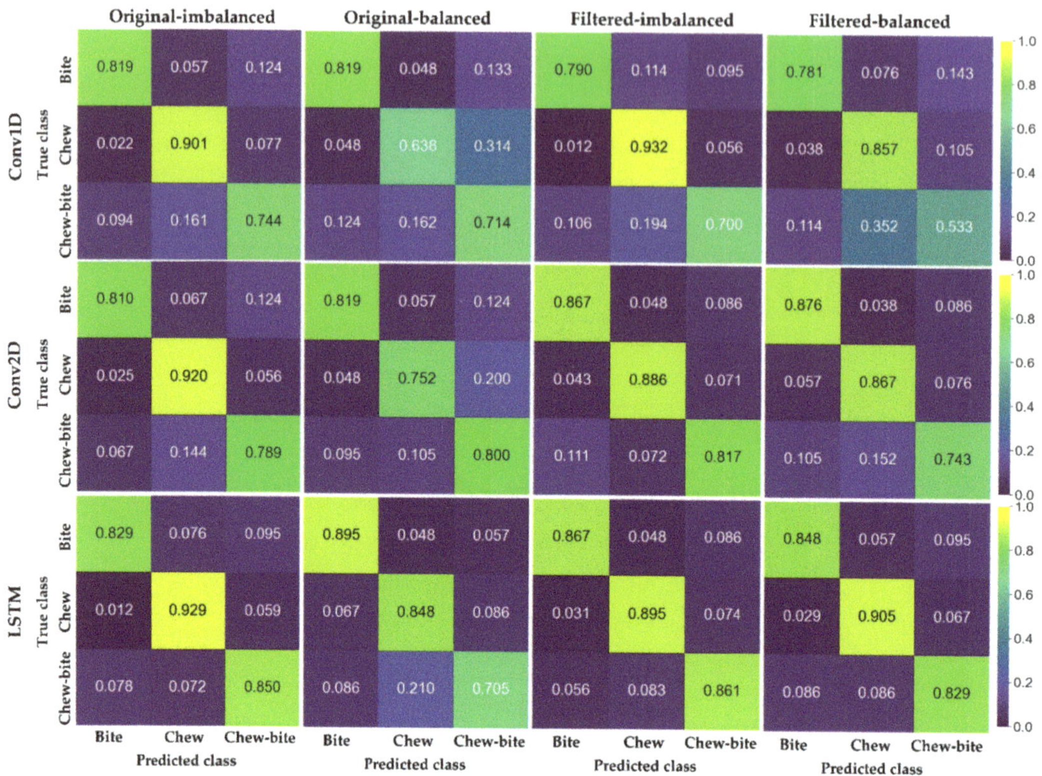

Figure 8. Confusion matrixes of the three deep learning models for classifying the ingestive behaviors in various datasets. Conv1D represents one-dimensional convolutional neural network; Conv2D represents two-dimensional convolutional neural network; and LSTM represents long short-term memory network. "Original" indicates the original dataset without any filtering; "Filtered" indicates the dataset was filtered with the bandstop filter to remove background beeping sounds; "imbalanced" indicates the dataset with unequal audio file sizes for the three ingestive behaviors; and "balanced" indicates the dataset with equal audio file sizes for the three ingestive behaviors.

3.3. Performance for Classifying the Ingestive Behaviors under Various Forage Conditions

The classification performance to identify particular ingestive behaviors associated with key forage characteristics was further investigated with the LSTM and filtered-imbalanced dataset (Table 4 and Figure 9). The overall *precision*, *recall*, and *F1 score* were 0.1–0.2 lower than those in Section 3.2. On average, the LSTM had similar classification performance for the two forage species (0.758 for alfalfa and 0.738 for tall fescue) and heights (0.620 for short and 0.620 for tall). By contrast, the overall classification performance of the ingestive behaviors for the forage species was approximately 0.1 higher than for the two forage heights. As for forage species (Figure 9), classifying biting behavior under tall fescue had the lowest accuracy (0.436) while classifying chewing under tall fescue had the highest accuracy (0.905). As for forage heights, the highest (0.804) and lowest (0.418) accuracies were observed when classifying chewing and biting for short forage.

Table 4. *Precision*, *recall*, and *F1 score* for classifying the ingestive behaviors under various forage conditions.

Behavior	Forage Species	*Precision*	*Recall*	*F1 Score*	Behavior	Forage Height	*Precision*	*Recall*	*F1 Score*
Bite	Alfalfa	0.742	0.697	0.719	Bite	Short	0.590	0.418	0.489
	Tall fescue	0.630	0.436	0.515		Tall	0.500	0.480	0.490
Chew	Alfalfa	0.720	0.753	0.736	Chew	Short	0.594	0.603	0.599
	Tall fescue	0.784	0.835	0.809		Tall	0.771	0.723	0.746
Chew-bite	Alfalfa	0.838	0.801	0.819	Chew-bite	Short	0.723	0.804	0.761
	Tall fescue	0.851	0.905	0.877		Tall	0.746	0.779	0.762
Overall		0.793	0.797	0.795	Overall		0.694	0.698	0.696

Notes: The model performance was investigated with the long short-term memory model and the filtered and imbalanced dataset.

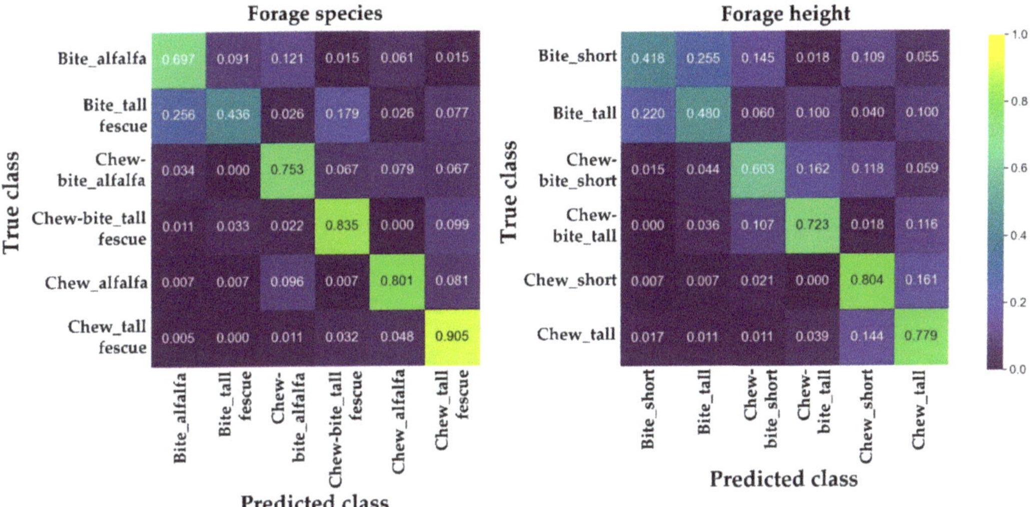

Figure 9. Confusion matrixes for classifying the ingestive behaviors under various forage characteristics. The model performance was investigated with the long short-term memory model and the filtered and imbalanced dataset.

4. Discussion

4.1. Effects of Forage on Ingestive Sound Characteristics

Dairy cows can generate different sounds when ingesting grass materials (e.g., different grass/hay species, multiple heights, etc.) based on this research and previous investigations [5]. Understanding such information, especially the relationship between the forage characteristics and ingestive behaviors, is important to provide good cattle grazing strategies with appropriate welfare status. The ingestive sounds can be further linked with forage intake for providing supplemental information on precision management of grass utilization and animal health status [5,20]. Variations in ingestive sound characteristics may be attributed to grass/feed bulk density [21], dry matter content [20], surface area [22], diurnal pattern of intake [10], and many other options e.g., relative sheer strength of the forage. However, due to a lack of detailed information about forage characteristics, individual information, and animal status in the open-access dataset, the actual reason for the variations remains unclear and should be researched in the future.

4.2. Overall Classification Performance

The overall positive classification performance of this study and previous literature is presented in Table 5. The performance of previous studies was mainly obtained with machine learning models or knowledge-based models. One possible reason for the slightly reduced performance of the current study was the inclusion of all performance cases in the analysis (i.e., Conv1D, Conv2D, balanced dataset, etc.). With the optimal case (LSTM with filtered and imbalanced dataset), the classification performance (0.820–0.867 for bites, 0.895–0.935 for chews, and 0.824–0.861 for chew-bites) outperformed or was comparable to previous studies. The successful classification could be attributed to obvious differences in the acoustic signals among bites, chews, and chew-bites, robust data cleaning, and appropriate design of model architectures. Current forage characteristics may have little influence on model performance improvement. Perhaps, more diverse forage characteristics should be included in future research for optimizing model performance.

Table 5. Performance comparison for classifying bites, chews, and chew-bites of dairy cows among different studies.

Positive Performance			Reference
Bites	**Chews**	**Chew-Bites**	
0.728–0.895	0.638–0.941	0.533–0.861	Current study
0.620–0.900	0.880–0.990	0.430–0.940	[11]
0.760–0.900	0.880–0.990	0.610–0.940	[14]
–	0.670–0.990	–	[23]

Notes: The positive performance includes accuracy, *precision*, *recall*, and *F1 score*. "–" indicates missing information.

4.3. Deep Learning Models

To the authors' knowledge, this paper is the first to assess the application of deep learning models for classifying dairy cow ingestive behavior sounds. During model development, key features for effective classification were learned directly from the dataset, and exhaustive labeling and dedicated manual design were not required for feature extractors that are typically required in machine learning or knowledge-based algorithms [24], enabling scientists in other domains without extensive computer science expertise applying deep learning techniques. Deep CNNs are good at handling acoustic signals because of their efficient computation and powerful learning ability [15,25]. In this study, the Conv1D and Conv2D did not perform as well as the LSTM, due to two possible reasons. Firstly, more efficient and accurate connection schemes (e.g., residual connection [26], inception connection [27], etc.) were not applied in the CNNs. Secondly, the LSTM can learn backward and feedforward features from acoustic signals, which is critical for dealing with sequential data [28]. Besides acceptable detection accuracy, decent processing speed (85.366–88.035 ms for processing 1-s acoustic data) was also achieved by the LSTM. Although powerful computing devices (with GPU of RTX 3080) are crucial components, the extremely light weight ($\leq$2 MB) of the network architectures was the primary factor for the fast processing speed [29]. Because current networks can balance detection accuracy and processing speed, they offer new opportunities for real-time monitoring of animal conditions [30], behaviors [31], etc. for cattle industry.

4.4. Other Factors Influencing Classification Performance

Other possible influencing factors should be considered as well for future model development of sound classification. Currently, deep learning experts are switching their attention from model-centric to data-centric to improve detection performance [32]. Data quality plays a crucial role in deep learning, where improving the model hits a bottleneck now. For instance, particularly for acoustic datasets, data challenges of noise, balance, and quantity must be addressed. Considerable data noise can downgrade model performance to varying degrees. The dataset was recorded in a controlled environment with minimal introduced noises, and therefore, there was no significant model performance improvement after noise filtering. However, such a controlled environment with minimal background noise is hard to achieve in on-farm or in-field conditions. Imbalanced datasets can result in biased inference for classes with the larger proportions [33]. However, in this study the balanced dataset with the same number of audio files but uneven durations had poorer performance than the imbalanced dataset. Thus, the length of audio files may play a more important role in classification improvement than audio file quantity when selected for balancing classes. Uneven duration of audio files also downgraded the performance for different forages. Sufficient data is necessary to explore the optimal performance of deep learning models [34]. The current dataset contained only 27.5 min of useful data, and such audio length became smaller when the dataset was split based on ingestive behaviors and forage characteristics. The relatively small ingestive sound dataset is typical of studies with dairy cows (i.e., 13 min in the study of Chelotti et al. [11]) because of their quick ingestive actions, challenging environment for data collection, laborious manual labeling, etc. A large ingestive sound dataset is recommended to be built in the future to improve model performance.

Forage characteristics also influenced the automatic classification of ingestive sounds. The best and poorest performance difference for classifying the ingestive behaviors was 0.4–0.5 among different forage characteristics. Chelotti et al. [11] reported a 0.11–0.41 performance difference and Milone et al. [14] demonstrated a 0.11–0.33 performance difference for classifying the ingestive behaviors under similar forage characteristics used in this study. Apart from uneven class balance and small datasets, similar acoustic features between the alfalfa and tall fescue or the two forage heights could decrease the classification performance. This may indicate that current techniques may not be sufficient or generalizable to differentiate the ingestive behaviors for various forage characteristics. When sound classification for specific forage is needed, the model may need re-development with custom datasets for robust classification.

5. Conclusions

Classification of the three ingestive behaviors (bites, chews, and chew-bites) of dairy cows using deep learning models was conducted in this study. The results showed that forage species (alfalfa vs. tall fescue) and heights (tall and short) significantly influenced the amplitude and duration of the ingestive sounds of dairy cows. The LSTM using a filtered dataset with balanced duration and imbalanced audio files had better performance than its counterparts. Currently, it is difficult to differentiate the bites, chews, and chew-bites between alfalfa and tall fescue under two different heights. In addition to training the LSTM with more temporal data, sophisticated feature extraction techniques will be considered in a future study.

Author Contributions: Conceptualization, G.L., Y.X., and R.S.G.; methodology, G.L.; software, G.L.; validation, G.L.; formal analysis, G.L.; investigation, G.L.; resources, G.L., Y.X., and R.S.G.; data curation, G.L.; writing—original draft preparation, G.L.; writing—review and editing, Y.X., Q.D., Z.S., and R.S.G.; visualization, G.L.; supervision, Y.X. and R.S.G.; project administration, G.L. and Y.X.; funding acquisition, Y.X. and R.S.G. All authors have read and agreed to the published version of the manuscript.

Funding: This research was funded by College of Agriculture and Life Sciences, Iowa State University, and the Institution of Agriculture and Natural Resources, the University of Nebraska-Lincoln.

Institutional Review Board Statement: Not applicable.

Informed Consent Statement: Not applicable.

Data Availability Statement: The acoustic dataset of dairy cows is available at https://github.com/sinc-lab/dataset-jaw-movements (accessed on 15 July 2021).

Acknowledgments: The authors acknowledge the research team of Sebastián R. Vanrell for releasing the acoustic dataset.

Conflicts of Interest: The authors declare no conflict of interest. The funders had no role in the design of the study; in the collection, analyses, or interpretation of data; in the writing of the manuscript, or in the decision to publish the results.

References

1. Li, G.; Huang, Y.; Chen, Z.; Chesser, G.D.; Purswell, J.L.; Linhoss, J.; Zhao, Y. Practices and applications of convolutional neural network-based computer vision systems in animal farming: A review. *Sensors* **2021**, *21*, 1492. [CrossRef] [PubMed]
2. U.S. Department of Agriculture. Milk Production. 2021. Available online: https://www.nass.usda.gov/Publications/Todays_Reports/reports/mkpr0321.pdf (accessed on 21 May 2021).
3. MacDonald, J.M.; Law, J.; Mosheim, R. *Consolidation in U.S. Dairy Farming*; U.S. Department of Agriculture: Washington, DC, USA, 2020.
4. Perdue, S.; Hamer, H. *Census of Agriculture*; U.S. Department of Agriculture: Washington, DC, USA, 2019.
5. Galli, J.R.; Cangiano, C.A.; Pece, M.A.; Larripa, M.J.; Milone, D.H.; Utsumi, S.A.; Laca, E.A. Monitoring and assessment of ingestive chewing sounds for prediction of herbage intake rate in grazing cattle. *Animal* **2018**, *12*, 973–982. [CrossRef]
6. Chelotti, J.O.; Vanrell, S.R.; Galli, J.R.; Giovanini, L.L.; Rufiner, H.L. A pattern recognition approach for detecting and classifying jaw movements in grazing cattle. *Comput. Electron. Agric.* **2018**, *145*, 83–91. [CrossRef]
7. Andriamandroso, A.; Bindelle, J.; Mercatoris, B.; Lebeau, F. A review on the use of sensors to monitor cattle jaw movements and behavior when grazing. *Biotechnol. Agron. Société Environ.* **2016**, *20*, 1–14. [CrossRef]
8. Forbes, T.D.A. Researching the Plant-Animal Interface: The investigation of Ingestive Behavior in Grazing Animals. *J. Anim. Sci.* **1988**, *66*, 2369–2379. [CrossRef]
9. Milone, D.H.; Rufiner, H.L.; Galli, J.R.; Laca, E.A.; Cangiano, C.A. Computational method for segmentation and classification of ingestive sounds in sheep. *Comput. Electron. Agric.* **2009**, *65*, 228–237. [CrossRef]
10. Barrett, P.D.; Laidlaw, A.S.; Mayne, C.S.; Christie, H. Pattern of herbage intake rate and bite dimensions of rotationally grazed dairy cows as sward height declines. *Grass Forage Sci.* **2001**, *56*, 362–373. [CrossRef]
11. Chelotti, J.O.; Vanrell, S.R.; Milone, D.H.; Utsumi, S.A.; Galli, J.R.; Rufiner, H.L.; Giovanini, L.L. A real-time algorithm for acoustic monitoring of ingestive behavior of grazing cattle. *Comput. Electron. Agric.* **2016**, *127*, 64–75. [CrossRef]
12. Chelotti, J.O.; Vanrell, S.R.; Rau, L.S.M.; Galli, J.R.; Planisich, A.M.; Utsumi, S.A.; Milone, D.H.; Giovanini, L.L.; Rufiner, H.L. An online method for estimating grazing and rumination bouts using acoustic signals in grazing cattle. *Comput. Electron. Agric.* **2020**, *173*, 105443. [CrossRef]
13. Clapham, W.M.; Fedders, J.M.; Beeman, K.; Neel, J.P. Acoustic monitoring system to quantify ingestive behavior of free-grazing cattle. *Comput. Electron. Agric.* **2011**, *76*, 96–104. [CrossRef]
14. Milone, D.H.; Galli, J.R.; Cangiano, C.A.; Rufiner, H.L.; Laca, E.A. Automatic recognition of ingestive sounds of cattle based on hidden Markov models. *Comput. Electron. Agric.* **2012**, *87*, 51–55. [CrossRef]
15. LeCun, Y.; Bengio, Y.; Hinton, G. Deep learning. *Nature* **2015**, *521*, 436–444. [CrossRef]
16. Vanrell, S.R.; Chelotti, J.O.; Bugnon, L.A.; Rufiner, H.L.; Milone, D.H.; Laca, E.A.; Galli, J.R. Audio recordings dataset of grazing jaw movements in dairy cattle. *Data Brief* **2020**, *30*, 105623. [CrossRef]
17. Saxton, A. A Macro for Converting Mean Separation Output to Letter Groupings in Proc Mixed. In Proceedings of the 23rd SAS Users Group International, Nashville, TN, USA, 22–25 March 1998; SAS Institute Inc.: Nashville, TN, USA, 1998; pp. 1243–1246.
18. Muda, L.; Begam, M.; Elamvazuthi, I. Voice recognition algorithms using mel frequency cepstral coefficient (MFCC) and dynamic time warping (DTW) techniques. *J. Comput.* **2010**, *2*, 138–143.
19. Seth18. Audio Classification. 2020. Available online: https://github.com/seth814/Audio-Classification (accessed on 21 May 2021).
20. Galli, J.R.; Cangiano, C.A.; Demment, M.W.; Laca, E.A. Acoustic monitoring of chewing and intake of fresh and dry forages in steers. *Anim. Feed Sci. Technol.* **2006**, *128*, 14–30. [CrossRef]
21. Rook, A.J.; Huckle, C.A.; Penning, P.D. Effects of sward height and concentrate supplementation on the ingestive behaviour of spring-calving dairy cows grazing grass-clover swards. *Appl. Anim. Behav. Sci.* **1994**, *40*, 101–112. [CrossRef]
22. Gibb, M.J.; Huckle, C.A.; Nuthall, R.; Rook, A.J. Effect of sward surface height on intake and grazing behaviour by lactating Holstein Friesian cows. *Grass Forage Sci.* **1997**, *52*, 309–321. [CrossRef]
23. Ungar, E.D.; Rutter, S.M. Classifying cattle jaw movements: Comparing IGER Behaviour Recorder and acoustic techniques. *Appl. Anim. Behav. Sci.* **2006**, *98*, 11–27. [CrossRef]
24. Khalid, S.; Khalil, T.; Nasreen, S. A survey of feature selection and feature extraction techniques in machine learning. In Proceedings of the 2014 Science and Information Conference, London, UK, 27–29 August 2014; IEEE Xplore: London, UK, 2014; pp. 372–378. [CrossRef]
25. Hinton, G.; Deng, L.; Yu, D.; Dahl, G.E.; Mohamed, A.; Jaitly, N.; Senior, A.; Vanhoucke, V.; Nguyen, P.; Sainath, T.N.; et al. Deep Neural Networks for Acoustic Modeling in Speech Recognition: The Shared Views of Four Research Groups. *ISPM* **2012**, *29*, 82–97. [CrossRef]
26. He, K.; Zhang, X.; Ren, S.; Sun, J. Deep residual learning for image recognition. In Proceedings of the IEEE Conference on Computer Vision and Pattern Recognition, Las Vegas, NV, USA, 26 June–1 July 2016; Caesars Palace: Las Vegas, NV, USA, 2016; pp. 770–778.
27. Szegedy, C.; Ioffe, S.; Vanhoucke, V.; Alemi, A. Inception-v4, Inception-ResNet and the Impact of Residual Connections on Learning. In Proceedings of the AAAI Conference on Artificial Intelligence, San Francisco, CA, USA, 4–9 February 2017; Volume 31. Available online: https://ojs.aaai.org/index.php/AAAI/article/view/11231 (accessed on 21 May 2021).

28. Sherstinsky, A. Fundamentals of Recurrent Neural Network (RNN) and Long Short-Term Memory (LSTM) network. *Phys. D Nonlinear Phenom.* **2020**, *404*, 132306. [CrossRef]
29. Huang, J.; Rathod, V.; Sun, C.; Zhu, M.; Korattikara, A.; Fathi, A.; Fischer, I.; Wojna, Z.; Song, Y.; Guadarrama, S. Speed/accuracy trade-offs for modern convolutional object detectors. In Proceedings of the IEEE conference on computer vision and pattern recognition, Hawai'i Convention Center, Honolulu, HI, USA, 21–26 July 2017; Hawai'i Convention Center: Honolulu, HI, USA, 2017; pp. 7310–7311.
30. Jung, D.-H.; Kim, N.Y.; Moon, S.H.; Jhin, C.; Kim, H.-J.; Yang, J.-S.; Kim, H.S.; Lee, T.S.; Lee, J.Y.; Park, S.H. Deep Learning-Based Cattle Vocal Classification Model and Real-Time Livestock Monitoring System with Noise Filtering. *Animals* **2021**, *11*, 357. [CrossRef] [PubMed]
31. Rau, L.M.; Chelotti, J.O.; Vanrell, S.R.; Giovanini, L.L. Developments on real-time monitoring of grazing cattle feeding behavior using sound. In Proceedings of the 2020 IEEE International Conference on Industrial Technology (ICIT), Buenos Aires, Argentina, 26–28 February 2020; IEEE Xplore: Buenos Aires, Argentina, 2020; pp. 771–776. [CrossRef]
32. Wu, A. A Chat with Andrew on MLOps: From Model-Centric to Data-Centric AI. 2021. Available online: https://www.youtube.com/watch?v=06-AZXmwHjo&t=1048s (accessed on 21 May 2021).
33. Mikołajczyk, A.; Grochowski, M. Data augmentation for improving deep learning in image classification problem. In Proceedings of the 2018 International Interdisciplinary PhD Workshop (IIPhDW), Świnoujście, Poland, 9–12 May 2018; IEEE Xplore: Świnoujście, Poland, 2018; pp. 117–122. [CrossRef]
34. Jan, B.; Farman, H.; Khan, M.; Imran, M.; Islam, I.U.; Ahmad, A.; Ali, S.; Jeon, G. Deep learning in big data Analytics: A comparative study. *Comput. Electr. Eng.* **2019**, *75*, 275–287. [CrossRef]

MDPI

Article

Automatic Detection and Segmentation for Group-Housed Pigs Based on PigMS R-CNN †

Shuqin Tu, Weijun Yuan, Yun Liang *, Fan Wang and Hua Wan

College of Mathematics and Informatics, South China Agricultural University, Guangzhou 510642, China; Tsq5_6@scau.edu.cn (S.T.); ywj@stu.scau.edu.cn (W.Y.); 201827010518@stu.scau.edu.cn (F.W.); wanhua@scau.edu.cn (H.W.)

* Correspondence: yliang@scau.edu.cn

† This manuscript is extension version of the conference paper: Tu, S.; Liu, H.; Li, J.; Huang, J.; Li, B.; Pang, J.; Xue, Y. Instance Segmentation Based on Mask Scoring R-CNN for Group-housed Pigs. In Proceedings of the 2020 International Conference on Computer Engineering and Application (ICCEA 2020), Guangzhou, China, 27–29 March 2020.

Abstract: Instance segmentation is an accurate and reliable method to segment adhesive pigs' images, and is critical for providing health and welfare information on individual pigs, such as body condition score, live weight, and activity behaviors in group-housed pig environments. In this paper, a PigMS R-CNN framework based on mask scoring R-CNN (MS R-CNN) is explored to segment adhesive pig areas in group-pig images, to separate the identification and location of group-housed pigs. The PigMS R-CNN consists of three processes. First, a residual network of 101-layers, combined with the feature pyramid network (FPN), is used as a feature extraction network to obtain feature maps for input images. Then, according to these feature maps, the region candidate network generates the regions of interest (RoIs). Finally, for each RoI, we can obtain the location, classification, and segmentation results of detected pigs through the regression and category, and mask three branches from the PigMS R-CNN head network. To avoid target pigs being missed and error detections in overlapping or stuck areas of group-housed pigs, the PigMS R-CNN framework uses soft non-maximum suppression (soft-NMS) by replacing the traditional NMS to conduct post-processing selected operation of pigs. The MS R-CNN framework with traditional NMS obtains results with an F1 of 0.9228. By setting the soft-NMS threshold to 0.7 on PigMS R-CNN, detection of the target pigs achieves an F1 of 0.9374. The work explores a new instance segmentation method for adhesive group-housed pig images, which provides valuable exploration for vision-based, real-time automatic pig monitoring and welfare evaluation.

Keywords: pig identification; mask scoring R-CNN; soft-NMS; group-housed pigs

Citation: Tu, S.; Yuan, W.; Liang, Y.; Wang, F.; Wan, H. Automatic Detection and Segmentation for Group-Housed Pigs Based on PigMS R-CNN . *Sensors* **2021**, *21*, 3251. https://doi.org/10.3390/s21093251

Academic Editors: Yongliang Qiao, Lilong Chai, Dongjian He and Daobilige Su

Received: 11 April 2021
Accepted: 2 May 2021
Published: 7 May 2021

Publisher's Note: MDPI stays neutral with regard to jurisdictional claims in published maps and institutional affiliations.

1. Introduction

With the development of artificial intelligence and automation technology, utilizing video cameras to monitor the health and welfare of pigs has become more important in the modern pig industry. In group-housed environments, instance segmentation of pigs includes detection, which automatically obtains the positions of all pigs, and segmentation, which distinguishes each pig in the images [1]. Many high-level and intelligent pig farming applications, such as pig weight estimation [2], pig tracking [3], and behavior recognition [4–7], require accurate detection and segmentation of pig objects in complex backgrounds. The premise and foundation of pig behavior analysis involves the accurate detection and segmentation of group pig images [8]. Therefore, detecting and segmenting group-housed pigs can help improve the efficiency of instance segmentation, complete high-level applications, and improve the welfare of pigs in pig farms.

Digital image processing combined with pattern recognition studies use automatic detection and segmentation techniques to group-housed pigs [9]. Pig detection and segmentation methods include two categories based on non-deep learning and deep learning algorithms. These non-deep learning approaches have more mature technologies and have been widely applied in video surveillance of pigs [1,10–12]. Guo et al. [10] proposed a pig's foreground detection method based on the combination of a mixture of Gaussians and threshold segmentation. This approach achieved an average pig object detection rate with approximately 92% in complex scenes. Guo et al. [11] also proposed an effective method for identifying individual group-housed pigs from a feeder and a drinker using a multilevel threshold segmentation. The method achieved a 92.5% average detection rate on test video data. An approach [5] was proposed to detect mounting events amongst pigs for pig video files. The method used the Euclidean distances of the different parts of pigs to automatically recognize a mounting event. This approach can obtain the results of sensitivity, specificity, and accuracy, with 94.5%, 88.6%, and 92.7%, respectively, for identifying mounting events. The algorithm [12] for the group-housed pig detection was developed, and it used Gabor and LBP features for feature extraction and classified each pig by SVM. A recognition rate with 91.86% was obtained by this algorithm. Li et al. [1] combined appearance features and the template matching framework to detect each pig under a group-housed environment. From these related studies, we found that these non-deep learning methods have some disadvantages, and that feature extraction and recognition of the pigs are separated. Meanwhile, feature extractions of these methods obtain features with handed-design features and cannot automatically learn features from the amount of data, which causes these methods to lack robustness.

The studies on pig detection and segmentation methods, based on deep learning models, have increased in recent years, since Girshick et al. [13] proposed the R-CNN method. These algorithms can automatically extract the pig's target features in an image and avoid the process of extracting the target features by artificial observation, so the obtained model has strong universality. Faster R-CNN [14] was used to detect pig targets and classify lactating sow postures, including standing, sitting, sternal recumbency, ventral recumbency, and lateral recumbency, using depth images [15]. Yang et al. [16] first used faster R-CNN to detect individual pigs and their heads. Then, a behavior identification algorithm was implemented for feeding behavior recognition from a group-housed pen. Finally, the results of a precision rate with 0.99 and recall rate with 0.8693 can be obtained for feeding behavior recognition of pigs. Xue et al. [17] proposed an approach based on a fully convolutional network (FCN) [18] and Otsu's thresholding to segment the lactating sow images. The approach achieved a 96.6% mean accuracy rate. He et al. [19] proposed an end-to-end framework named mask R-CNN for object detection and segmentation. The mask R-CNN achieved good results for the challenging instance segmentation dataset COCO [13], and was used in cattle segmentation [20]. These recent research results show that pig detection and segmentation approaches, based on deep learning models, were effective and widely used under complex scene environments. However, when pigs are under heavy overlap and adhesion, pig instance segmentation methods, based on non-deep learning and deep learning algorithms, have some difficulty accurately detecting individual pigs with a low miss rate.

Segmenting the adhesive pig images is important to the next extraction of pig herd behavioral characteristic parameters. The adhesion segmentation methods mainly include ellipse fitting, watershed transformation, and concave point analysis. In [4,9], the least square method was used to conduct ellipse fitting for pigs, and then it separated the adhered pig bodies according to the length of the major and minor axis, and the position of the center point, but this method did not evaluate the performance of the ellipse fitting method when several pigs adhered. Xiong et al. [21] divided piglet adhesion into four cases, as follows: no adhesion, slight, mild, and severe adhesion. Firstly, the contour line of the adhered part was extracted. Then, the contour line, according to the concave point, was segmented. The ellipse fitting for the contour lines was carried out after segmentation. Finally, five ellipse-screening rules were developed. The ellipses, which did not comply with the rules, were integrated. The recognition accuracy of this method for pigs was over 86%. A Kinect-based segmentation of touching-pigs was used for segmentation of touching-pigs, by applying YOLO and CPs [8], and this method was effective at separating touching-pigs with an accuracy of 91.96%. Mask scoring R-CNN (MS R-CNN) [22] was explored for instance segmentation of standing posture images of group-housed pigs, from the top view and front view pig video sequences, which achieved a best F1 score with 0.9405 on pig test datasets in our previous work [23]. However, severe adhesion was not analyzed, but is further researched in these studies.

Although these works have been performed effectively on pig detection and segmentation, most approaches are designed under fewer pigs/one pig environments or top view images; it is difficult for these algorithms to detect each pig in a group-housed condition when the pigs are closely grouped. Based on the previous work of our team [23], we propose a PigMS R-CNN model, which can efficiently obtain the instance segmentation mask for each pig, while simultaneously decrease the faulty detection results for group-housed pigs. To reduce the missed rate of pigs in group-housed environments, soft-NMS [24] was used in the PigMS R-CNN model. The algorithm gains improvements in precision measured over multiple overlap thresholds, which are especially suitable for group-housed pig detection. The PigMS R-CNN model, combining the MS R-CNN and soft-NMS network, is simple to implement and does not need extra resources. We can easily use the improved approach for a pig instance segmentation application.

2. Materials and Methods

2.1. Data Acquisition

The experiment's dataset was collected in "Lejiazhuang Pig Farm" of live pigs in Foshan city, Guangdong Province, China. The settings of the cameras of the experimental pigsty are shown in Figure 1. In Figure 1a, the pigsty was 3 m high, 7 m long, and 5 m wide. Figure 1b,c present the images, in video surveillance, of top view and front view. A camera was placed in the middle of the pigsty, 2 m above the ground. Another camera was placed in front of the pigsty, 1.35 m above the ground. We used the FL3-U3-88S2C-C camera, which obtained images at 1920 × 1080 pixels.

Figure 1. (**a**) Video surveillance system for data acquisition, (**b**) top view image, and (**c**) front view image.

To generate sufficient images in the experiments, we focused on 5 pigsties, where the number of pigs included is 3–20; video data were randomly obtained during a course of 10 days, each day containing over 7 h of video, from 9 a.m. to 4 p.m. The video dataset was saved in AVI format and the frame frequency of the video was 25 fps. We chose 420 images as the dataset. Among 420 images, 290 images that included 2838 pigs were used for the training set, and 130 images that included 1147 pigs were used for the test set. We marked pigs manually with VIA software for these images. The typical augmentation methods, including left-right flipping, rotation, and re-scaling were automatically used to enlarge the training dataset. With no external light used, all images showed uneven illumination. Therefore, the dataset reflected the common characteristics of pig monitoring, and it can objectively estimate a pig's instance segmentation performance in a group-housed condition.

2.2. Data Labeling

To evaluate the proposed algorithm, the frames with the standing status of pigs in videos were labeled to illustrate the detection and segmentation results of the algorithm. VIA (http://www.robots.ox.ac.uk/~vgg/software/via/ accessed on 7 May 2021) is an open-source image annotation tool that can be used online or offline. VIA software can label rectangles, circles, ellipses, polygons, points, and lines, set area properties, and save the annotation information as CSV or JSON file formats.

In this study, we used it to label the contour of the group-house pigs with the shape of the polygon area, extracted the annotation information of the pig's contour, and saved as a JSON file format. These JSON files included image name, image size, label name, anchor coordinate information of each pig object in each image, and area attribute name. Labeling the 420 images with each, including the 3–20 pig objects, cost about 180 person-hours. In the 420 images, the total number of pigs was 3985, including some incomplete pigs (with over half of their bodies visible) in the image borders.

3. The Proposed Approach

3.1. The PigMS R-CNN Model

The PigMS R-CNN model (as shown in Figure 2) based on MS R-CNN [22] included three stages. In the first stage, a residual network of 101 (or 50) layers, combining the feature pyramid network (FPN), was used as a feature extraction network to obtain feature maps for input images. FPN can obtain different levels of the feature maps according to three scales. In the second stage, the region proposal network (RPN) extracted the regions of interest (RoIs), according to the feature maps. In the third stage, for each RoI, we can get the location, classification, and segmentation results of detected pig targets in the group-housed scenes through the category, and regression, and mask three branches in the PigMS R-CNN head network. In the segmentation branch, the Maskiou head in FCN was used to regress between the predicted mask and the true ground mask to improve the segmentation accuracy.

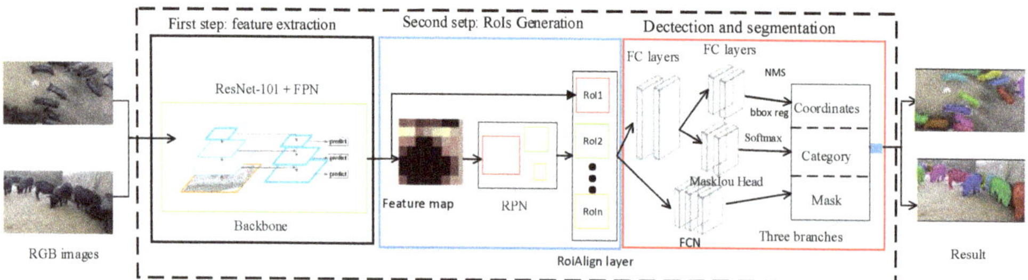

Figure 2. Flowchart of the detection and segmentation algorithm based on PigMS R-CNN.

In addition, the third stage extracted features using RoIAlign from each candidate RoI, and performed BB regression, classification with softmax, and a binary mask prediction for each potentially detected pig by FCN. During the process of BB regression, NMS was applied to these potentially detected pigs to remove highly overlapping BB and obtain the ultimate locations of the pigs. The steps are detailed in the following sections.

3.2. The Feature Extraction

The feature extraction uses backbone architecture to extract different levels of features over an entire image. The backbone architecture includes two parts, named ResNet [25] and FPN [26]. The deep ResNets are easy to optimize for training and obtain high accuracy from greatly increased depth compared with other networks. ResNets with 50-layer, 101-layer, and 152-layer depths are the most commonly used residual structures on detection and segmentation. In this paper, with comprehensive consideration of accuracy and operation time, the ResNet-101 network of a depth of 101 layers was used in the implementation of the mask R-CNN, and primarily extracted features from the three convolutional layers of the third, fourth, and fifth stages, as shown in Figure 3, left, which fed to the next multiscale backbone architecture.

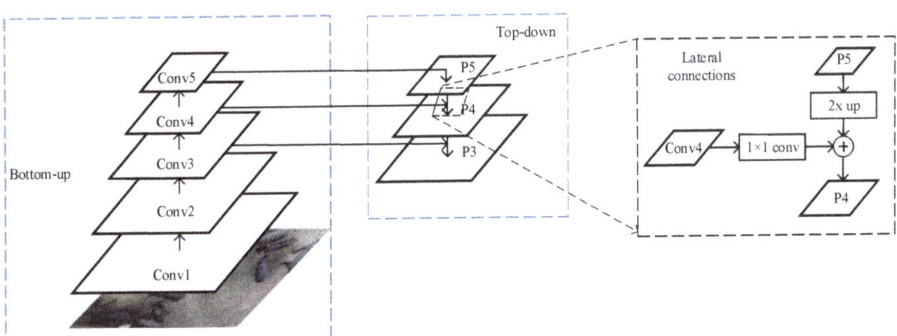

Figure 3. The architecture of feature pyramid network.

Another part of effective backbone architecture was FPN, proposed by Tsung Yi Lin [26]. FPN can construct a multi-output feature pyramid from a single-scale input by using top-down architecture and lateral connections, as shown in Figure 3, middle and right. The detailed process of Figure 3 is described in the following.

The conv1, conv2, conv3, conv4, and conv5 outputs are obtained from the last residual blocks in the ResNet101 network. First, the top-layer feature map (P5) can be achieved by performing a 1×1 convolutional layer on conv5. Then, the upsampled feature map was generated by a factor of 2 on P_5, and the feature map P_4 (Figure 3, right lateral connections) can be obtained by element-wise addition operation for the upsampled feature map, and the result, which is obtained by performing a 1×1 convolutional operation on C_4. Finally, according to $\{C_3, C_4, C_5\}$, this final set of feature maps is $\{P_3, P_4, P_5\}$.

The backbone architecture of PigMS R-CNN in this study uses ResNe-101+FPN. The feature maps of backbone architecture for a housed-pig image are shown in Figure 4. Figure 4a shows source image, Figure 4b–d extract low-level features, including texture and edge contour information. Figure 4e represents high-level abstract features.

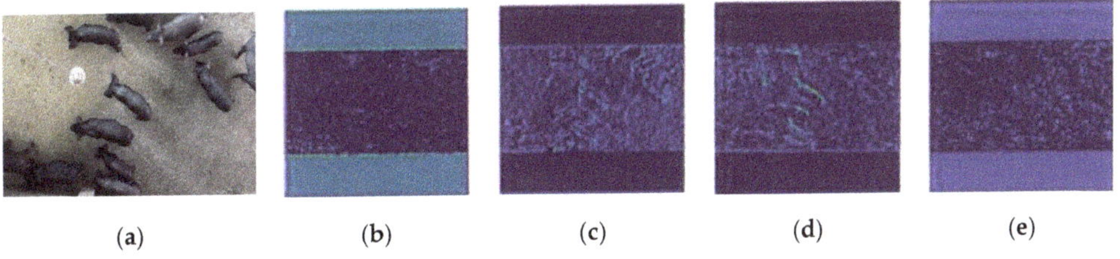

Figure 4. The feature map of backbone architecture based on ResNe101+FPN. (**a**) shows source image, (**b–d**) extract low-level features, and (**e**) represents high-level abstract features.

3.3. RoIs Generation Based on RPN

The candidate RoIs will be produced by RPN using the feature maps from the first stage. The convolution layers of a pre-trained network are followed by a 3×3 convolutional layer in the RPN. The function of this operation is to map a large spatial window or receptive field in the input image to a low-dimensional feature vector at a window location. Then two 1×1 convolutional layers are used for classification and regression operations of all spatial windows [14].

In the RPN, the algorithm introduces anchors to manage different scales and aspect ratios of objects. An anchor is located at each sliding location of the convolutional maps, and lies at the center of each spatial window, which is associated with a scale and an aspect ratio. Following the default setting of [26], five scales (32^2, 64^2, 128^2, 256^2, and 512^2 pixels) and three aspect ratios (1:1, 1:2, and 2:1) are used, and k = 15 anchors at each location are created and used for each sliding window. These anchors go through a classification layer (cls) and a regression layer (reg). The RPN then completes the following two tasks: (1) determining whether the anchors are targets or non-targets (2k); (2) performing coordinate correction on the target anchors (4k). In the classification layer branch, two scores (target and non-target) are generated for each anchor; in the regression layer branch, the parameterizations of the four coordinates are corrected, and shown as Equation (1). For each target anchor:

$$\begin{aligned} m_x &= (x - x_a)/w_a, m_y = (y - y_a)/h_a, m_w = \log(w/w_a), m_h = \log(h/h_a), \\ m^*_x &= (x^* - x_a)/w_a, m^*_y = (y^* - y_a)/h_a, m^*_w = \log(w^*/w_a), m^*_h = \log(h^*/h_a), \end{aligned} \tag{1}$$

where x, y denote the two coordinates of the box center, w, h is the width and height of the box. The x, x_a, and x^* variables denote the predicted box, anchor box, and ground-truth box respectively (likewise for y, w, h). Finally, after sorting scores of the target anchors in descending order, the first n anchors are selected for the next stage of detection.

The loss function of training RPN is as follows:

$$L(\{p_i\},\{t_i\}) = \frac{1}{Ncls}\sum_i L_{cls}(p_i, p^*_i) + \lambda\frac{1}{Nreg}\sum_i p^*_i L_{reg}(t_i, t^*_i) \tag{2}$$

where i is the index of an anchor and p_i is the predicted probability of anchor i, which is taken as a positive object. If the anchor is positive sample, the ground-truth label p^*_i is 1, otherwise it is 0. The m_i vector includes the 4 parameterized coordinates of the predicted bounding box, and m^*_i is a vector of the ground-truth box associated with a positive anchor. L_{cls} is the classification loss to express log loss for two classes (object vs. not object). $L_{reg}(m_i, m^*_i) = R(m_i - m^*_i)$ is the regression loss, where R is the robust loss function (smooth L1). The term $p^*_i L_{reg}$ denotes the regression loss which is activated only for positive anchors (p^*_i = 1) and is disabled otherwise (p^*_i = 0). The outputs of the *cls* and *reg* layers consist of $\{p_i\}$ and $\{m_i\}$, respectively. The two terms are normalized with $Ncls$, $Nreg$, and a balancing weight λ.

Some RPN proposals highly overlap with each other. To reduce redundancy, NMS, on the proposal regions based on their cls scores, is adopted. The IoU threshold for NMS is set 0.7, which leaves us about 2k proposal regions per image. NMS does not harm the ultimate detection accuracy, but substantially reduces the number of proposals. After NMS, the top-N ranked proposal regions are used for further detection and segmentation.

The process of bounding box generation based on RPN is shown in Figure 5. Figure 5a shows the source image, Figure 5b shows the results that the anchors are targets or non-targets; Figure 5c shows the results of the coordinate corrections on the target anchors.

(**a**)

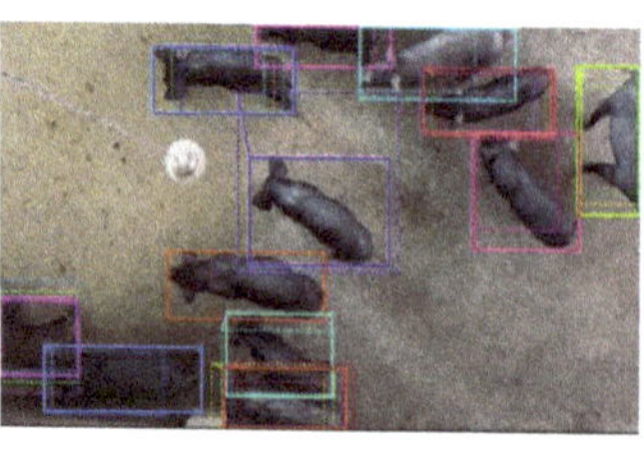

(**b**)

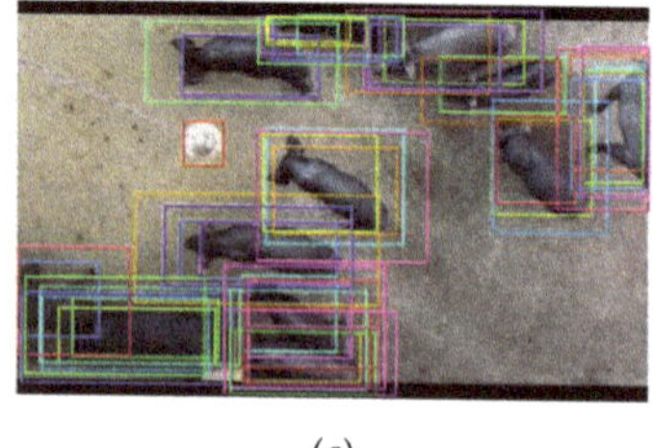

(**c**)

Figure 5. The process of bounding box generation based on RPN, (**a**) shows the source image, (**b**) shows the results that the anchors are targets or non-targets, and (**c**) shows the results of the coordinate corrections on the target anchors.

3.4. The Three Branches of Detection and Segmentation

The MS R-CNN outputs the three branches, including performing the proposal classification, regression, and a binary mask for each RoI in parallel. The first two branch structures use two fully connected (FC) layers for the region of interest (RoI) performing classification prediction and BB regression. The last branch is instance segmentation, which uses a fully convolutional network (FCN) [18] for predicting a mask from each BB.

Formally, during training, a multi-task loss (L_{total}) on each BB is defined:

$$L_{\text{total}} = L_{box} + L_{\text{cls}} + L_{mask} \tag{3}$$

where L_{box} and L_{cls} are the bounding-box and classification loss, L_{mask} was defined as the average binary cross-entropy loss. For a BB associated with ground-truth class k, L_{mask} is only defined on the k-th mask in MS R-CNN model, L_{box}, L_{cls} and L_{mask} are identical as those defined in [23].

The output of the three branches is shown in Figure 6. Figure 6a shows the BB's positions can be adjusted using a regression network. Figure 6b shows the BBs that are assigned a score for each class label using a classification network after NMS. Figure 6c shows the detection and segmentation final results of group-housed pigs.

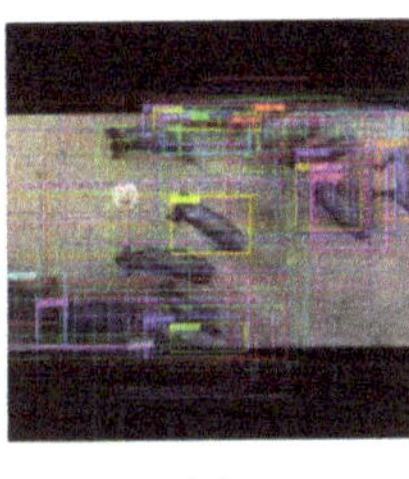

(a)

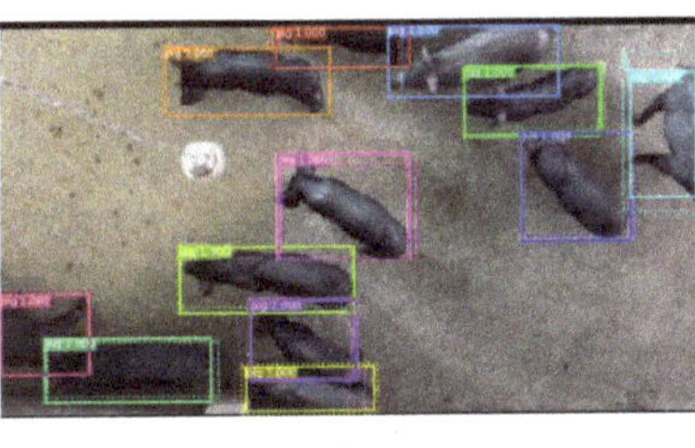

(b)

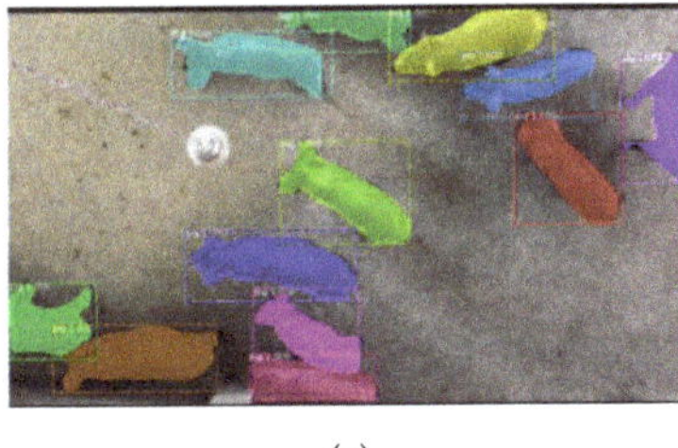

(c)

Figure 6. The output of three branches. (**a**) The BB positions, which can be updated using a regression network, (**b**) the classification prediction of each BB, (**c**) the detection and segmentation mask result of each BB.

3.5. Improving Non-Maximum Suppression

For detection tasks, NMS is a necessary component after BB regression, which is a post-processing algorithm for redundancy removal of detection results. It is a handcrafted algorithm, to greedily select high-scoring detections, and remove their overlapping, low confidence neighbors. The NMS algorithm first sorts the proposal boxes according to the classification scores from high to low, then the detection box m with the highest score(s_m) is selected, and the other boxes with obvious overlap (Intersection over Union (IoU) > threshold used N_t) are suppressed. This process is recursive until all proposal boxes are traversed. Traditional NMS processing methods can be expressed by the following fraction resetting function:

$$s_i = \begin{cases} s_i, iou(b_m, b_i) < N_t \\ 0, iou(b_m, b_i) >= N_t \end{cases} \tag{4}$$

where b_m, b_i, s_i and N_t denote the detection box with the maximum score, the i-th of detection box, the score of the i-th detection box, and threshold value. NMS sets a hard threshold N_t while deciding what should be kept or removed from the neighborhood of b_m.

NMS performs well in generic object detection, adopting different thresholds; however, due to local maximum suppression, there is missing detection for the easily overlapped target objects. In the group-housed pig-breeding environment, there are serious adhesions and high overlaps in group pigs. To improve the detection performance of group-housed pigs, an improved NMS algorithm named soft-NMS [24] is used.

The soft-NMS algorithm process is as follows in Figure 7:

Input: $B=\{b_1,\ldots,b_n\}$, $S=\{s_1,\ldots,s_n\}$, N_t

B is a set of detected bounding boxes.

S contains corresponding detection scores.

N_t is the NMS threshold.

Output: D, the set of detected bounding boxes with classification scores.

$D \leftarrow \phi$

while $B \neq \phi$ do

$m \leftarrow \arg\max S$

$M \leftarrow \{b_m, s_m\}$

$D \leftarrow D \cup M$; $B \leftarrow B - b_m$

for b_i in B do

if $IoU(b_m, b_i) \geq N_t$ then

$B \leftarrow B - b_m$; $S \leftarrow S - s_i$

end NMS

$s_i \leftarrow s_i f(IoU(b_m, b_i))$

soft-NMS

end for

end while

return D

Figure 7. The soft-NMS algorithm.

where $f(IoU(b_m, b_i))$ is the overlap based weighting function with a Gaussian penalty function in soft-NMS as follows,

$$f(IoU(b_m, b_i)) = e^{-\frac{IoU(b_m, b_i)^2}{\sigma}}, \forall b_i \notin D \quad (5)$$

This rule is applied in the algorithm for each iteration and scores of all remaining detection boxes are updated. In the soft-NMS algorithm, the computational complexity of each step is $O(N)$, where N is the number of detection boxes. Moreover, the algorithm updates the scores for all detection boxes that overlap with b_m. Therefore, for N detection boxes, the computational complexity of soft-NMS is $O(N^2)$, and it is the same as traditional NMS. Soft-NMS hence does not require any additional training and uses the same running time with NMS's detectors.

3.6. Experiment Detail

The experimental environment is described below as follows:

PC: CPU, Intel® Xeon(R) CPU E5-2620 v4 @ 2.10 GHz × 8; Memory, 64 GB; Graphics, Tesla K40c.

OS: Ubuntu 16.04, CUDA10.1, Python3, Pytorch1.0, PyCharm, Jupyter Notebook.

The procedure of the PigMS R-CNN model mainly involves three steps: image data annotation, model training, and verification, as shown in Figure 8. Firstly, the training set and test set are labeled via the annotation tool, and the corresponding annotation files with JSON format are obtained. The train images, test images, and the corresponding annotation JSON files form the pig object dataset for training. Then, the pig.py of the training file adopts the PigMS R-CNN algorithm to set up an instance segmentation model using python3.6 on the pig object dataset, which includes 290 training and 130 test images. Finally, the validated pig file inspect_pig_model.ipynb written in jupyter notebook software is used to test set and get the pig object detection and segmentation result by calling the built model file of pig.py.

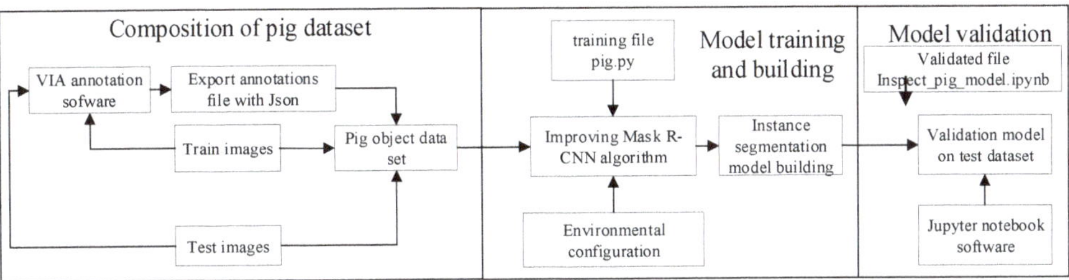

Figure 8. Establishment process of the pig instance segmentation model based on the PigMS R-CNN.

Back-propagation and stochastic gradient descent (SGD) are used to train the PigMS R-CNN model. For RPN networks, each mini batch comes from a single image containing many positive and negative example anchors. The input image is resized to 800 pixels on its shorter side. Synchronized SGD is used to train the model on eight GPUs. Each mini batch includes two images per GPU and 512 anchors in each image with a weight decay of 0.001 and a momentum of 0.9. The learning rate is 0.01 for the first 6000 mini-batches and 0.001 for the next 1000, and 0.0001 for the next 1000. NMS is performed over the proposals with an IoU threshold of 0.7. N_t and σ in soft-NMS are set to 0.7 and 0.5, respectively. The pre-trained model of ResNet-101 was available at https://dl.fbaipublicfiles.com/detectron/ImageNetPretrained/MSRA/R-101.pkl (accessed on 7 May 2021). The experimental results, models, and images are obtained by the Baidu network disk's address (https://pan.baidu.com/s/1_BOpAJ8trdjhBZO1fe4hWA, accessed on 23 April 2021), where the extracted code is a5h6. The precision–recall curve with the corresponding F1 score is used as the evaluation metric for pig detection [23]. The F1 score is computed as:

$$F1 = \frac{2 * Precision * Recall}{Precison + Recall} \tag{6}$$

4. Results and Discussion

4.1. Experimental Results of the MS R-CNN Model

We verified the reliability and effectiveness of the algorithm using the test dataset, which includes 130 images with 1920 × 1080 pixels. The precision–recall curve (PRC) of MS R-CNN is shown in Figure 9. The black dotted line expresses the points where recall and precision are identical. The more convex the purple line PRC is, the better the result is. The results of the red line PRC show that this method is suitable for pig detection and segmentation in a group-housed pig environment.

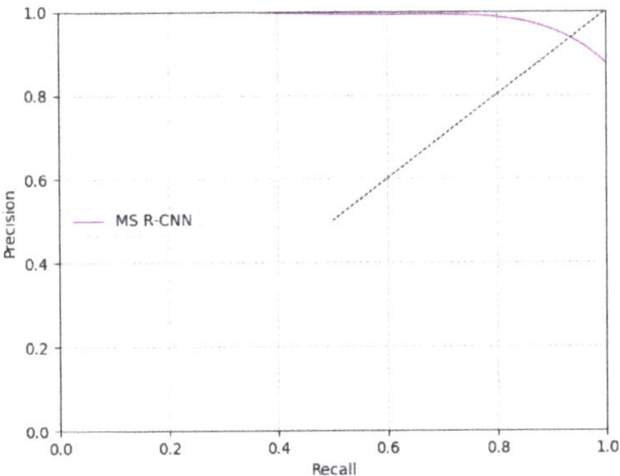

Figure 9. Precision–recall curves based on MS R-CNN.

The pig segmentation recall, and precision, are also shown in Table 1. Among 130 images, 50 images were taken from the front view images, including 352 pig objects. Moreover, 80 images were taken from the top view images, including 795 pig objects. Among a total of 1147 test pig objects, there were 1159 pigs detected, and 1064 pigs were correctly detected. The MS R-CNN model achieved a recall(R) rate of 92.76%, a precision (P) rate of 91.80%, and an F1 value of 0.9228. In the front view, 296 pig objects were correctly detected among a total of 369 pigs, the recall rate was 84.09%, and the precision rate was 80.22%. In the top view, 768 pig objects were correctly detected among a total of 790 pigs, the recall rate was 96.60%, and the precision rate was 97.22%. The average detection time for each image was 0.5233 s—favorable for the actual production requirements.

Table 1. The detection results of 130 images based on MS R-CNN.

Image Type	The Total of Pig Objects	The Detected Number	The Correct Detected Number	Recall (%)	Precision (%)	F1
Front view	352	369	296	84.09	80.22	0.8211
Top view	795	790	768	96.60	97.22	0.9691
Total number	1147	1159	1064	92.76	91.80	0.9228

As can be seen from Table 1, the result of the top view are significantly better than that of the front view. There are two main reasons: (1) by comparison with the top view, the overlap and adhesion between pigs in the front view were more serious. At the same time, the morphology and features were more complicated. (2) The number of verified images in the front view was close to that from top view. However, due to the limited range from the front view and serious occlusion of pigs, the number of pig objects in the front view was far lower than that in top view, accounting for only 30.69% of the total number of pig objects. Due to the influence of these two factors, the detection results of pig objects in the front view was lower than that of pig objects from the top view.

Figure 10 shows some experimental results of the detection and segmentation of pig objects. The dotted lines in the figure indicate all pig objects detected; the yellow solid line is marked as a pig object detected by mistake, and the red solid line is marked as a pig object detected by omission. The MS R-CNN model achieved good detection and segmentation accuracy for pig objects. However, it had some defects, mainly due to imprecise segmentation of the mask, incomplete detection of pig ears, legs, and tails, and a few cases of false or missed detection.

The results of the top and front view image detections are analyzed below. For the front view images, the main reasons for failure in the detection results are as follows: (1) the front view images included small pig objects, which caused false and missed detection. (2) The front view images included an overlap of pig bodies and disturbance of the light shadow, resulting in missed pigs (as shown with a solid line red box in Figure 10f,h, and false detection of the pig objects (as shown with the solid line yellow box in Figure 10f,h. For the top view images, the main reason for failure in detection and segmentation results was the serious adhesion of the pigs. If the distance between pigs in group-housed pigs is relatively close, it is prone to missed pig objects (as shown in the solid line red box in Figure 10d).

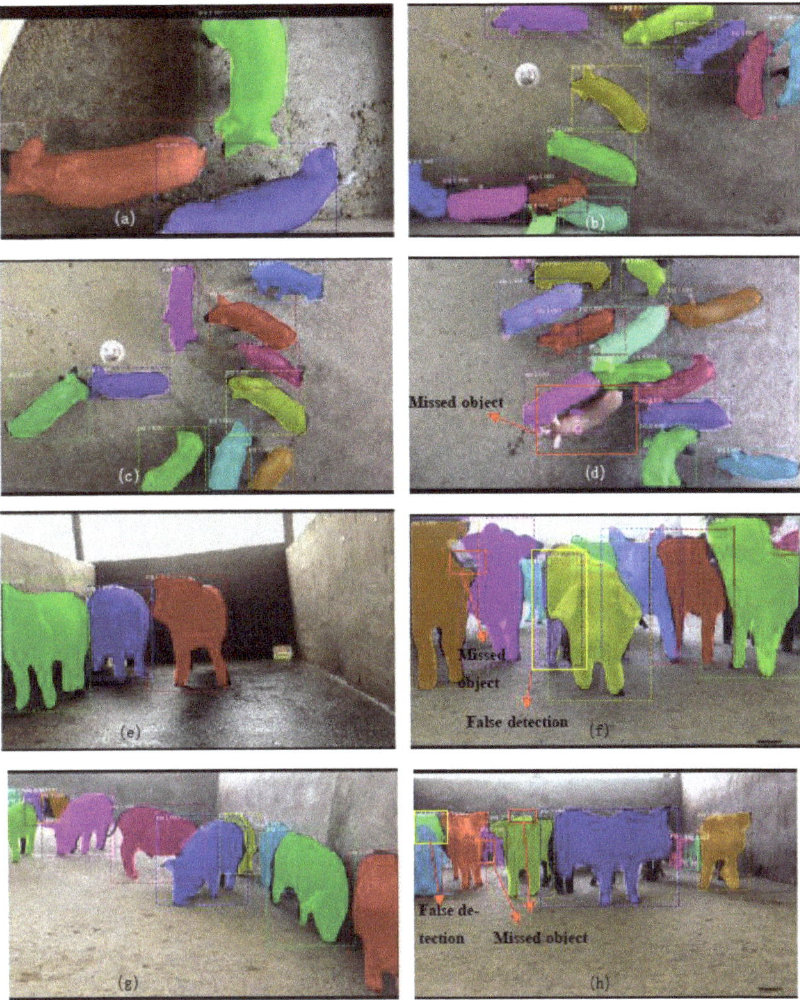

Figure 10. Detection and instance segmentation of the pig objects. Results include a small number of pigs (**a**), a medium number of pigs (**b**,**c**), a large number of pigs (**d**) in the top view images. Results include a small number of pigs (**e**), a medium number of pigs (**f**,**g**), a large number of pigs (**h**) in the front view images.

In addition, for the missed detection caused by adhesion, the results of the top view images are more serious than that of the front view images. The missed detection in two pig objects caused from top view images can be divided into two cases, where the two pigs were mistaken as a pig object (solid line orange box marked in Figure 11); one pig was correctly detected while another pig was missed, as shown with the solid line blue box marked in Figure 11.

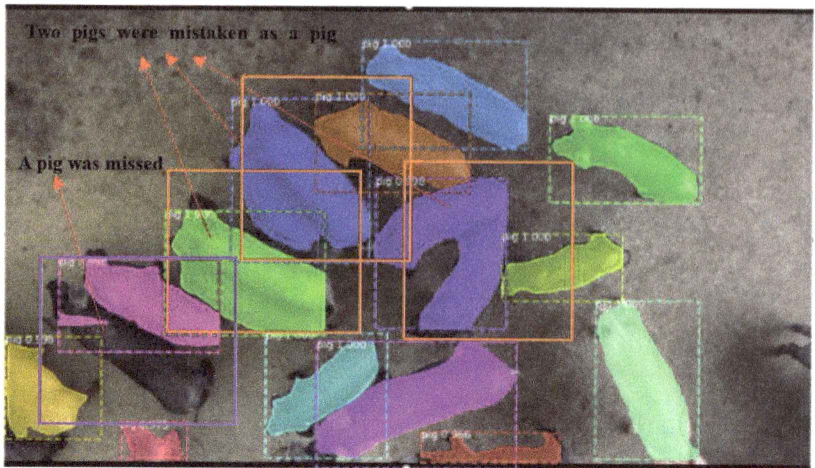

Figure 11. The results of missed detection from the top view image.

4.2. The Instance Segmentation and Detection Result of the PigMS R-CNN

To improve the missed and wrong detection of target pigs caused by overlapping, adhesion, and other complex environmental issues in a crowded room, the paper used the soft-NMS method instead of the traditional NMS in the MS R-CNN model, without adding extra times. The same 130 images of the test were used for validation, using soft-NMS with a threshold value of 0.5 and a traditional NMS algorithm.

The results of PigMS R-CNN are illustrated in Table 2. Among a total of 1147 test pig objects, 1138 pigs were detected, and 1071 pigs were correctly detected. The PigMS R-CNN model achieved a recall rate of 93.37% and a precision rate of 94.11%, which is better than the MS R-CNN model (Table 2, columns 6 and 7). The PigMS R-CNN based on soft-NMS leads to less missed and incorrectly detected objects.

Table 2. Detection and instance segmentation result of PigMS R-CNN.

Image Type	The Total of Pig Objects	The Detected Number	The Correct Detected Number	Recall (%)	Precision (%)	F1
Front view	352	350	294	83.52	84.00	0.8376
Top view	395	788	777	97.74	98.60	0.9817
Total number	1147	1138	1071	93.37	94.11	0.9374

4.3. Discussion

A PigMS R-CNN method is developed for group-housed pig detection and instance segmentation under the natural scene. Previous studies demonstrated that the MS R-CNN method can achieve good detection accuracies with a recall rate of 0.9276 and a precision rate of 0.918. However, the method would not work well to separate the touching-pigs under the overlapping backgrounds of group-housed pigs. To solve the problems, the improved approach was developed by soft-NMS.

First, the comparison of the touching-pigs detection between the MS R-CNN method with NMS and soft-NMS is discussed. Figure 12 presents the comparisons of some examples of detection results between NMS and soft-NMS, with the same threshold of 0.5. The different solid lines in the figure indicate all pig objects detected, the red dotted line is marked as the pig object, detected successfully with the red arrows, and the blue arrows point to the missed detection.

NMS | Soft-NMS

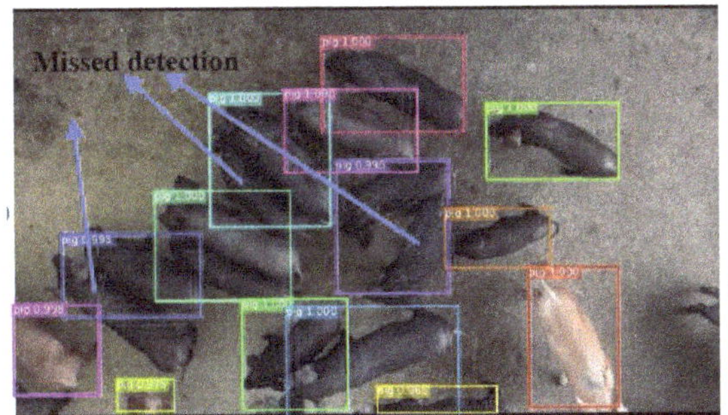

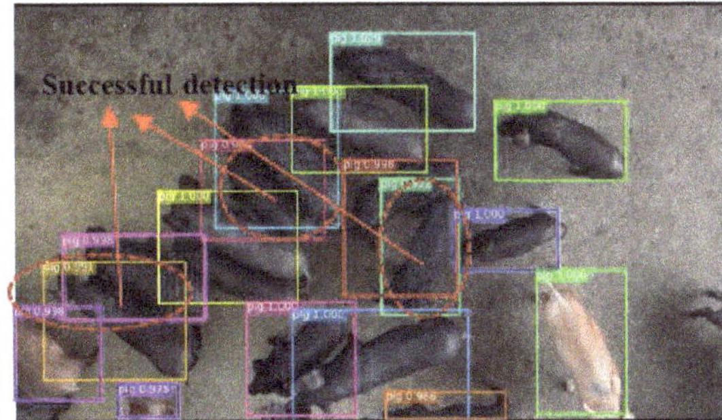

(**a**)

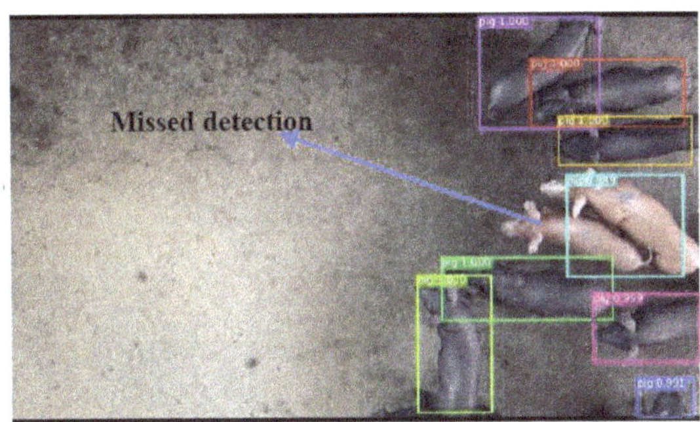

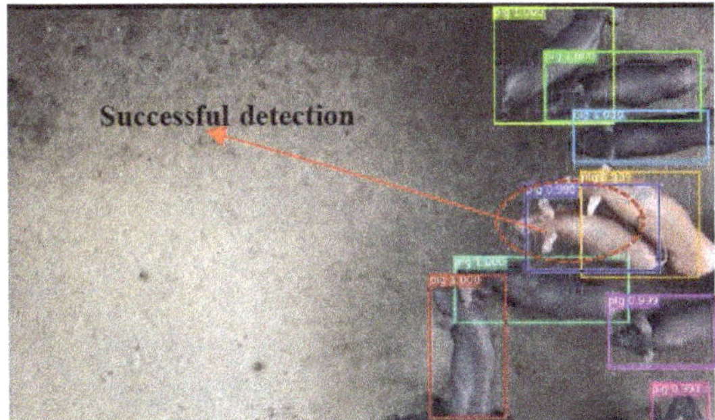

(**b**)

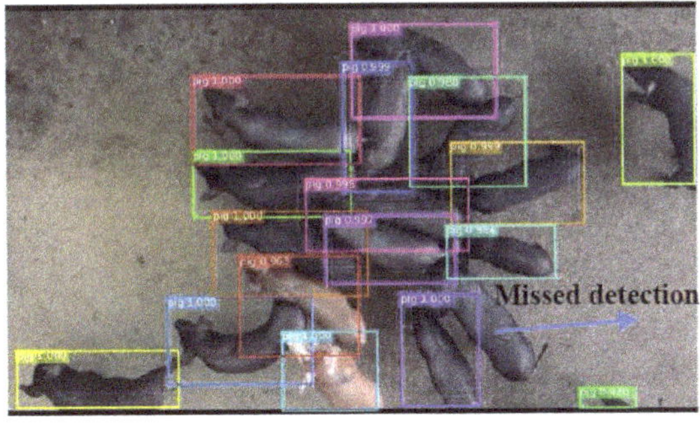

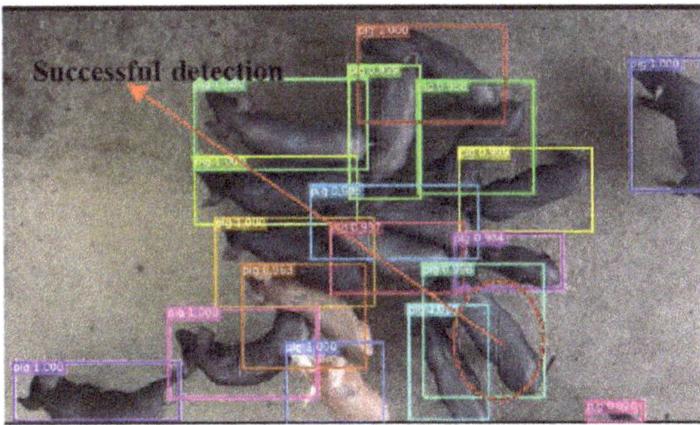

(**c**)

Figure 12. *Cont.*

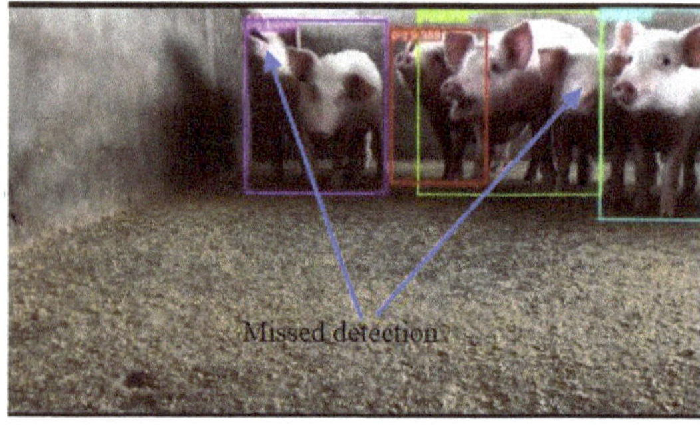

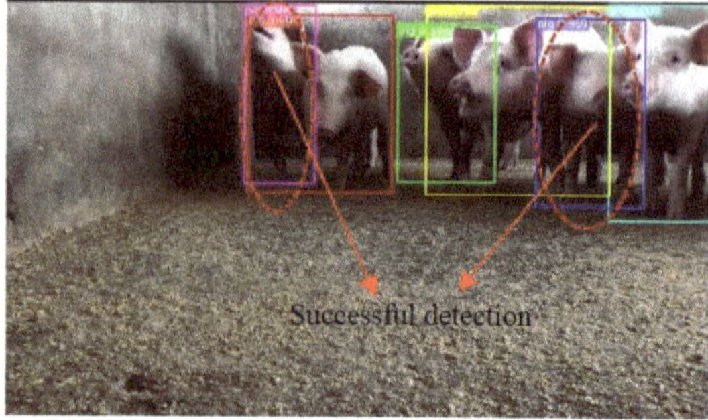

(d)

Figure 12. The comparison of detection between NMS and soft-NMS. The left column of (**a–d**) shows that the MS R-CNN method with NMS produces serious missed detection results under dense touching-pig conditions. The right column of (**a–d**) shows that the PigMS R-CNN approach with the soft-NMS algorithm overcomes the influence of the connected pigs by a Gaussian penalty function.

The MS R-CNN method with NMS and soft-NMS can successfully detect the group-housed pig under a "non-close together: environment. However, the MS R-CNN method with NMS (e.g., Figure 12 left column) produces serious missed detection results under dense touching-pig conditions. For example, the pigs in Figure 12a,c,d are too close together to distinguish individual pigs. The reason is that NMS selects the BB with the max score and sets the scores for neighboring detections to zero. Therefore, if an object actually existed in the overlap threshold, it would be missed, and this would reduce the recall rate. By contrast, the soft-NMS algorithm can overcome the influence of the connected pigs by decaying the scores of neighboring detection boxes, which have an overlap by a Gaussian penalty function, and obtain reliable pig detection results (e.g., Figure 12 right column). The soft-NMS method proved significantly effective at improving the missed detection of touching-pigs.

Table 3 shows the comparison of results between MS R-CNN and PigMS R-CNN when the IoU thresholds were set at different values (0, 0.3, 0.5, 0.8, and 0.9). It is observed that the PigMS R-CNN model is stable with an average recall of 94% and average precision of 93%. By contrast, the MS R-CNN model achieves an average recall of 93.5% and average precision of 90%. In particular, when the IoU threshold value is set to zero, the precision of MS R-CNN is 86.33%, which is lowest in other IoU threshold value situations, and that of the PigMS R-CNN, is 92.77%. Therefore, PigMS R-CNN can obtain better results than MS R-CNN in precision when the two methods achieve almost equal the recall rates.

Table 3. The comparison of results of different IoU thresholds between MS R-CNN and PigMS R-CNN.

Method	0			0.3			0.5			0.8		
	R	P	F1	R	P	F1	R	P	F1	R	P	F1
MS R-CNN	94.68%	86.33%	0.9031	93.55%	91.59%	0.9256	93.27%	90.19%	0.9170	92.68%	92.6%	0.9264
PigMS R-CNN	94.85%	92.77%	0.9380	93.85%	92.99%	0.9342	93.95%	92.69%	0.9332	93.33%	93.80%	0.9356

Next, some detection errors of the soft-NMS algorithm are analyzed and given in Figure 13. In our experiments, most of the false detections occur under two situations. First, false detections can occur when pigs from top view show incomplete appearances and severe shape deformation when compared with the pigs in the training images, such as the pigs show part appearances, as shown with the green box in Figure 13a. Second, for pigs from the front view, due to the heavy overlapping, false detections in the soft-NMS algorithm can occur, as shown in Figure 13b. Therefore, inferior environments will reduce the detection performance for front view images. To overcome these limitations, more advanced detection and segmentation techniques can be used to improve image quality. A pig key-point detection algorithm based on human posture [27,28] can be studied and utilized to enhance the reliability of the detection results.

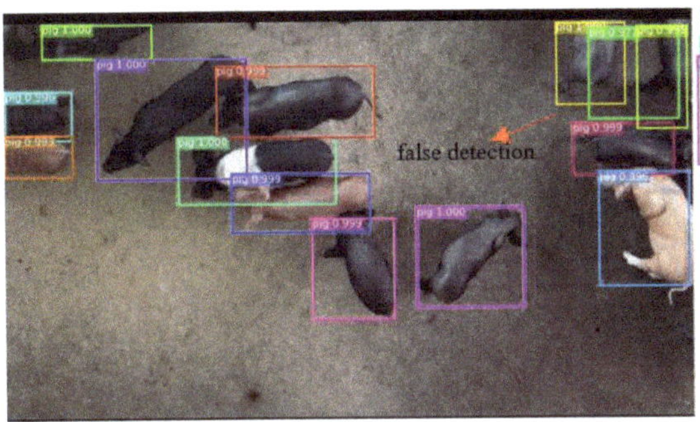

(**a**)

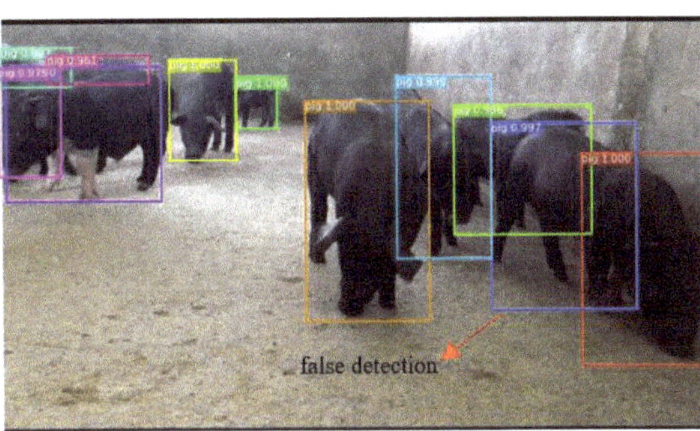

(**b**)

Figure 13. Example of detection failures of the proposed algorithm. (**a**) shows that false detections can occur when pigs from top view show incomplete appearances and severe shape deformation. For pigs from the front view, (**b**) shows that due to the heavy overlapping, false detections in the soft-NMS algorithm can occur.

5. Conclusions and Future Work

The group-housed instance segmentation of pigs in a natural environment is a significant operation to efficiently manage pig farms. However, using a traditional method, group-housed pigs cannot be separated accurately in real-time for heavily overlapped pigs in complex backgrounds.

In this paper, an improved MS R-CNN framework with soft-NMS was proposed to obtain the locations and segmentation of each pig in group-housed pig images. To prevent the missed and wrong detection of target pigs, caused by overlapping, adhesion, and other complex environmental issues in a crowded room, this paper employs the soft-NMS method instead of the traditional NMS in the MS R-CNN model, without adding extra times. All boxes, which have an overlap greater than a threshold N_t in traditional NMS, are given a zero score. Compared with NMS, soft-NMS rescores neighboring boxes instead of suppressing them altogether, which obtains improvement in precision and recall values.

Based on the pig detection and segmentation results for 130 images, with 1147 for top view and front view, the basic MS R-CNN framework obtained results with an F1 of 0.9228, while the target pigs using PigMS R-CNN had an F1 of 0.9374 in complex scenes. This algorithm can achieve good performance in terms of F1 without adding extra time. Our work on pig instance segmentation supports the foundation for pig behavior monitoring, posture recognition, and other related applications, such as body size and weight measurement estimations of pigs. Furthermore, it provides a deep learning framework for detecting and segmenting animals using overhead and front view cameras in a natural environment. In the future, we will develop pig behavior applications, such as pig monitoring of drinking water and fighting behavior under group-housed pigs' scenes.

Author Contributions: S.T. conceived the idea of the study, analysed the data, contributed significantly to the writing of the initial draft and the finalizing of this paper. W.Y. performed the experiment, carried out additional analyses of data. Y.L. contributed to the sponsorship of this paper. F.W. collected the data, analysed the data and revised the manuscript. H.W. helped perform the analysis with constructive discussions and revised the manuscript. All authors have read and agreed to the published version of the manuscript.

Funding: This research was supported by the National Natural Science Foundation of China (61772209 and 31600591), the Science and Technology Planning Project of Guangdong Province (Grant No. 2015A020224038, 2014A020208108, and 2019A050510034), and College Students' Innovation and Entrepreneurship Competition (202010564031).

Institutional Review Board Statement: Not applicable.

Informed Consent Statement: Not applicable.

Data Availability Statement: Not applicable.

Acknowledgments: We would like to thank Lejiazhuang Pig Farm for allowing us to capture data.

Conflicts of Interest: The authors declare no conflict of interest.

References

1. Li, B.; Liu, L.; Shen, M.; Sun, Y.; Lu, M. Group-housed pig detection in video surveillance of overhead views using multi-feature template matching. *Biosyst. Eng.* **2019**, *181*, 28–39. [CrossRef]
2. Wongsriworaphon, A.; Arnonkijpanich, B.; Pathumnakul, S. An approach based on digital image analysis to estimate the live weights of pigs in farm environments. *Comput. Electron. Agric.* **2015**, *115*, 26–33. [CrossRef]
3. Zhang, L.; Gray, H.; Ye, X.; Collins, L.; Allinson, N. Automatic Individual Pig Detection and Tracking in Pig Farms. *Sensors* **2019**, *19*, 1188. [CrossRef] [PubMed]
4. Nasirahmadi, A.; Richter, U.; Hensel, O.; Edwards, S.; Sturm, B. Using machine vision for investigation of changes in pig group lying patterns. *Comput. Electron. Agric.* **2015**, *119*, 184–190. [CrossRef]
5. Nasirahmadi, A.; Hensel, O.; Edwards, S.A.; Sturm, B. Automatic detection of mounting behaviours among pigs using image analysis. *Comput. Electron. Agric.* **2016**, *124*, 295–302. [CrossRef]
6. Chen, C.; Zhu, W.; Guo, Y.; Ma, C.; Huang, W.; Ruan, C. A kinetic energy model based on machine vision for recognition of aggressive behaviours among group-housed pigs. *Livest. Sci.* **2018**, *218*, 70–78. [CrossRef]
7. Yang, A.; Huang, H.; Zhu, X.; Yang, X.; Chen, P.; Li, S.; Xue, Y. Automatic recognition of sow nursing behaviour using deep learning-based segmentation and spatial and temporal features. *Biosyst. Eng.* **2018**, *175*, 133–145. [CrossRef]
8. Ju, M.; Choi, Y.; Seo, J.; Sa, J.; Lee, S.; Chung, Y.; Park, D. A Kinect-Based Segmentation of Touching-Pigs for Real-Time Monitoring. *Sensors* **2018**, *18*, 1746. [CrossRef] [PubMed]
9. Kashiha, M.; Bahr, C.; Ott, S.; Moons, C.P.H.; Niewold, T.A.; Ödberg, F.O.; Berckmans, D. Automatic identification of marked pigs in a pen using image pattern recognition. *Comput. Electron. Agric.* **2013**, *93*, 111–120. [CrossRef]
10. Guo, Y.; Zhu, W.; Jiao, P.; Chen, J. Foreground detection of group-housed pigs based on the combination of Mixture of Gaussians using prediction mechanism and threshold segmentation. *Biosyst. Eng.* **2014**, *125*, 98–104. [CrossRef]
11. Guo, Y.; Zhu, W.; Jiao, P.; Ma, C.; Yang, J. Multi-object extraction from topview group-housed pig images based on adaptive partitioning and multilevel thresholding segmentation. *Biosyst. Eng.* **2015**, *135*, 54–60. [CrossRef]
12. Huang, W.; Zhu, W.; Ma, C.; Guo, Y.; Chen, C. Identification of group-housed pigs based on Gabor and Local Binary Pattern features. *Biosyst. Eng.* **2018**, *166*, 90–100. [CrossRef]
13. Lin, T.-Y.; Serge Belongie, M.M.; Bourdev, L.; Girshick, R.; James Hays, P.P.; Deva Ramanan, C.; Lawrence, Z.; Piotr, D. Microsoft COCOCommon Objects in Context. In Proceedings of the European Conference on Computer Vision, Zurich, Switzerland, 6–12 September 2014; pp. 740–755.
14. Ren, S.; He, K.; Girshick, R.; Sun, J. Faster R-CNN towards real-time object detection with region proposal networks. *IEEE Trans. Pattern Anal. Mach. Intell.* **2015**, *39*, 1137–1149. [CrossRef]
15. Zheng, C.; Zhu, X.; Yang, X.; Wang, L.; Tu, S.; Xue, Y. Automatic recognition of lactating sow postures from depth images by deep learning detector. *Comput. Electron. Agric.* **2018**, *147*, 51–63. [CrossRef]
16. Yang, Q.; Xiao, D.; Lin, S. Feeding behavior recognition for group-housed pigs with the Faster R-CNN. *Comput. Electron. Agric.* **2018**, *155*, 453–460. [CrossRef]
17. Yang, A.; Huang, H.; Zheng, C.; Zhu, X.; Yang, X.; Chen, P.; Xue, Y.J.B.E. High-accuracy image segmentation for lactating sows using a fully convolutional network. *Biosyst. Eng.* **2018**, *176*, 36–47. [CrossRef]
18. Long, J.; Shelhamer, E.; Darrell, T. Fully Convolutional Network for Semantic Segmentation. In Proceedings of the Conference on Computer Vision and Pattern Recognition, Boston, MA, USA, 7–12 June 2015; pp. 3431–3440.
19. He, K.; Gkioxari, G.; Dollár, P.; Girshick, R. Mask R-CNN. In Proceedings of the IEEE International Conference on Computer Vision, Venice, Italy, 22–29 October 2017; pp. 2980–2988.
20. Qiao, Y.; Truman, M.; Sukkarieh, S. Cattle segmentation and contour extraction based on Mask R-CNN for precision livestock farming. *Comput. Electron. Agric.* **2019**, *165*. [CrossRef]
21. Lu, M.; Xiong, Y.; Li, K.; Liu, L.; Yan, L.; Ding, Y.; Lin, X.; Yang, X.; Shen, M. An automatic splitting method for the adhesive piglets' gray scale image based on the ellipse shape feature. *Comput. Electron. Agric.* **2016**, *120*, 53–62. [CrossRef]
22. Huang, Z.; Huang, L.; Gong, Y.; Huang, C.; Wang, X. Mask Scoring R-CNN. In Proceedings of the IEEE Conference on Computer Vision and Pattern Recognition CVPR, Long Beach, CA, USA, 16–20 June 2019.

23. Tu, S.; Liu, H.; Li, J.; Huang, J.; Xue, Y. Instance Segmentation Based on Mask Scoring R-CNN for Group-housed Pigs. In Proceedings of the 2020 International Conference on Computer Engineering and Application (ICCEA), Guangzhou, China, 27–29 March 2020; pp. 458–462.
24. Bodla, N.; Singh, B.; Chellappa, R.; Davis, L.S. Improving Object Detection with One Line of Code. In Proceedings of the IEEE International Conference on Computer Vision, Venice, Italy, 22–29 October 2017; pp. 5562–5570.
25. He, K.; Zhang, X.; Ren, S.; Sun, J. Deep Residual Learning for Image Recognition. In Proceedings of the 2016 IEEE Conference on Computer Vision and Pattern Recognition (CVPR), Las Vegas, NV, USA, 27–30 June 2016; pp. 770–778. [CrossRef]
26. Lin, T.Y.; Dollár, P.; Girshick, R.; He, K.; Hariharan, B.; Belongie, S. Feature Pyramid Networks for Object Detection. In Proceedings of the IEEE Conference on Computer Vision and Pattern Recognition, Honolulu, HI, USA, 21–26 July 2017; pp. 936–944.
27. Toshev, A.; Szegedy, C. DeepPose: Human Pose Estimation via Deep Neural Networks. In Proceedings of the Computer Vision and Pattern Recognition, Columbus, OH, USA, 24–27 June 2014; pp. 1653–1660.
28. Papandreou, G.; Zhu, T.; Kanazawa, N.; Toshev, A.; Tompson, J.; Bregler, C.; Murphy, K. Towards Accurate Multi-person Pose Estimation in the Wild. In Proceedings of the IEEE Conference on Computer Vision and Pattern Recognition, Honolulu, HI, USA, 21–26 July 2017; pp. 3711–3719.

 sensors

MDPI

Article

Pig Weight and Body Size Estimation Using a Multiple Output Regression Convolutional Neural Network: A Fast and Fully Automatic Method

Jianlong Zhang [1,2], Yanrong Zhuang [1,2], Hengyi Ji [1,2] and Guanghui Teng [1,2,3,*]

1 College of Water Resources & Civil Engineering, China Agricultural University, Beijing 100083, China; zhangjianlong@cau.edu.cn (J.Z.); zyr123@cau.edu.cn (Y.Z.); s20203091757@cau.edu.cn (H.J.)
2 Key Laboratory of Agricultural Engineering in Structure and Environment, Ministry of Agriculture and Rural Affairs, Beijing 100083, China
3 Beijing Engineering Research Center on Animal Healthy Environment, Beijing 100083, China
* Correspondence: futong@cau.edu.cn

Abstract: Pig weight and body size are important indicators for producers. Due to the increasing scale of pig farms, it is increasingly difficult for farmers to quickly and automatically obtain pig weight and body size. Due to this problem, we focused on a multiple output regression convolutional neural network (CNN) to estimate pig weight and body size. DenseNet201, ResNet152 V2, Xception and MobileNet V2 were modified into multiple output regression CNNs and trained on modeling data. By comparing the estimated performance of each model on test data, modified Xception was selected as the optimal estimation model. Based on pig height, body shape, and contour, the mean absolute error (MAE) of the model to estimate body weight (BW), shoulder width (SW), shoulder height (SH), hip width (HW), hip width (HH), and body length (BL) were 1.16 kg, 0.33 cm, 1.23 cm, 0.38 cm, 0.66 cm, and 0.75 cm, respectively. The coefficient of determination (R^2) value between the estimated and measured results was in the range of 0.9879–0.9973. Combined with the LabVIEW software development platform, this method can estimate pig weight and body size accurately, quickly, and automatically. This work contributes to the automatic management of pig farms.

Keywords: pig weight; body size; estimation; deep learning; convolutional neural network

Citation: Zhang, J.; Zhuang, Y.; Ji, H.; Teng, G. Pig Weight and Body Size Estimation Using a Multiple Output Regression Convolutional Neural Network: A Fast and Fully Automatic Method. *Sensors* **2021**, *21*, 3218. https://doi.org/10.3390/s21093218

Academic Editor: Son-Lam Phung

Received: 1 March 2021
Accepted: 27 April 2021
Published: 6 May 2021

1. Introduction

Animal husbandry is shifting toward automation, intelligence, and precision [1,2]. Pig weight and body size, two of the most important indicators for pig producers, provide information about feed conversion ratio (FCR), growth rate, uniformity, and health conditions [3,4]. Weight and body size also provide important references to regulate nutrition and the environment [5,6]. Precisely and automatically weighing pigs and measuring their body size can improve the feeding, breeding management, and selling, as well as preventing raisers from incurring unnecessary costs, humanpower, and materials, consequently improving the economic benefits [7,8].

Pig weight and body size are traditionally measured using ground scales and measuring sticks. This process causes stress to the animals and requires tremendous effort on behalf of the farm workers [9]. With the development of machine vision technology over the last 30 years, several researchers have searched for methods to estimate pig weight and body size using images to avoid direct measurements [10–13]. The estimation methods can be divided into four categories:

(i) Projection method. Project a slide with grids onto the back of a pig, then calculate the pig shoulder height and area according to the principle of stereo projection to estimate pig weight [14]. This method is difficult to automate.

(ii) Two-dimensional image method. Extract the pig body size, back area size, and other parameters from 2D images of pig backs, and use the model of the relationship between the pig weight and these parameters to achieve weight estimation. The average error of the weight estimation using this method is 3.38–5.3% [15–18].

(iii) Three-dimensional image method. After acquiring 3D images of pig backs using a depth camera, extract the pig back height, body size, back area size, and other parameters from the 3D image and use these parameters to estimate the pig weight. The 2D image mainly shows color, texture and contour information of the pig back, but the color and texture information are not related to the pig weight and body size. The 3D image shows outline and height information on the pig back; these parameters are highly correlated with the body size and pig weight. In addition, it was impossible to estimate the pig height using the 2D image. Therefore, this method is more promising than the 2D image method. The mean absolute error (MAE) of estimating pig body size for this method is 1.44–5.81% [9,19–25].

(iv) Ellipse fitting method. The ellipse fitting method is used to fit the area of a pig back image and estimate the weight of the pig based on the relationship model between the pig weight and center of mass, the length of the long axis and the short axis, the area, and the regional eccentricity of the fitted ellipse. The average relative error when using the ellipse fitting method to estimate pig weight is 3–3.8% [26–29].

In most of the aforementioned studies, the pig body images generally need to be processed as follows: background removal, image enhancement, image binarization, filtering and denoising, and head and tail removal, followed by the extraction of body size, volume, back area, and other parameters. The entire image process is cumbersome and time-consuming, and there is a chance of failure, all of which pose obstacles to automation.

The convolutional neural network (CNN) is one of the representative algorithms of deep learning. It is a type of feed-forward neural network that includes convolution calculation and has a deep structure. A CNN generally includes convolutional layers, pooling layers, fully connected layers, and an output layer, using the back propagation algorithm for the model training process [30]. Trained CNNs can extract information from images in an end-to-end manner with fast processing speed, and have been widely used in animal farming [31], clinical diagnosis [32], industrial production [33], and other aspects. In some equipment such as sorting systems for fattening pigs and breeding stations, there are strict requirements on the speed of pig weight and body size acquisition to improve operating efficiency. Due to the cumbersome and time-consuming process of the existing weight estimation methods and the real-time processing of images by CNN, a multiple output regression CNN model may be able to extract body shape features and estimate pig weight and body size quickly and accurately.

Given the above rationale, we aimed to develop a pig weight and body size estimation method using 3D images and a multiple output regression CNN. The study objectives were: (i) to train and select a pig weight and body size estimation model, (ii) to test the accuracy of this model, and (iii) to apply the method.

2. Materials and Methods

2.1. Design of the Pig Weight and Back Image Acquisition System

To train and evaluate the pig weight and body size estimation model, pig weight data, body size data, and 3D images of pig backs were needed. For pig weight data and 3D images of pig backs, the pig weight and back image acquisition system was designed (Figure 1). The size of the system is 1.5 mL $\times$ 0.5 mW $\times$ 0.9 mH. In the top of the system, there is an Intel RealSense D435 depth camera with a resolution of 1280 $\times$ 720 pixels to acquire 3D and 2D images simultaneously. There are 4 weighing sensors with a measurement range of 0–500 kg at the bottom of system, and the measurement accuracy after calibration was $\pm$0.1 kg. The limit bars on both sides of the system ensure that the whole pig is on the scale when weighing. The acquisition system could easily move when necessary as it is on wheels.

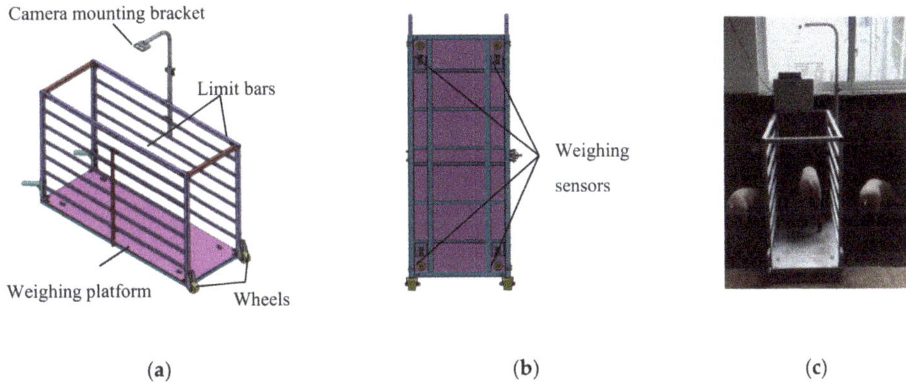

Figure 1. Pig weight and back image acquisition system: (**a**) three-dimensional diagram of system; (**b**) distribution of weighing sensors; (**c**) photo of system.

The system control program was developed based on the Internet of Things and the LabVIEW V18.0 software development platform, using Client-Server (C/S) architecture. The specific control scheme and program interface used are shown in Figure 2. The depth camera was connected to the server through a USB interface, and the pig weight data obtained by the weighing sensors were converted into a network signal by the USR-TCP232 and transmitted to the server through a switchboard. All software and hardware were controlled by the server, and MySQL version 5.5 database was installed in the server to store pig weight and image data. To ensure the quality of the acquired images, the running time of the system for this study was 8:00–17:00. When the program started, the depth camera and weighing sensors were initialized. Then, the system read the weighing data every 0.2 s. When 4 consecutive weighing data points were within the range of the pig population, and the difference between the maximum and minimum of the 4 data points was less than 0.2 kg, it was assumed that there was a pig on the weighing platform and that the pig was relatively quiet. When these requirements were met, the camera acquired the 3D and 2D images of the pig back simultaneously and the pig weight data and back images were saved in the database, so pig weight and back image data were able to be acquired continuously. Pig back images could also be manually acquired by using the capture button. The acquired 3D and 2D images were in PNG and APD format, respectively, and the file name of the 3D and 2D images was the acquisition time (accurate to milliseconds).

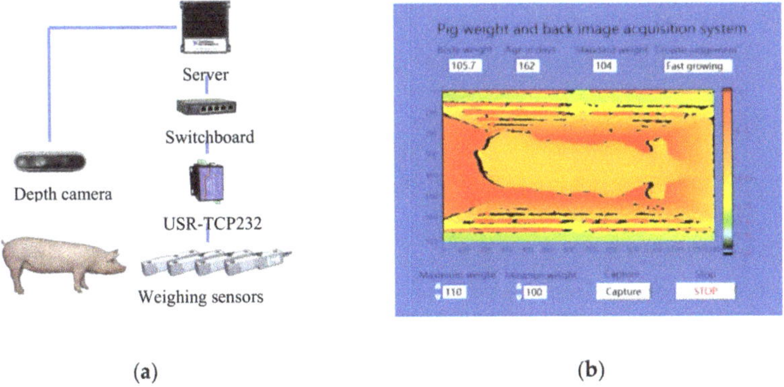

Figure 2. Pig weight and back image acquisition system: (**a**) data acquisition scheme; (**b**) software interface.

2.2. Acquisition Method of Body Size

As shown in Figure 3a, the body size data includes body length (BL), shoulder width (SW), shoulder height (SH), hip width (HW), and hip height (HH). BL is the length of line L_1–L_2, which is the straight-line distance from the root of the ears to the root of the tail. The SW is the length of line S_1–S_2, which is the transverse horizontal straight-line distance at the widest part of the shoulder. SH is the height of point M, which is the highest point of the shoulder along the line S_1–S_2. HW is the length of line H_1-H_2, which is the transverse horizontal straight-line distance at the widest part of the hip. HH is the height of point N, which is the highest point of the hip along the line H_1–H_2. Each body size data point was measured using a measuring stick (Figure 3b). To match the body size data to each pig, different marks were used to identify different pigs.

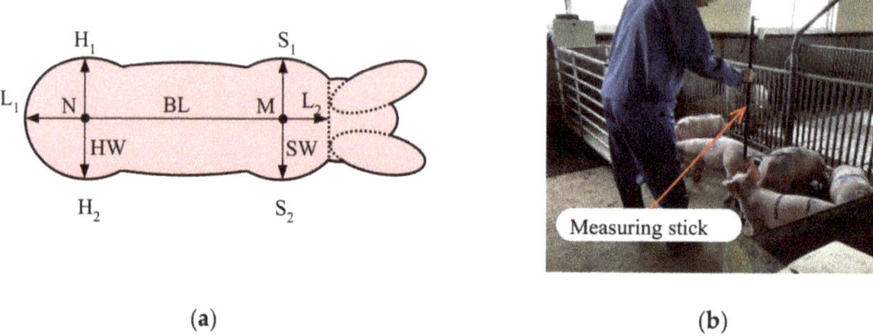

Figure 3. Specific locations of body size parameters and measurement of body size: (**a**) specific locations of body size parameters; (**b**) body size measurement using a measuring stick.

2.3. Data Collection and Preprocessing

In this study, two types of data were collected: modeling data and test data. The modeling data were used to train the models, and the test data were used as unknown data to test the generalization ability of the trained models. The data collection process in this study was in compliance with European Union legislation concerning the protection of animals for scientific purposes (European Parliament, 2010).

The modeling data were collected in a pig house at the Rongchang Experimental Station, Chongqing, China. There were 50 pens in this pig house and 5 pigs per pen. The area of each pen was 4.2 mL × 2.5 mW × 1.0 mH, and it was equipped with a duckbill drinker. The pig weight and back image data were obtained with the acquisition system. The acquisition system was placed in front of the duckbill drinker. Pigs would enter the system and stand on the weighing platform every time they drank. Therefore, the acquisition system could obtain the back images and body weight (BW) data automatically whenever a pig come to drink throughout the running time of the acquisition system. The body size data of each pig were measured manually once in the morning and once in the afternoon. Each body size data point was measured 5 times. The maximum and minimum were removed and the average of the remaining 3 values was taken as the final result. The result was accurate to within millimeters. Combined with the marks on the pig back, the measured body size data point could be matched to 3D images by the 2D images. The system was cleaned and disinfected every night and put into the following pig pen the next day to start a new collection. Since the pigs had been living in the pig house for some time before the experiment began, they were familiar with the drinking methods, so the pigs were not trained to go to the weighing platform to drink. The data collection period lasted for 88 days. During the experiment, 8 pigs were sold, and 3 pigs died of illness, and a total of 38,112 pig back images and corresponding weight and body size data in various postures from 239 Duroc × Landrace × Yorkshire growing and finishing pigs (121 castrated boars and 129 gilts) were collected (159 images per pig). Pig weight data were in the range of 16.5–117.0 kg, and the number of data points in the weight categories of 16.5–40 kg, >40–65 kg, >65–90 kg, and >90–117 kg was 8094, 10,340, 11,330, and 8348, respectively.

The test data were collected at a commercial pig farm belonging to the Shandong Rongchang Breeding Company, Binzhou, China. At this farm, pig houses were divided into 10 pens and 20 Duroc × Landrace × Yorkshire growing and finishing pigs were reared in each pen. The area of each pen was 7.5 mL × 4.0 mW × 0.95 mH. The collection method of the test data was the same as that used for the modeling data. The data collection period lasted for 60 days. During the experiment, 4 pigs died of illness and 8 pigs were eliminated, and a total of 20,026 test data in various postures (Figure 4) from 188 pigs in the weight range of 22.0–105.4 kg were collected (106 images per pig). The number of data points in the weight categories of 22.0–42.0 kg, >42.0–62.0 kg, >62.0–82.0 kg, and >82.0–105.0 kg was 6506, 5180, 4154, and 4186, respectively.

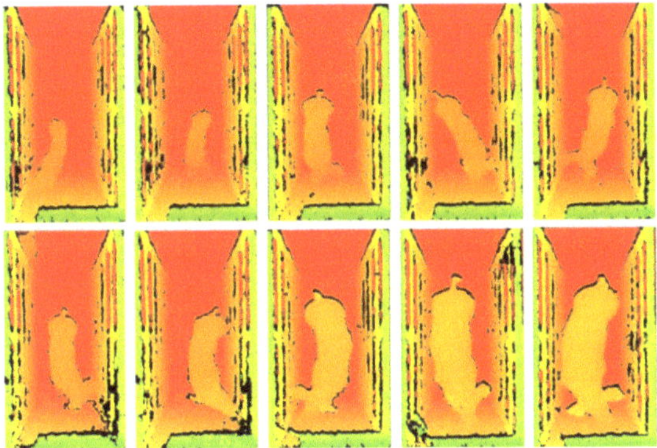

Figure 4. Samples of pig images in various postures.

The size of the original 3D images was 1280 × 720 pixels. To improve training speed, all images were preprocessed in the same way (Figure 5). The distance from the depth camera to the weighing platform was 1650 mm, and the pixel value of each point in the original image was the distance in millimeters from the point to the depth camera. To convert this distance to true height, each pixel value in the images was inverted as

$$P_{\mathrm{i}} = 1650 - P_{\mathrm{o}} \tag{1}$$

where P_{i} represents the pixel's value in the inverted image, and P_{o} represents the pixel's value in the original image. After inversion, the pixel value of each point was the distance from the point to the weighing platform in the range of 0–1650 mm. Then, the pixel value of the inverted image was scaled into 0–255 and converted into a gray scale image, where the lighter the color, the greater the height. The gray scale image was then resized into 2 different sizes (299 × 299 pixels and 224 × 224 pixels) as the inputs for different models. Since all images were processed in the same way, the process did not change the relative position and size of the pigs in the images, so it had little impact on the final estimation. Finally, each image was tagged with 6 labels in the order of BW, SW, SH, HW, HH, and BL.

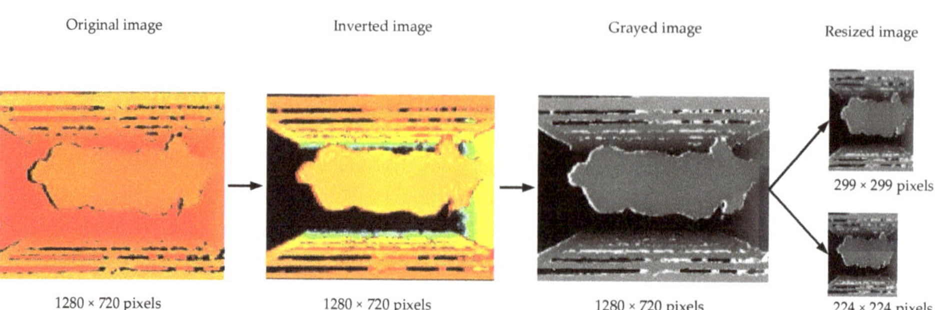

Figure 5. Image preprocessing process.

2.4. Construction, Training and Testing of Pig Weight and Body Size Estimation Models

As many CNNs have achieved excellent results in the ImageNet Large Scale Visual Recognition Challenge, 4 state-of-the-art classification CNNs (DenseNet201 [34], ResNet152 V2 [35], Xception [36], and MobileNet V2 [37]) were used as base models and transformed into multiple output regression CNNs for pig weight and body size estimation. DenseNet201 uses dense blocks. In a dense block, the inputs of each layer contain the output of all previous layers and this mechanism can reduce the disappearance of gradients and make the network much deeper. ResNet152 V2 is built based on VGGNet and residual block. The core idea of a residual block is to apply an identity shortcut connection to skip one or more layers directly. This operation can also deepen the network depth. Xception uses depthwise separable convolutions to reduce model size and uses an extreme inception module to fuse features extracted from different convolution kernels. MobileNet V2 is characterized by the use of depthwise separable convolutions and inverted residual structure. The depthwise separable convolution can reduce the parameters of the model and the inverted residual structure can reduce the information loss caused by activation function. The specific transformation process was: (i) the last classification layer of each model was removed; (ii) 6 dense layers (DLs) with only one node and no activation function were added to each model in parallel (Figure 6) to output BW, SW, SH, HW, HH, and BL, separately.

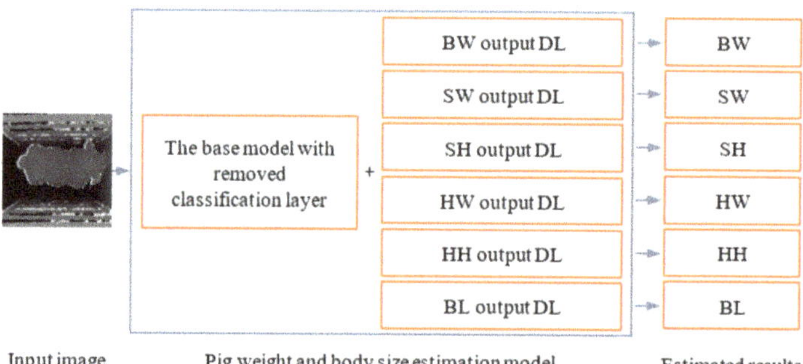

Figure 6. Pig weight and body size estimation model and estimate process. DL: dense layer; BW: body weight; SW: shoulder width; SH: shoulder height; HW: hip width; HH: hip height; BL: body length.

All models were written with the available libraries in Python 3.7.0 and tensorflow-gpu-2.2.0. All code was run on a desktop computer with an Intel i7-9700 processor, 32 GB RAM, Windows 10 (64 bit), and a NVidia GeForce GTX 1660 Ti 6 GB graphics card with Turing^{TM} architecture. The developed computer code was available in GitHub: https://github.com/18801389568/Pig-weight-and-body-size-estimation (accessed on 11 April 2021). Model training is the process of continuously changing model parameters to make the estimation results more accurate. The data used to train the models were modeling data. In all models, the modeling data were randomly divided into training sets and validation sets in a ratio of 7:3 after the order was shuffled. The preprocessed 3D images were used as input during model training, and the output was the corresponding pig weight and body size. As the quantity of data across the weight ranges was similar, the mean square error (MSE) was used as the loss function to evaluate the estimation ability of the models. The MSE was calculated as

$$\mathrm{MSE} = \mathrm{MSE_{BW}} + \mathrm{MSE_{SW}} + \mathrm{MSE_{SH}} + \mathrm{MSE_{HW}} + \mathrm{MSE_{HH}} + \mathrm{MSE_{BL}} \tag{2}$$

where $\mathrm{MSE_{BW}}$, $\mathrm{MSE_{SW}}$, $\mathrm{MSE_{SH}}$, $\mathrm{MSE_{HW}}$, $\mathrm{MSE_{HH}}$, and $\mathrm{MSE_{BL}}$ are the MSE generated by estimating BW, SW, SH, HW, HH, and BL, respectively. The calculation methods of $\mathrm{MSE_{BW}}$, $\mathrm{MSE_{SW}}$, $\mathrm{MSE_{SH}}$, $\mathrm{MSE_{HW}}$, $\mathrm{MSE_{HH}}$, and $\mathrm{MSE_{BL}}$ are similar and can be calculated as

$$\mathrm{MSE_V} = \frac{1}{M}\sum_{m=1}^{M}\left(y_m^{\mathrm{V}} - \hat{y}_m^{\mathrm{V}}\right)^2 \tag{3}$$

where V can be any one of BW, SW, SH, HW, HH, and BL; M is the total number of data points in the validation set; m is the sample number of the data in the validation set; y_m^{V} is the measured value for V of the mth sample; and $\hat{y}_m^{\mathrm{V}}$ is the estimated result for V of the mth sample. In order to compare the performance of each model under the same condition, the configuration of the hyper-parameters used in each model was the same, as shown in Table 1. The loss of each model on the validation set was used as the evaluation standard to retain the best parameters in the training process for each model.

Table 1. Hyper-parameters of models.

Optimization Function	Learning Rate	Loss Function	Batch Size	Iterations
Adam	0.001	MSE	16	150

Information about the trained models is shown in Table 2. The number of parameters is the number of all parameters in the model and the number of trainable parameters is the number of parameters except the parameter in batch-normalization layers and global-average-pooling layers. Among the 4 models, the input image size for the modified Xception model is 299 × 299 pixels, while the image size for the other 3 models is 224 × 224 pixels. The model size and number of parameters for modified ResNet152 V2 were largest, while the training time for modified Xception was longest due to the big input image size. Due to modified MobileNet V2 having the smallest model size, the lowest number of parameters, and the smallest input image size, the training time for this model was the shortest.

Table 2. Model information.

Model	Size of Input Image (pixels)	Model Size (MB)	Number of Parameters	Number of Trainable Parameters	Training Time (h)
Modified DenseNet201	224 × 224	229	18,333,510	18,104,454	29.1
Modified MobileNet V2	224 × 224	31	2,265,670	2,231,558	12.9
Modified ResNet152 V2	224 × 224	683	58,343,942	58,200,198	35.7
Modified Xception	299 × 299	243	20,873,774	20,819,246	54.0

After model training, test data were used to examine the generalization capability of each model. The models were investigated from the aspect of the estimated root mean square error (RMSE), MAE, mean relative error (MRE), and mean estimation time (MET) of an image.

3. Results and Discussion

3.1. Model Training Results

The change in loss (MSE) of each model on the validation set during the training steps is shown in Figure 7. During the training process, modified MobileNet V2 was observed to experience a larger fluctuation on the validation set. This may be due to the fact that the model has fewer parameters and cannot estimate pig weight and body size well. When the 80th iteration was reached, the other three models had converged and achieved good estimation results. Finally, the lowest MSE obtained by modified DenseNet201, modified MobileNet V2, modified ResNet152 V2, and modified Xception on the validation set were 0.132, 1.243, 0.221, and 0.092, respectively. The modified Xception achieved the highest estimation accuracy.

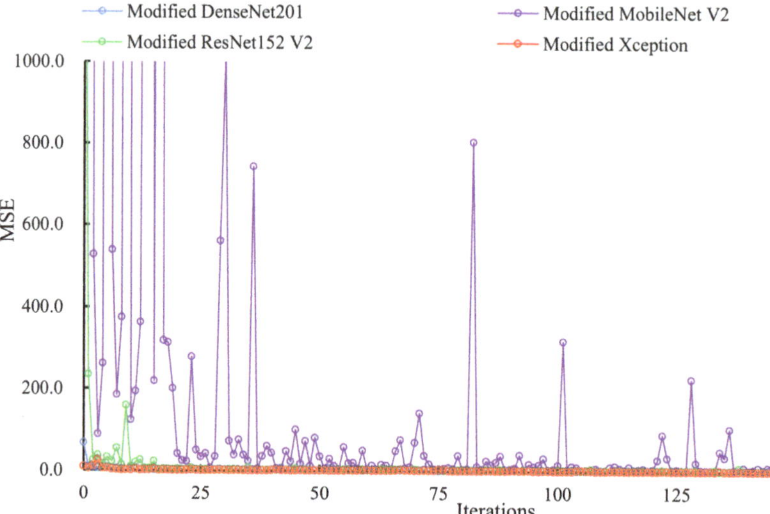

Figure 7. Loss change on validation set of each model.

3.2. Model Test Results

Table 3 presents the results of investigating the generalization performance of the models using the test data collected from a commercial pig farm. Similar to the results for the validation set, the four trained models also had good estimation performance. The lowest errors when estimating BW, SW, SH, HW, HH, and BL were obtained by modified Xception, Xception, ResNet152, MobileNet V2, Xception, and ResNet152, respectively. As for the validation set, modified Xception produced the most accurate estimation performance among the four models. This may be because the Xception module in the model can more effectively synthesize information. Although modified MobileNet V2 fluctuated during the training process, it performed well on the test set after training. This is because the task of pig weight and body size estimation is not as complicated as object classification, as it does not need to extract complex textures and edge information from images. When the four models were tasked with estimating body size, the largest MRE was generated by estimating SH. This is because the movement of a pig head when drinking can cause a change in SH. According to observations, when the head of an 80 kg pig moves up and down, it will cause an SH change of about 4 cm. The MET of the four models were all within 27.1 ms, which could meet the requirement of real-time operation.

Table 3. Performance of the models on the test set. BW: body weight; SW: shoulder width; SH: shoulder height; HW: hip width; HH: hip height; BL: body length; RMSE: root mean square error; MAE: mean absolute error; MRE: mean relative error; MET: mean estimation time; MSE: total mean square error.

Items		Modified DenseNet201	Modified MobileNet V2	Modified ResNet152 V2	Modified Xception
BW	RMSE (kg)	2.51	1.84	1.73	1.53
	MAE (kg)	2.03	1.49	1.31	1.16
	MRE	3.44%	2.54%	2.26%	1.99%
SW	RMSE (cm)	0.48	0.44	0.46	0.43
	MAE (cm)	0.38	0.34	0.37	0.33
	MRE	1.49%	1.35%	1.47%	1.31%
SH	RMSE (cm)	1.53	1.38	1.31	1.36
	MAE (cm)	1.42	1.22	1.17	1.23
	MRE	2.79%	2.38%	2.30%	2.40%
HW	RMSE (cm)	0.50	0.40	0.47	0.47
	MAE (cm)	0.45	0.31	0.38	0.38
	MRE	1.84%	1.29%	1.55%	1.58%
HH	RMSE (cm)	1.11	0.96	1.10	0.87
	MAE (cm)	0.90	0.76	0.89	0.66
	MRE	1.59%	1.34%	1.58%	1.16%
BL	RMSE (cm)	1.16	0.89	0.84	0.94
	MAE (cm)	0.97	0.69	0.63	0.75
	MRE	1.05%	0.74%	0.69%	0.82%
MET (ms)		17.98	5.99	27.10	12.32
MSE (kg^2)		11.699	7.357	7.057	6.236

Considering the total MSE of each model, modified Xception was selected as the final pig weight and body size estimation model. Measured and estimated pig weights and body sizes are shown in Figure 8. The coefficient of determination (R^2) value between the measured and estimated BW, SW, SH, HW, HH, and BL were as high as 0.9973, 0.9922, 0.9911, 0.9937, 0.9879, and 0.9971, respectively. Even if the pig body is not straight, high estimation accuracy can still be obtained. The estimation accuracy of this model is higher than the projection method [14], the 2D image method [15–18], and the ellipse fitting method [26–29], as this model estimates pig weight and body size based on the height and distance of all points in a 3D image rather than the individual information points extracted by these other methods. The accuracy is same when using the 3D image method [9,19–25], but the processing operation of the model is simpler. The estimation accuracy of pig weight and body size cannot be further improved because pig weight changes with eating, drinking, and excretion. Pig weight is also affected by the lean meat ratio. Such changes are difficult to see in images of a pig back, and thus the model cannot tell the difference in pig weight.

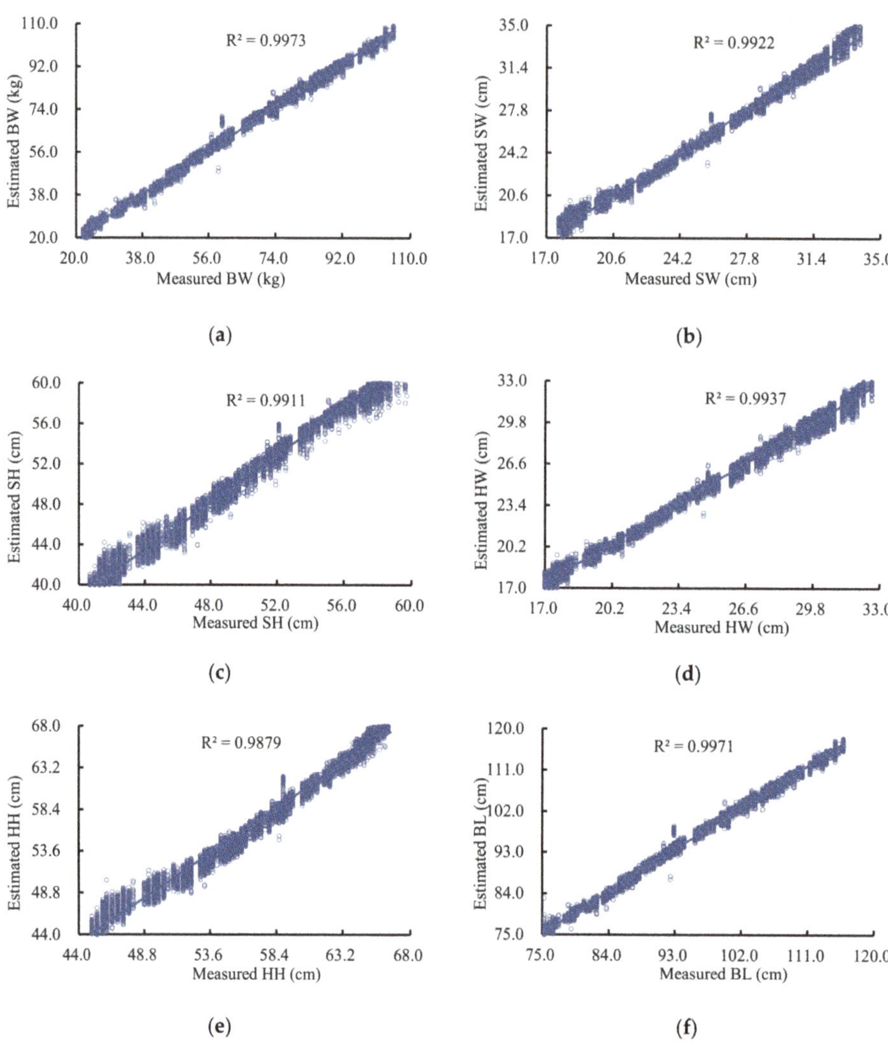

Figure 8. Comparison between measured and estimated BW (**a**), SW (**b**), SH (**c**), HW (**d**), HH € and BL (**f**).

3.3. Feature Maps

Models can detect elementary features, such as texture and outline, in their shallow convolutional layers and learn to detect more comprehensive features in their deeper layers. To determine what information had been learned and on what basis the modified Xception estimates pig weight and body size, the feature maps that were output by the first convolutional layer were examined. After the original image (Figure 9a) was input to the first convolutional layer of modified Xception, a total of 32 feature maps were output (Figure 9b). When comparing the input image with the feature maps, we found that the input image was smoothed after the first convolutional layer, the background interference was eliminated, the contour, edge, and depth features of the pig body were extracted. Therefore, it was demonstrated that the model estimated pig weight and body size based on a pig height and body shape characteristics. Notably, the model is not necessarily based on the distance between specific points to estimate the body size: it could be based on the overall body physique of the pig, but nevertheless, the performance on the test set showed that the method still produced accurate results. Compared with the method of estimating body size based on the distance of the points in an image, this method might reduce the estimation error caused by posture changes of the pig.

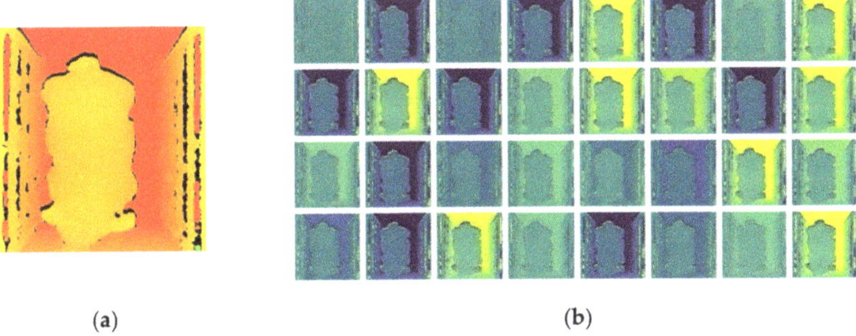

Figure 9. Original image (**a**) and feature maps (**b**) output from the first convolutional layer of modified Xception.

3.4. Application Prospect

Benefitting from the powerful development capability and Python Integration Toolkit provided by LabVIEW, this method can be used to measure pig weight and body size in a fully automated way (Figure 10). Pig weight and body size can be quickly estimated without a complex operation after the preprocessed 3D image was input into the model. Such a simple and convenient operation will reduce the workload and technical requirements for farm breeders. In addition, this non-contact measurement method can also avoid stress or injury to pigs. It is also feasible for the model to be integrated into control programs and be applied to commercial farms.

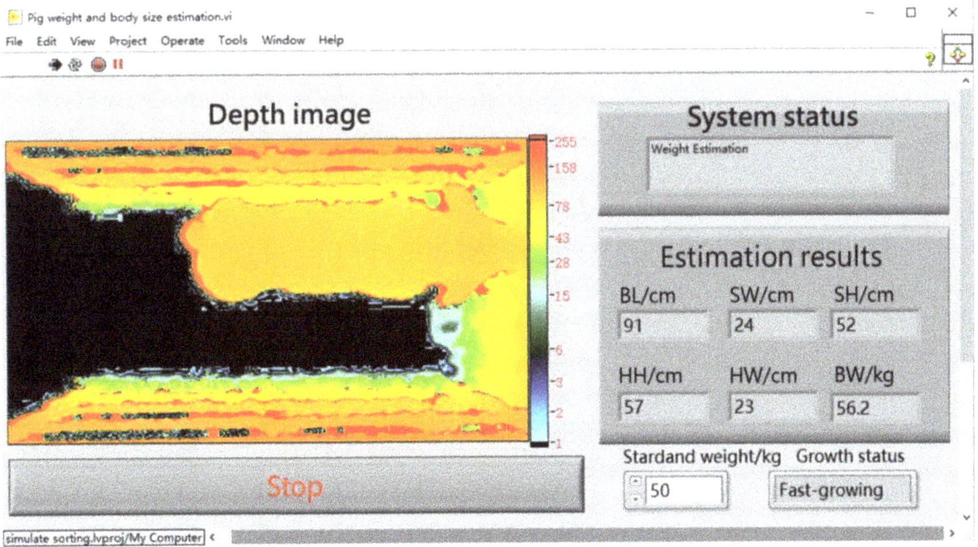

Figure 10. LabVIEW panel of the pig weight and body size estimation system.

In conclusion, multiple output regression CNN can be used to accurately estimate pig weight and body size. The estimation process only requires the simple preprocessing of acquired 3D images and can be automated. The high estimation speed of this method can ensure real-time operation in commercial farms. The influence of light on the estimation accuracy can be reduced by using 3D images. Even when we are estimating the weight of pigs of different breeds, only a small amount of pig data need to be collected and corrected on the basis of the original mode output.

4. Conclusions

We propose an innovative method of estimating pig weight and body size using a multiple output regression CNN. After training the modified DenseNet201, ResNet152 V2, Xception, and MobileNet V2 on modeling data and comparing the estimation results on test data, modified Xception was finally selected as the optimal pig weight and body size estimation model. This method estimates pig weight and body size based on a pig height, contour, and body shape, and yielded a MAE of 1.16 kg, 0.33 cm, 1.23 cm, 0.38 cm, 0.66 cm, and 0.75 cm when estimating BW, SW, SH, HW, HH, and BL, respectively. The MAT for modified Xception was 0.012 s. This method can successfully estimate pig weight and body size on the LabVIEW platform in a fully automated way. It is feasible to apply this method to sorting systems for fattening pigs, breeding stations and other occasions where there are strict requirements on the speed of pig weight and body size acquisition. This method can also be used to estimate the weight and body size of other animals such as cattle and sheep. Future work should combine modified Xception with object detection technology to realize pig weight and body size estimation through the depth camera installed on the top of the pig house.

Author Contributions: Data curation, J.Z. and H.J.; Investigation, G.T.; Methodology, J.Z. and Y.Z.; Project administration, G.T.; Software, J.Z. and Y.Z.; Supervision, G.T.; Writing—original draft, J.Z.; Writing—review & editing, G.T., Y.Z. and H.J. All authors have read and agreed to the published version of the manuscript.

Funding: This work is funded by the National Key Research and Development Program of China (2016YFD0700204).

Informed Consent Statement: Not applicable.

Data Availability Statement: Data is contained within the article.

Conflicts of Interest: The authors declare no conflict of interest.

References

1. Li, D.; Li, Z. System Analysis and Development Prospect of Unmanned Farming. *Trans. Chin. Soc. Agric. Mach.* **2020**, *51*, 1–12. [CrossRef]
2. Lee, S.; Ahn, H.; Seo, J.; Chung, Y.; Park, D.; Pan, S. Practical Monitoring of Undergrown Pigs for IoT-Based Large-Scale Smart Farm. *IEEE Access* **2019**, *7*, 173796–173810. [CrossRef]
3. Amraei, S.; Abdanan Mehdizadeh, S.; Salari, S. Broiler weight estimation based on machine vision and artificial neural network. *Brit. Poult. Sci.* **2017**, *58*, 200–205. [CrossRef]
4. Wang, Y.; Yang, W.; Walker, L.T.; Rababah, T.M. Enhancing the accuracy of area extraction in machine vision-based pig weighing through edge detection. *Int. J. Agric. Biol. Eng.* **2008**, *1*, 37–42. [CrossRef]
5. Matthews, S.G.; Miller, A.L.; Clapp, J.; Plötz, T.; Kyriazakis, I. Early detection of health and welfare compromises through automated detection of behavioural changes in pigs. *Vet. J.* **2016**, *217*, 43–51. [CrossRef] [PubMed]
6. Liu, T.H.; Teng, G.H.; Fu, W.S. Research and development of pig weight estimation system based on image. In Proceedings of the 2011 International Conference on Electronics, Communications and Control (ICECC), Ningbo, China, 9–11 September 2011.
7. McFarlane, N.J.B.; Wu, J.; Tillett, R.D.; Schofield, C.P.; Siebert, J.P.; Ju, X. Shape measurements of live pigs using 3-D image capture. *Anim. Sci.* **2005**, *81*, 383–391. [CrossRef]
8. Doeschl-Wilson, A.B.; Whittemore, C.T.; Knap, P.W.; Schofield, C.P. Using visual image analysis to describe pig growth in terms of size and shape. *Anim. Sci.* **2004**, *79*, 415–427. [CrossRef]
9. Kongsro, J. Estimation of pig weight using a Microsoft Kinect prototype imaging system. *Comput. Electron. Agric.* **2014**, *109*, 32–35. [CrossRef]
10. Menesatti, P.; Costa, C.; Antonucci, F.; Steri, R.; Pallottino, F.; Catillo, G. A low-cost stereovision system to estimate size and weight of live sheep. *Comput. Electron. Agric.* **2014**, *103*, 33–38. [CrossRef]
11. Frost, A.R.; Schofield, C.P.; Beaulah, S.A.; Mottram, T.T.; Lines, J.A.; Wathes, C.M. A review of livestock monitoring and the need for integrated systems. *Comput. Electron. Agric.* **1997**, *17*, 139–159. [CrossRef]

12. Itoh, T.; Kawabe, M.; Nagase, T.; Matsushita, H.; Kato, M.; Miyoshi, M.; Miyahara, K. Body surface area measurement in juvenile miniature pigs using a computed tomography scanner. *Exp. Anim.* **2017**, *66*, 229–233. [CrossRef]
13. Wu, J.; Tillett, R.; McFarlane, N.; Ju, X.; Siebert, J.P.; Schofield, P. Extracting the three-dimensional shape of live pigs using stereo photogrammetry. *Comput. Electron. Agric.* **2004**, *44*, 203–222. [CrossRef]
14. Minagawa, H.; Hosono, D. A light projection method to estimate pig height. In *Swine Housing, Proceedings of the First International Conference, Des Moines, IA, USA, 9–11 October 2000*; American Society of Agricultural Engineers: St. Joseph, MO, USA, 2000; pp. 120–125.
15. Yang, Y.; Teng, G.; Li, B.; Shi, Z. Measurement of pig weight based on computer vision. *Trans. Chin. Soc. Agric. Eng.* **2006**, *2*, 135–139. [CrossRef]
16. Schofield, C.P.; Marchant, J.A.; White, R.P.; Brandl, N.; Wilson, M. Monitoring Pig Growth using a Prototype Imaging System. *J. Agric. Eng. Res.* **1999**, *72*, 205–210. [CrossRef]
17. Liu, T.; Li, Z.; Teng, G.; Luo, C. Prediction of Pig Weight Based on Radical Basis Function Neural Network. *Trans. Chin. Soc. Agric. Mach.* **2013**, *44*. [CrossRef]
18. Kollis, K.; Phang, C.S.; Banhazi, T.M.; Searle, S.J. Weight estimation using image analysis and statistical modelling: A preliminary study. *Appl. Eng. Agric.* **2007**, *23*, 91–96. [CrossRef]
19. Fernandes, A.; Rea, J.D.O.; Fitzgerald, R.; Herring, W.; Rosa, G. A novel automated system to acquire biometric and morphological measurements and predict body weight of pigs via 3D computer vision. *J. Anim. Sci.* **2019**, *97*, 496–508. [CrossRef] [PubMed]
20. Ke, W.; Hao, G.; Qin, M.; Wei, S.; Zhu, D. A portable and automatic Xtion-based measurement system for pig body size. *Comput. Electron. Agric.* **2018**, *148*, 291–298. [CrossRef]
21. Shi, C.; Teng, G.; Li, Z. An approach of pig weight estimation using binocular stereo system based on LabVIEW. *Comput. Electron. Agric.* **2016**, *129*, 37–43. [CrossRef]
22. Li, Z.; Mao, T.; Liu, T.; Teng, G. Comparison and optimization of pig mass estimation models based on machine vision. *Trans. Chin. Soc. Agric. Eng.* **2015**, *31*, 155–161. [CrossRef]
23. Shi, C.; Zhang, J.; Teng, G. Mobile measuring system based on LabVIEW for pig body components estimation in a large-scale farm. *Comput. Electron. Agric.* **2019**, *156*, 399–405. [CrossRef]
24. Pezzuolo, A.; Milani, V.; Zhu, D.; Guo, H.; Guercini, S.; Marinello, F. On-Barn Pig Weight Estimation Based on Body Measurements by Structure-from-Motion (SfM). *Sensors* **2018**, *18*, 3603. [CrossRef] [PubMed]
25. Fu, W.; Teng, G.; Yang, Y. Research on three-dimensional model of pig's weight estimating. *Trans. Chin. Soc. Agric. Eng.* **2006**, *22*, 84–87.
26. Kashiha, M.; Bahr, C.; Ott, S.; Moons, C.P.H.; Niewold, T.A.; Ödberg, F.O.; Berckmans, D. Automatic weight estimation of individual pigs using image analysis. *Comput. Electron. Agric.* **2014**, *107*, 38–44. [CrossRef]
27. Suwannakhun, S.; Daungmala, P. Estimating Pig Weight with Digital Image Processing using Deep Learning. In Proceedings of the 2018 14th International Conference on Signal-Image Technology & Internet-Based Systems (SITIS), Beijing, China, 26–29 November 2018.
28. Wang, Y.; Yang, W.; Winter, P.; Walker, L. Walk-through weighing of pigs using machine vision and an artificial neural network. *Biosyst. Eng.* **2008**, *100*, 117–125. [CrossRef]
29. Kashiha, M.; Bahr, C.; Ott, S.; Moons, C.P.H.; Niewold, T.A.; Ödberg, F.O.; Berckmans, D. Weight Estimation of Pigs Using Top-View Image Processing. In Proceedings of the International Conference Image Analysis and Recognition (ICIAR), Vilamoura, Portugal, 22–24 October 2014.
30. LeCun, Y.; Bengio, Y.; Hinton, G. Deep learning. *Nature* **2015**, *521*, 436–444. [CrossRef]
31. Kamilaris, A.; Prenafeta-Boldú, F.X. Deep learning in agriculture: A survey. *Comput. Electron. Agric.* **2018**, *147*, 70–90. [CrossRef]
32. Li, G.; Bai, L.; Zhu, C.; Wu, E.; Ma, R. A Novel Method of Synthetic CT Generation from MR Images Based on Convolutional Neural Networks. In Proceedings of the 2018 11th International Congress on Image and Signal Processing, BioMedical Engineering and Informatics (CISP-BMEI), Beijing, China, 13–15 October 2018.
33. Chen, D.; Li, S.; Wu, Q. A Novel Supertwisting Zeroing Neural Network with Application to Mobile Robot Manipulators. *IEEE Trans. Neural. Netw. Learn. Syst.* **2020**, 1–12. [CrossRef]
34. Huang, G.; Liu, Z.; van der Maaten, L.; Weinberger, K.Q. Densely Connected Convolutional Networks. In Proceedings of the 2017 IEEE Conference on Computer Vision and Pattern Recognition (CVPR), Honolulu, HI, USA, 21–26 July 2017.
35. Yu, X.; Yu, Z.; Ramalingam, S. Learning strict identity mappings in deep residual networks. In Proceedings of the 2018 IEEE/CVF Conference on Computer Vision and Pattern Recognition, Salt Lake City, UT, USA, 18–22 June 2018.
36. Chollet, F. Xception: Deep Learning with Depthwise Separable Convolutions. In Proceedings of the 2017 IEEE Conference on Computer Vision and Pattern Recognition (CVPR), Honolulu, HI, USA, 21–26 July 2017.
37. Sandler, M.; Howard, A.; Zhu, M.; Zhmoginov, A.; Chen, L. MobileNetV2: Inverted Residuals and Linear Bottlenecks. In Proceedings of the 2018 IEEE/CVF Conference on Computer Vision and Pattern Recognition, Salt Lake City, UT, USA, 18–22 June 2018.

MDPI
St. Alban-Anlage 66
4052 Basel
Switzerland
Tel. +41 61 683 77 34
Fax +41 61 302 89 18
www.mdpi.com

Sensors Editorial Office
E-mail: sensors@mdpi.com
www.mdpi.com/journal/sensors

www.ingramcontent.com/pod-product-compliance
Lightning Source LLC
LaVergne TN
LVHW071258170726
843515LV00006BA/1427

* 9 7 8 3 0 3 6 5 4 0 3 5 1 *